AF358815

Building's Vulnerability Assessment against Natural Hazards by Using Modern Computational Techniques

Building's Vulnerability Assessment against Natural Hazards by Using Modern Computational Techniques

Editors

Tom Lahmer
Ehsan Harirchian
Viviana Novelli

Basel • Beijing • Wuhan • Barcelona • Belgrade • Novi Sad • Cluj • Manchester

Editors

Tom Lahmer
Institut für Strukturmechanik
(ISM), Bauhaus-Universität
Weimar
Weimar
Germany

Ehsan Harirchian
Institut für Strukturmechanik
(ISM), Bauhaus-Universität
Weimar
Weimar
Germany

Viviana Novelli
School of Engineering,
Cardiff University
Cardiff
UK

Editorial Office
MDPI
St. Alban-Anlage 66
4052 Basel, Switzerland

This is a reprint of articles from the Topical Collection published online in the open access journal *Buildings* (ISSN 2075-5309) (available at: https://www.mdpi.com/journal/buildings/topical_collections/Build_Vulner_Asses_Natur_Haz_Moder_Comput_Tech).

For citation purposes, cite each article independently as indicated on the article page online and as indicated below:

Lastname, A.A.; Lastname, B.B. Article Title. *Journal Name* **Year**, *Volume Number*, Page Range.

ISBN 978-3-7258-0485-6 (Hbk)
ISBN 978-3-7258-0486-3 (PDF)
doi.org/10.3390/books978-3-7258-0486-3

Contents

Preface

In a world marked by the relentless force of natural hazards, the imperative to understand and fortify our built environment against their impact has never been more urgent. Recent destructive events across the globe serve as poignant reminders of the vulnerability of structures to the whims of nature. It is against this backdrop that we present this collection, a concerted effort to delve into the intricate realm of assessing buildings' resilience against natural hazards.

The emphasis on research in this domain is not merely a scholarly pursuit but a vital necessity for the safety and well-being of communities worldwide.

This compilation is more than just a repository of insights; it is an invitation to the ingenious minds and dedicated researchers who have made strides in understanding the complex interplay between buildings and natural forces. Their work, showcased within these pages, reflects a commitment to advancing our knowledge and fostering innovative solutions for a safer future.

We extend our heartfelt appreciation to each author who has contributed their expertise, insights, and time to this endeavor. Their dedication to the pursuit of knowledge and the dissemination of valuable findings is instrumental in steering the course of scientific inquiry.

Furthermore, our gratitude extends to the visionary publishers who recognize the significance of disseminating cutting-edge research. Their unwavering support has been indispensable in bringing this collection to fruition and ensuring its accessibility to a wider audience. The rigorous evaluation and refinement of the manuscripts owes much to the diligent efforts of our esteemed reviewers. Their expertise and constructive feedback have played a pivotal role in maintaining the quality and credibility of the content presented herein.

Last but certainly not least, we express our appreciation to everyone involved in this collaborative effort, from the contributors to the editorial and production teams. It is through the collective dedication of these individuals that we take a significant step forward in the realm of science, contributing to the ongoing dialogue on building resilience and natural hazard safety.

As we navigate the intricate landscape of evaluating buildings' vulnerability to natural hazards, may this collection serve as a beacon for future research and inspire a collective commitment to building a safer and more resilient world.

Tom Lahmer, Ehsan Harirchian, and Viviana Novelli

Editors

Article

Response Spectra-Based Post-Earthquake Rapid Structural Damage Estimation Approach Aided with Remote Sensing Data: 2020 Samos Earthquake

Onur Kaplan * and Gordana Kaplan

Institute of Earth and Space Sciences, Eskisehir Technical University, 26555 Eskisehir, Turkey; kaplangorde@gmail.com
* Correspondence: onur_kaplan@eskisehir.edu.tr; Tel.: +90-532-694-7905

Abstract: Effective post-event emergency management contributes substantially to communities' earthquake resilience, and one of the most crucial actions following an earthquake is building damage assessment. On-site inspections are dangerous, expensive, and time-consuming. Remote sensing techniques have shown great potential in localizing the most damaged regions and thus guiding aid and rescue operations in recent earthquakes. Furthermore, to prevent post-earthquake casualties, heavily damaged, unsafe buildings must be identified immediately since in most earthquakes, strong aftershocks can cause such buildings to collapse. The potential of the response spectrum concept for being associated with satellite-based remote sensing data for post-earthquake structural damage estimation was investigated in this study. In this respect, a response spectra-based post-earthquake structural damage estimation method aided by satellite-based remote sensing data was proposed to classify the buildings after an earthquake by prioritizing them based on their expected damage levels, in order to speed up the damage assessment process of critical buildings that can cause casualties in a possible strong aftershock. A case study application was implemented in the Bayrakli region in Izmir, Turkey, the most affected area by the Samos earthquake, on 30 October 2020. The damage estimations made in this research were compared with the in situ damage assessment reports prepared by the Republic of Turkey Ministry of Environment and Urbanization experts. According to the accuracy assessment results, the sensitivity of the method is high (91%), and the necessary time spent by the in situ damage assessment teams to detect the critical buildings would have been significantly reduced for the study area.

Keywords: response spectrum; rapid damage assessment; remote sensing; deep learning

Citation: Kaplan, O.; Kaplan, G. Response Spectra-Based Post-Earthquake Rapid Structural Damage Estimation Approach Aided with Remote Sensing Data: 2020 Samos Earthquake. *Buildings* **2022**, *12*, 14. https://doi.org/10.3390/buildings12010014

Academic Editors: Tom Lahmer, Ehsan Harirchian and Viviana Novelli

Received: 28 November 2021
Accepted: 23 December 2021
Published: 26 December 2021

Publisher's Note: MDPI stays neutral with regard to jurisdictional claims in published maps and institutional affiliations.

1. Introduction

A magnitude Mw 7.0 earthquake struck offshores of Samos Island (Greece) in the eastern Aegean Sea on 30 October 2020. Two people lost their lives, and many buildings were damaged or collapsed due to the earthquake in Karlovasi–Samos Island, Greece. However, the main effect of the earthquake was on the city of Izmir, Turkey, located about 65 km from the earthquake epicenter. Twelve buildings suffered an immediate collapse, and many buildings experienced heavy damage. One hundred and seventeen lives were lost, according to Turkey's Ministry of Interior Disaster and Emergency Management Presidency (AFAD) reports. The majority of the damages were located in the Bayrakli region because of amplified ground motion, despite the long distance from the source [1] The presence of soft stories, lack of proper detailing, poor construction quality, the presence of heavy overhangs, and lack of code compliance were attributed as the primary causes of damage. The earthquake also demonstrated the effect of infill walls on the seismic performance of deficient and inadequate buildings [2].

Rapid urbanization poses a significant challenge to earthquake resilience in cities. The effective management of the post-event emergency contributes significantly to resilience, and one of the most critical activities following an earthquake is assessing building damage [3]. Earthquake-induced building damage and collapse is a major cause of human casualties. In situ expert inspections are risky, expensive, and time-consuming. Therefore, rapid and reliable identification of areas affected by an earthquake is vital in activating appropriate aid and rescue missions after the disaster. In addition, to prevent post-earthquake casualties, the risky, heavily damaged buildings must be detected immediately because, in most earthquakes, strong aftershocks can cause the collapse of those buildings. In recent earthquakes, the use of satellite imagery has demonstrated great promise in localizing the most damaged areas and thus guiding rescue and reconstruction operations. Recent studies have reported that high-resolution satellite imagery can be used to detect post-seismic building damage [4–7].

Inspired by the success of deep learning methods for semantic segmentation in computer vision fields, many researchers have been focusing on Convolutional Neural Networks (CNN) for object detection from remote sensing data. CNN, a deep learning supervised neural network which uses labeled data, has been recognized as one of the most successful and widely used deep learning approaches [8,9]. For example, CNN has been used for object detection of trees [10–15], buildings extraction [16,17], ship detection [18,19], etc.

The potential of the response spectrum concept for being associated with satellite-based remote sensing data for post-earthquake structural damage estimation was investigated in this study. The response spectrum concept was first introduced in the mid-1930s. By the late-1960s, the elastic response spectrum was well developed and understood. This concept is very well integrated with both earthquake engineering practice and research [20]. The elastic response spectrum is plotted between the maximum response of single-degree-of-freedom (SDOF) systems subjected to a particular component of a ground motion and natural periods of SDOF systems. Each response spectrum has a fixed damping ratio ξ; the damping ratio is usually expressed as a proportion of the critical damping, which returns a displaced oscillator to rest without any vibrations, called equivalent viscous damping [21]. In earthquake engineering, 5% of critical damping is generally used for reinforced concrete buildings. Response spectra can be plotted for the natural period T_n vs. acceleration, velocity, or displacement. For example, an acceleration response spectrum of a ground motion component gives the maximum acceleration response of an SDOF system for a given period. In other words, investigating the spectrum, it can be said that for which natural periods the SDOF system will be exposed to more immense accelerations or vice versa. It should be noted that most of the buildings in practice are multi-degree-of-freedom (MDOF) systems, and different modes contribute to their behavior. However, it can be assumed that the fundamental vibration mode controls the seismic response, at least for the buildings whose response is not affected by the contribution of higher modes of vibration (mid-rise buildings) and the buildings without torsional irregularities.

Regarding this, the fundamental vibration period of a building is a crucial parameter for earthquake engineering in both earthquake-resistant design and seismic performance assessment. The equivalent seismic lateral force is determined from a design spectrum in most static design methods. Consequently, the earthquake force is a function of the fundamental vibration period of the building [22,23]. The fundamental period can be computed based on modeling or simplified empirical relationships defined in seismic design codes. For rapid assessment applications, empirical equations are preferable. Kaplan et al. (2021) [24] developed an empirical relationship to predict elastic fundamental vibration periods of reinforced concrete buildings with masonry infill walls. The equation estimates the elastic fundamental period (T_0, in seconds) with respect to building height (H, in meters) (Equation (1)). This equation was established by performing ambient vibration measurements on residential mid-rise reinforced concrete buildings in Eskisehir, Turkey. As being such, it may represent the typical dynamic characteristics of Turkish RC building stock.

$$T_0 = 0.0195H \tag{1}$$

The fundamental period can also be used to estimate the seismic demand during and after an earthquake using the response spectrum of the ground motion. This research aims to propose a response spectra-based post-earthquake structural damage estimation method aided by satellite-based remote sensing data. The research question that motivates this study is, "Can we classify buildings after an earthquake by prioritizing them based on their expected damage levels using the response spectra and remote sensing data?", in order to speed up the damage assessment process of critical buildings that can cause casualties in a possible strong aftershock. For this purpose, a case study application was implemented in the Bayrakli region in Izmir, Turkey, the most affected area by the Samos earthquake on 30 October 2020. A CNN model was applied to extract buildings from high-resolution satellite imagery to detect the buildings in the study area affected by the earthquake. In addition, satellite images were used to determine the height of the buildings. Furthermore, Equation (1) was used to predict the fundamental vibration periods of the buildings based on the calculated heights. The obtained data were imported into a geographic information system (GIS) environment and further used for seismic damage estimation of the detected buildings. The seismic demand for the buildings was estimated by comparing fundamental vibration periods of the buildings with the response spectra of the ground motion recorded at the critical station near the most affected area. The produced information was used to prioritize the in situ damage assessment interventions in the affected region.

The main contributions of the study are as follows:

(i)   Using a CNN to extract buildings from high-resolution remote sensing imagery pre- and post-earthquake and thus, detect collapsed buildings;

(ii)   Using shadow information from high-resolution satellite imagery to calculate building heights;

(iii) Estimation of fundamental vibration periods of detected buildings using the calculated building heights;

(iv) Comparing the fundamental periods with the response spectra of ground shaking gathered from the critical accelerometer station for predicting damaged buildings;

(v)   Proposing a model for rapid damage estimation and prediction for post-earthquake analyses.

The paper has been organized into five sections. Information about the methods and materials on both remote sensing and the response spectra-based damage estimation and the study area details are given in Section 2. The results are presented in Section 3, while Sections 4 and 5 present the discussion and conclusion.

2. Materials and Methods

2.1. Study Area

The earthquake on 30 October 2020, on Samos Island caused significant damage to the city of Izmir. Izmir is the third-largest city of Turkey with a provincial population of 4.5 million, located approximately 65 km from the earthquake epicenter [25]. The city of Izmir is located on the Inner Bay of Izmir. Inner Izmir Bay is a shallow marine basin that is actively growing and is controlled by active east–west trending extensional faults (normal faults) in the Aegean Extensional Province. The uppermost sediments in the basin are young alluvium and fan-delta to shallow marine deposits confined and controlled by the Izmir Fault to the south and the Karsiyaka–Bornova fault to the north [26]. Soft alluvial soils amplified the spectral accelerations, especially in the Bayrakli region, and consequently, most of the heavy damage in buildings occurred in Bayrakli. The study area was chosen from this region, where five buildings collapsed, and seven buildings were heavily damaged (demolished after the earthquake) (Figure 1). The majority of the heavily damaged buildings were constructed in the 1990s with poor code compliance and construction quality. Moreover, some of the buildings' structural systems were suspected to be modified to accommodate commercial use on the first floors [26]. Some of the substandard buildings in the region experienced gravity (pancake) type, and some of them sideways collapse [27] (Figure 2).

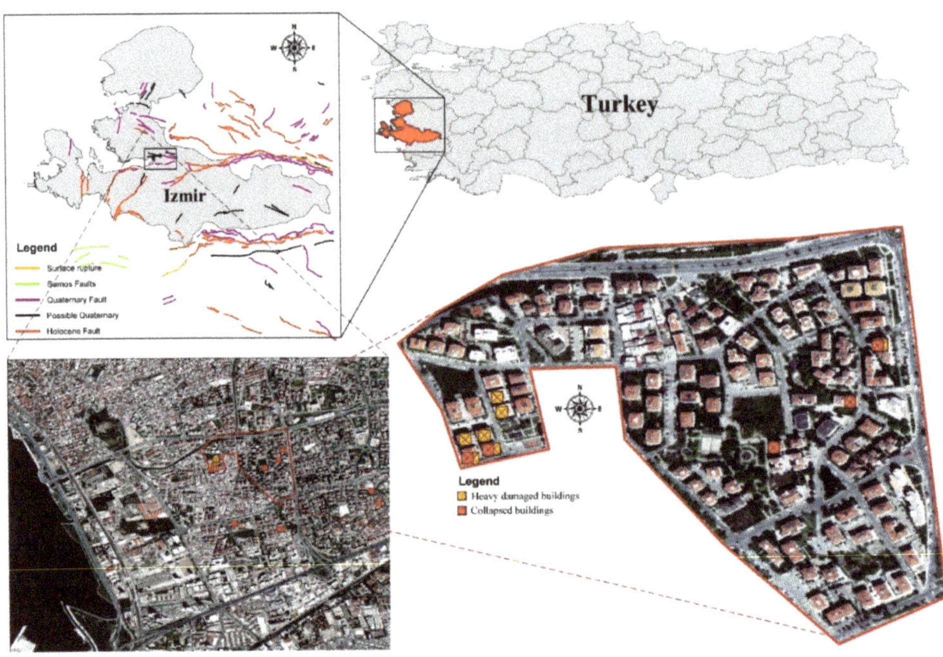

Figure 1. Study area.

Figure 2. Images from collapsed buildings in the Bayrakli region of Izmir [28,29].

2.2. Methodology

The presented study was carried out in two parts; remote sensing and structural damage estimation; then, the results of the two parts were compiled and assessed. The remote sensing part consists of four steps: (i) preparing the satellite data; (ii) implementing the CNN model for building extraction; (iii) assessment of the classification results; (iv) estimating the building heights using shadow lengths of the buildings. The structural damage estimation part also consists of six steps: (i) gathering real-time acceleration data from the stations located in the most affected area; (ii) deciding the critical station and computing response spectra of its both horizontal components; (iii) determining the critical period intervals of the selected critical station's response spectra; (iv) estimating fundamental vibration periods of the surrounding buildings according to the building heights retrieved from remote sensing data; (v) comparing the fundamental periods with the response spectra of ground motion gathered from the critical station for estimating damaged buildings; (vi) the accuracy assessment of the proposed approach has been conducted comparing the estimation results with the in situ damage assessment reports. The details about the methodology implemented in the study are presented in the flowchart shown in Figure 3.

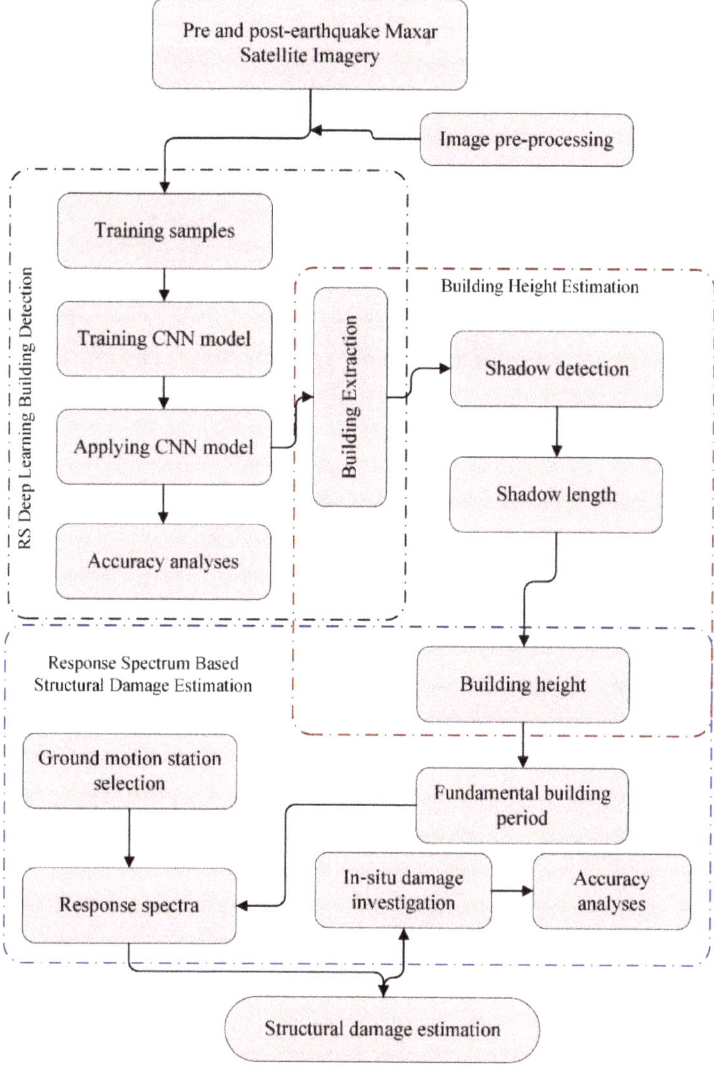

Figure 3. Flowchart of the methodology.

2.2.1. Remote Sensing Data and Techniques

Two satellite images, pre-earthquake (27 April 2020) and post-earthquake (3 November 2020), were provided by Maxar (DigitalGlobe company, Westminster, CO,

USA) with high-resolution (30 cm) and three bands (red, green, and blue, RGB) were used for the analyses.

The CNN workflow using Trimble's eCognition Developer 9.01 was applied to detect the buildings in the study area, based on the Google TensorFlow [30]. The analyses were done in a computer system with a 64-bit operating system, 16 GB RAM, and an Intel (R) Xeon (R) CPU E3-1535M v5 @ 3.60 GHz processor. The CNN model in eCognition was constructed of three different steps: (i) derivation of training samples, (ii) training the CNN model; (iii) applying the trained CNN model. For the training, 90 × 90 pixel samples were collected using the satellite imagery obtained from Maxar (Figure 4). A simple CNN model was created with one hidden layer with the highest correlation within the sample template.

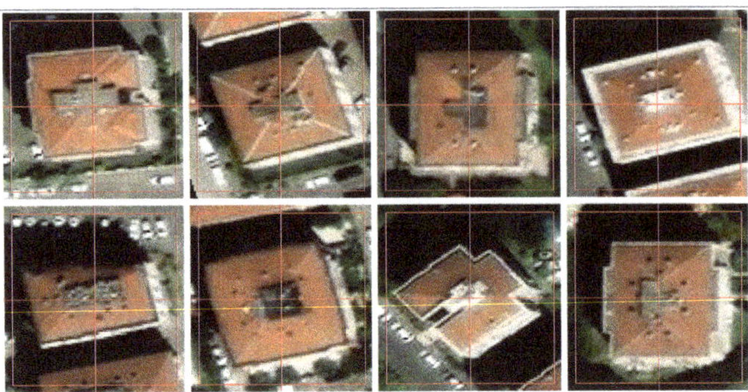

Figure 4. Example of 90 × 90 pixels samples generated from the CNN.

For the accuracy analyses, common evaluation statistics for binary classification were used. Namely, True Positives (TP) (a building is correctly identified), False Positives (FP) (a building is incorrectly identified; a commission error), and False Negatives (FN) (a building is missed; an omission error) parameters were taken into consideration. TP, FN, and FP indicate perfect identification, under-identification, and over-identification, respectively. Then the Precision (P), Recall (R), and F-score (F) were calculated. Precision (i.e., positive predictive value) describes the correctness of detected buildings and how well the algorithm dealt with FP (Equation (2)), Recall (i.e., sensitivity) describes the building detection rate and how well the algorithm dealt with FN (Equation (3)), and the F-score is the harmonic mean of Recall and Precision and reports the overall accuracy considering both commission and omission errors (Equation (4)) [30].

$$P = TP/(TP + FP) \tag{2}$$

$$R = TP/(TP + FN) \tag{3}$$

$$\text{F-score} = 2 \times ((P \times R)/(P + R)) \tag{4}$$

For the building heights, the length of the shadow of the buildings has been measured. Using the sun elevation angle of the first image (62°34′4.35″) of the time of the passing of the satellite over the study area has been used for calculating the buildings' heights. The estimated heights of the buildings were calculated using Equation (5).

$$\text{(Height of the object)} = \tan(\text{Sun Elevation Angle}) \times \text{(Length of the shadow)} \tag{5}$$

2.2.2. Response Spectra-Based Damage Estimation

This study aims to propose a response spectra-based post-earthquake structural damage estimation approach. Thus, response spectra of stations located in the neighborhood of primarily affected urban areas (Bayrakli region; Adalet and Manavkuyu neighborhoods in Izmir) from the October 30 Samos earthquake were investigated. The strong ground motion was recorded by the extensive network of AFAD. In the Bayrakli region, there are three seismic recording stations. Two of them are located on stiff soils (#3514 and #3520) having shear wave velocities of 836 and 875 m/s, respectively, and one (#3513) is on soft soil with 196 m/s. The study area is also on soft soil deposits similar to station #3513's location (Figure 5).

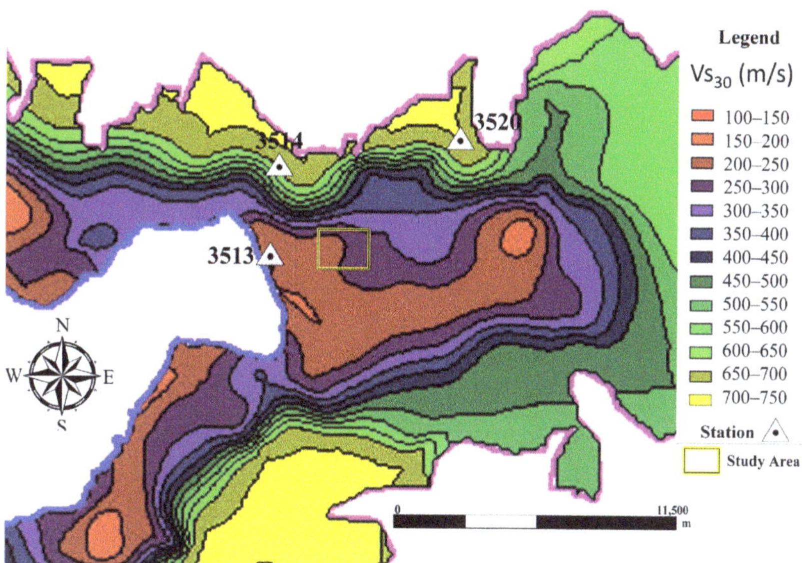

Figure 5. The distribution of Vs$_{30}$ (average shear wave velocity in the upper 30 m) map of the Inner Izmir Bay Basin modified from Izmir Earthquake Master Plan [31] and seismic recording stations around the study area in the Bayrakli region.

The elastic acceleration response spectra of stations #3513, #3514, and #3520 were plotted for their horizontal directions in Figure 6, strong ground motion records were provided from the Turkish Accelerometric Database and Analysis System (TADAS) [32]. Figure 6 also contains the elastic design spectra of the location of station #3513 according to Turkey Building Earthquake Code 2018 (TBEC-2018) [33] for soft and stiff soils. Additionally, the spectra for soft soils defined in the previous versions of the Turkish earthquake code (TEC-2007 [34], TEC-1997 [35], and TEC-1975 [36]) are shown in Figure 6, considering most of the collapsed buildings were constructed in the 1990s and were located on soft alluvial soils of the Bayrakli region. The design and response spectra were plotted with a damping ratio of 5%.

Comparing the design and response spectra, since the design spectra of the current and previous versions of Turkish code for both types of soil are well above the response spectra, and it can be noted that all versions of the code have well defined the seismic hazards of the region. Regardless of the hazard level for the collapsed buildings that were designed, they would not have collapsed if they had been designed following the seismic design principles of the current or previous versions of the governing building code. Moreover, there are buildings of similar typology nearby that did not collapse and were immediately reoccupied. It is very likely that the collapsed buildings, particularly those that experienced gravity (pancake) type failure, were not designed by the version of the code in force at the time of their construction and/or had structural and material-related deficiencies [27]. Figure 6 also shows that soft soil deposits in the Bayrakli region amplified the spectral accelerations. As a result of this amplification, the spectral accelerations on soft soils were approximately 2.5 times larger than those on the rock sites. Table 1 contains the information on recorded strong ground motions of the Samos earthquake for Bayrakli region stations.

In addition to spectral acceleration differences between the stations, Housner and Arias intensity and PGV values are significantly greater for #3513 (Table 1). The study area is also on soft soil deposits similar to station #3513's location (Figure 5). Therefore, station #3513 was selected as the critical station for applying the proposed rapid structural damage estimation approach in this study.

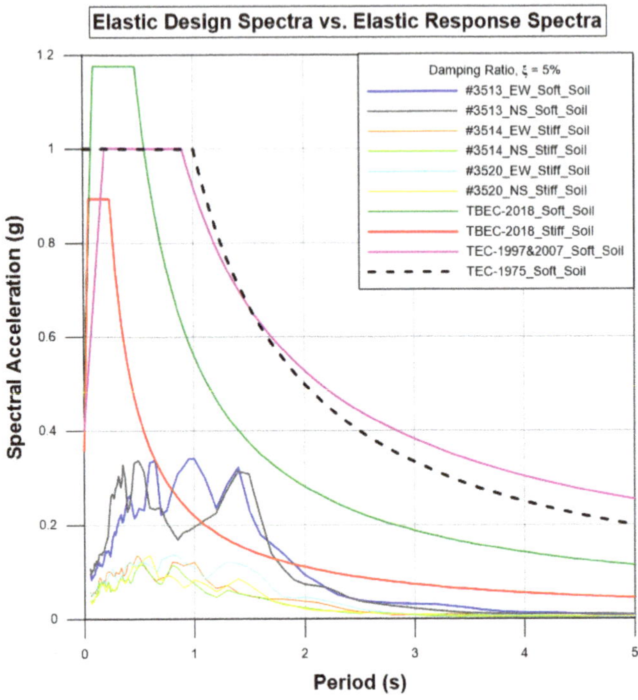

Figure 6. Comparison of the elastic acceleration response spectra of the recorded ground motion at the three stations of AFAD in the Bayrakli region against the design spectra of current and earlier versions of Turkish Seismic Codes at these locations.

Table 1. Information on recorded strong ground motions of the Samos earthquake for Bayrakli region stations [37].

Station Code	Rrup (km)	Repi (km)	Vs30 (m/s)	Comp.	PGA (cm/s^2)	PGV (cm/s)	PGD (cm)	Significant Duration (s)	Arias Intensity (cm/s)	Housner Intensity (cm)
				E-W	94.67	14.42	3.15	20.16	35.30	84.81
3513	65.05	72.00	196	N-S	106.28	17.11	2.90	20.59	33.17	79.76
				U-D	44.19	4.48	0.80	30.84	6.42	23.29
				E-W	56.02	6.41	1.31	23.75	4.58	28.26
3514	66.62	73.39	836	N-S	39.42	4.23	1.44	25.90	3.52	22.39
				U-D	25.15	1.94	0.73	27.17	2.10	10.91
				E-W	58.55	8.37	2.04	5.21	5.21	36.08
3520	68.46	75.78	875	N-S	36.11	4.65	1.13	3.60	3.60	24.21
				U-D	19.37	2.68	0.70	1.33	1.33	13.63

The response spectra of the ground motion recorded at station #3513 were thoroughly investigated. Figure 7 depicts response spectra of both horizontal components as well as their geometric mean. In addition, the fundamental vibration periods of various building types in and around the study area are demonstrated in Figure 7 to determine the critical period intervals in the response spectra. Especially the collapsed and heavily damaged mid-rise buildings were selected to investigate their fundamental periods and corresponding seismic demand based on the response spectra. During the earthquake, five mid-rise buildings in the study area and four nearby collapsed. The fundamental periods of those nine collapsed buildings range from 0.41 to 0.56 s. In the study area, there were four heavily damaged mid-rise buildings with fundamental periods ranging from 0.45 s to 0.49 s, and one mid-rise building with a period of 0.67 s did not sustain any damage. Low-rise buildings in the region have fundamental periods ranging from 0.09 s to 0.21 s, and a nearby high-rise building has a period of 1.07 s. Figure 8 shows their locations in the region. The fundamental vibration periods of the buildings were predicted based on their height using Equation (1). The building heights were calculated by implementing Equation (5) in the satellite images using remote sensing techniques.

Figure 7 shows three peak regions in the spectra where seismic demand is high: between 0.30 and 0.65 s, 0.85 and 1.15 s, and around 1.4 s. Notably, the elastic fundamental periods of collapsed and heavily damaged buildings are in the range of 0.40 and 0.65 s. Despite the presence of some substandard low-rise buildings, which have periods between 0.10 and 0.20 s in the study area, they were unaffected by the earthquake due to relatively low seismic demand. The other peaks of the spectra are around 1.00 and 1.40 s. There is only one building near the study area with a fundamental period of around 1.00 s (1.07 s), with a height of 55.12 m; it is the tallest building in the vicinity of the study area, and it suffered no damage as a result of the earthquake. Despite being severely shaken, no structural damage was discovered in the tall buildings in the Bayrakli region, about 30 buildings with heights between 100 m and 240 m [26] with periods approximately ranging between 2.00 to 4.50 s. Since there were no buildings with elastic fundamental periods between 1.00 and 1.40 s in the study area, the critical period interval in terms of seismic demand was set at 0.40 and 0.65 s. Buildings with fundamental periods between this range were labeled as "high-risk". Assuming that those buildings are most likely to be damaged, they should be inspected primarily by the in situ damage assessment teams. Moreover, as seen in Figure 6, there is no such region as the peaks of response spectra of the ground motion is above the design spectra for the selected study area. Indeed, if there are any peaks above the design spectra, such regions would be chosen as the critical period intervals.

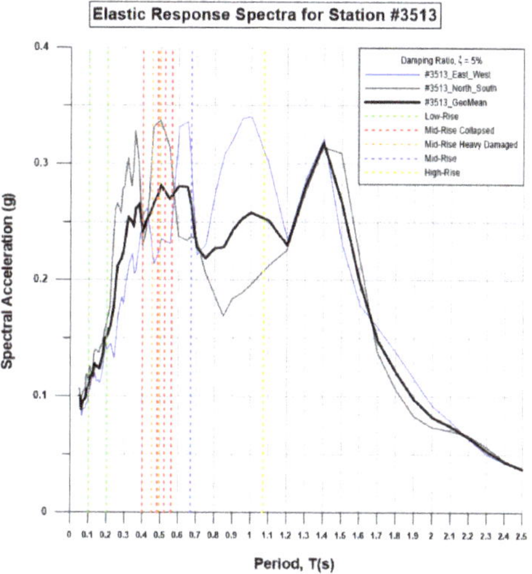

Figure 7. Response spectra of station #3513 and the fundamental periods of various building types in and around the study area.

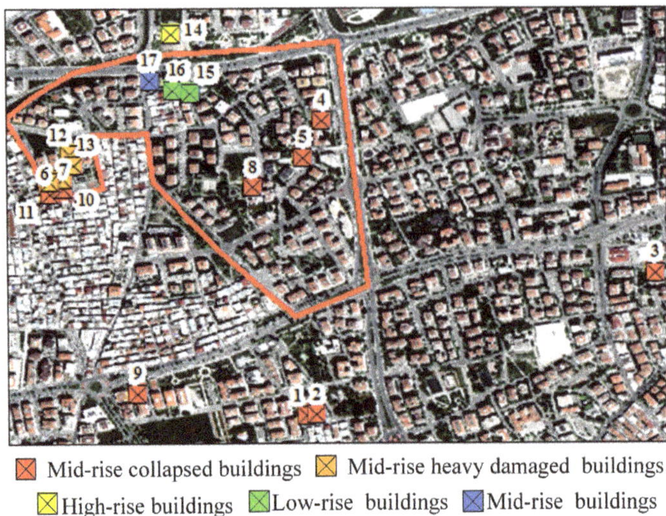

Mid-rise collapsed buildings   Mid-rise heavy damaged buildings
High-rise buildings   Low-rise buildings   Mid-rise buildings

Figure 8. Locations of various building types (consistently color-coded with the fundamental periods of the buildings in Figure 7) in and around the study area.

3. Results

3.1. Remote Sensing

The building detection using CNN performed well with accuracy higher than 95% in both satellite images (Table 2). However, the results also showed some confusion between discrete and adjacent buildings. For example, some adjacent buildings were classified as one building in the first image, while in the second image, some of the adjacent buildings were classified as separate buildings. Since there is no visible boundary between adjacent buildings and usually have the same characteristics, the adjacent buildings classified as discrete buildings were considered correctly classified. Additionally, it should be mentioned that a group of low-rise adjacent buildings were not taken into consideration and are shown in the results as NA (Not Applicable).

Table 2. Accuracy assessment for the CNN building classification.

	Pre-Earthquake Image	Post-Earthquake Image
P	97.6	93.5
R	98.4	98.5
F-score	98.0	95.9

The classified images are shown in Figure 9. The CNN approach accurately detected most of the buildings in the study area. A total of 127 buildings were detected in the first image and 140 in the second image. This difference comes from the difference in the classification of the attached buildings. The number of TP (truly detected buildings) was 122 for the first image and 129 for the second image, FN (missed buildings) was two in both images. The visual inspection concluded that these buildings are significantly smaller than the other detected buildings. The three for the first, and nine for the second image buildings were FP (falsely classified buildings). In the first image, the falsely classified buildings were objects with similar characteristics as the buildings, such as red-colored basketball court. In contrast, in the second image, the falsely classified objects were generally tents set for the earthquake victims in open areas like parks and sports areas in the study area. Additionally, compared to the pre-earthquake image, the accuracy of the post-earthquake image has been negatively affected by the poor image quality, which resulted in slightly lower P values, and thus the overall F-score of the classification.

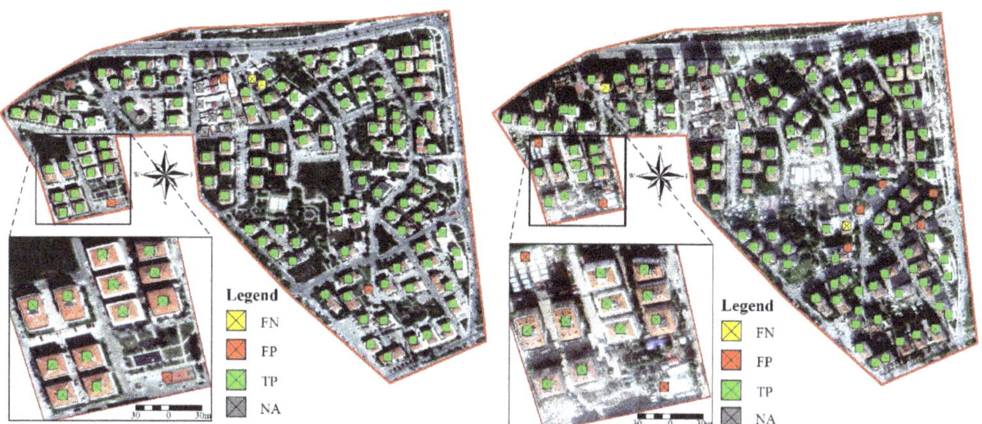

Figure 9. Pre (**left**) and post (**right**) earthquake building detection from Maxar satellite imagery.

3.2. Structural Damage Estimation Results

As stated in the "Response Spectra-Based Damage Estimation" section, buildings with fundamental periods between 0.40 and 0.65 s were labeled as "high risk" based on investigations conducted on the fundamental periods of collapsed and heavily damaged buildings and response spectra of the recorded ground motion at the critical station. As a result, those buildings are the most likely to be damaged and should be inspected first by in situ damage assessment teams. High-risk buildings are labeled with red color in Figure 10. Green and orange indicate low-risk buildings; green represents buildings with fundamental periods of less than 0.40 s, and orange represents buildings greater than 0.65 s. Estimation results were compared with the in situ damage assessment reports [38] prepared by the Republic of Turkey Ministry of Environment and Urbanization experts.

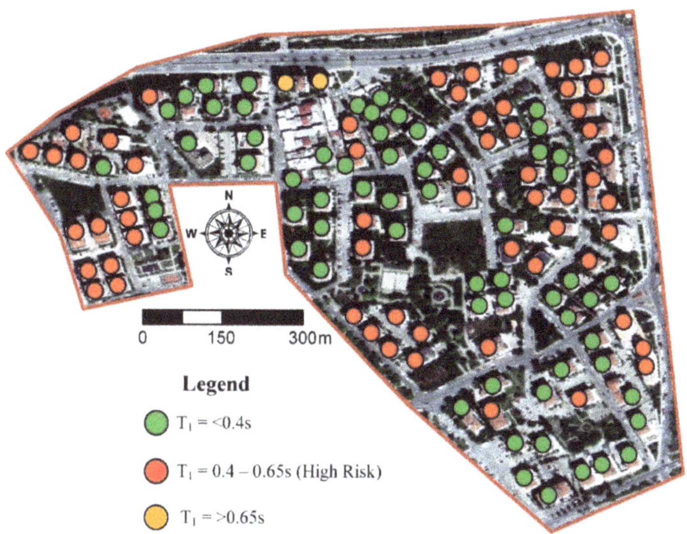

Figure 10. Structural damage estimation in the study area.

In order to evaluate the results, an accuracy assessment has been made using Equations (2)–(4). In the assessment, the predictions made in this research were compared with the on-site damage reports obtained from the Republic of Turkey Ministry of

Environment and Urbanization [38]. From 127 buildings, we were able to access damage reports of 117 buildings. In the comparison, collapsed, severely damaged, and moderately damaged buildings in the damage assessment reports were labeled as high-risk in this study. Buildings with minor damage and those with no damage were designated as low-risk buildings. TP represents the buildings that were correctly predicted to be high-risk and low-risk (69 buildings), FN represents the buildings that were incorrectly predicted to be low-risk (7 buildings). FP represents the buildings that were falsely predicted to be high-risk but were actually classified as low-risk based on the damage reports (39 buildings) (Figure 11). The accuracy assessment showed that the Recall of the method is 91%, the Precision is 64%, and the overall accuracy or the F-score is 75%.

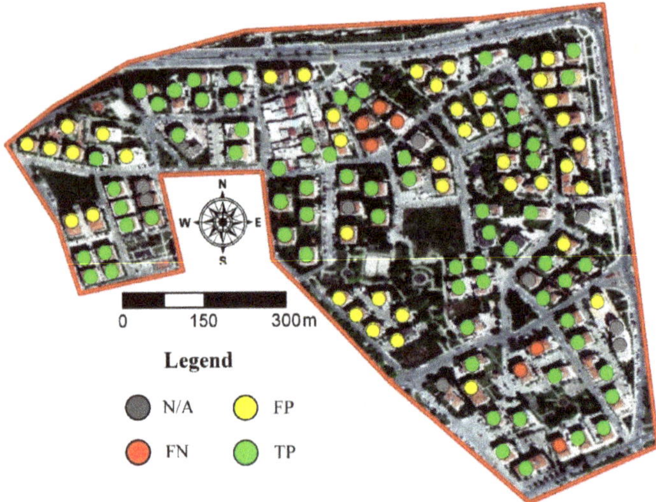

Figure 11. Accuracy assessment of structural damage estimation in the study area.

4. Discussion

The study aims to speed up the damage assessment process of critical buildings that can cause casualties in a possible strong aftershock; from this perspective, conventionally, the number of buildings that must have been investigated after the earthquake by in situ damage assessment experts were 127. If the proposed approach had been used, this number would have been reduced to 60. Because the proposed methodology suggests primarily investigating the potential high-risk buildings (Figure 10) due to their relatively high seismic demand, assuming that their seismic behavior is similar to those collapsed or severely damaged buildings in the region. Even if 39 of those 60 were classified as low-risk based on the in situ inspections, the necessary time spent by the damage assessment teams to detect the critical buildings would have been significantly reduced (more than 50%). It should be noted that in the proposed approach, the predicted seismic demand of the buildings is not compared with the capacity. Thus, the 39 buildings, classified as low-risk according to on-site inspections, can be considered as having higher seismic capacity than their collapsed and severely damaged counterparts. However, according to the on-site damage assessment reports, seven of the buildings classified as low-risk based on the proposed model were severely and/or moderately damaged; this number is meager in the total number of predictions. It corresponds to a 91% success rate in dealing with the false-negative predictions. In order to understand the reasons for the false-negative predictions, the photos of the seven buildings in the damage assessment reports were investigated in detail, and it was concluded that the decisions made by assessors for most of those seven buildings were questionable, even though, prior to this earthquake, the damage assessors were pre-trained for the first time in Turkey [37] by the Ministry of

Environment and Urbanization according to the methodologies developed by Ilki et al. [39] and Boduroglu et al. [40]. However, the damage assessment process is not an easy task due to many issues like varying experience and knowledge of the assessors. According to previous experiences in Turkey, in some cases, especially when the assessors have trouble deciding about the damage state of the building, they tend to choose heavy damage to avoid risks of a potential collapse of the building during strong aftershocks or future earthquakes. Sometimes the building owners also try to influence the assessors' decisions to be benefited from post-disaster help campaigns of the government. The reasons mentioned above may be argued to have influenced damage state decisions for some of the seven buildings. The methods developed by Ilki et al. [39] and Boduroglu et al. [40] aim to prevent such subjective decisions by providing a systematic, objective, and reliable seismic damage assessment system that results in the same structural damage level regardless of who makes the damage assessment.

In addition, it should be stated that, in the end, all of the buildings in the region must be inspected by in situ damage assessment teams. The proposed prioritization aims to provide an optimum allocation of time and resources after a catastrophic event for the civil protection units and decision-makers responsible for managing the consequences of the disaster. Instead of starting from the first building in the region, having a map produced by the proposed methodology would allow the decision-makers to divert the assessment teams primarily to the buildings expected to be severely damaged.

It is worth noting that some issues may be considered shortcomings of the approach, but they can be improved in future studies. For instance, in the proposed methodology, the classification of the expected damage is applied only based on the seismic demand, estimated from satellite-based remote sensing data and response spectra of the selected critical station in the region. The predicted seismic demand of the buildings is not compared with the capacity. When the rapid nature of the procedure and the idea of using remotely accessible data are taken into account, determining the seismic capacity of buildings is not consistent with the philosophy of the approach because it requires entering the buildings. The seismic capacity of the buildings or at least some indicators, like existing structural deficiencies and/or irregularities, can be provided from GIS-based building inventory databases if they exist for the investigated region. According to that information, a damage index can be defined. In some cases, the construction year can also be a good indicator if a milestone year can be designated for the region [41].

Another issue could be using the fundamental vibration period to estimate the seismic demand from response spectra which are designed to represent the SDOF system's response. It should be noted that most of the buildings in practice are MDOF systems, and different modes contribute to their behavior. However, it can be assumed that the fundamental vibration mode controls the seismic response, at least for the buildings whose response is not affected by the contribution of higher modes of vibration (mid-rise buildings) and the buildings without torsional irregularities. In addition, instead of using elastic values of fundamental periods, effective (yield) building periods can provide better damage predictions. However, it is not easy to identify the yield periods by empirical equations to the extent of usage in urban-level applications.

In the proposed methodology, the response spectra of nearby stations in the most affected area are used to estimate the seismic demand of the buildings, and it requires an extensive seismic recording network. For the regions with a limited strong ground motion network, some approaches such as empirical Ground Motion Prediction Models (GMPM) [42] or simulated datasets [25,43] could provide estimations on the hazard levels at locations where the actual motions are not recorded. In this way, the proposed approach can also be utilized for risk mitigation due to future events.

Without building inventory data or calculations, the collapsed and severely damaged reinforced concrete buildings' (7–10 floors) period range was considered between 0.60 and 1.50 s in some reports and papers published after the earthquake [2,37,44], assuming that the building damages correlated with the second and third peaks of station #3513's elastic response spectra (Figure 7). However, in this study, the building heights were calculated by implementing Equation (5) over the satellite images using remote sensing techniques. Then, the elastic fundamental vibration periods of the buildings were predicted based on their heights using Equation (1). According to the predictions made in this study, the elastic fundamental periods of the collapsed and severely damaged buildings range between 0.40 and 0.65 s and this period interval correspond to the first peak of the response spectra of station #3513 that correlates similar demand with the second and third peaks (Figure 7). The practitioners in Turkey generally use the number of stories (N) above the ground to predict the elastic fundamental periods of RC buildings (T) using an empirical equation (T = 0.1 N) which was used in the 1975 version of the Turkish earthquake code TEC-1975 [36], NEHRP-94 and the National Building Code of Canada [45,46]. That correlates well with the equations derived from measurements conducted on moment-resisting RC frame buildings during the 1971 San Fernando earthquake, infilled with drywall, thus more flexible than Turkish counterparts. This equation overestimates the fundamental periods compared to Equation (1), which was developed by performing ambient vibration measurements on residential mid-rise reinforced concrete buildings with masonry infill walls in Eskisehir, Turkey. As being such, Equation (1) may represent the typical elastic dynamic characteristics of Turkish RC building stock.

Ordaz et al. [47] proposed an early earthquake damage assessment system for Mexico City. In their study, the number of stories gathered from the existing building inventory database of Mexico City has been used to predict the fundamental vibration period of the buildings, which makes the model dependent on a building inventory database. Similarly, the Izmir Metropolitan Municipality has a GIS building inventory database for the study area, and the building height information exists in the inventory. Nevertheless, the building heights were retrieved from the story numbers assuming the height of each story is the same, and it equals three meters. This assumption can lead to a significant difference from the actual building heights, especially for buildings with commercial units on the ground floors, which are very common in the study area, due to their relatively high story heights. Therefore, in this study, the calculated building heights from the satellite images were used instead of using existing building height information in the inventory. The proposed approach uses remote sensing techniques to estimate the building heights from satellite images that can be provided anywhere on the earth, so the proposed methodology can be applied regardless of having an inventory database that includes building height information for the region. The reliability of retrieving building heights from remote sensing data using shadows has been successfully proven in various studies with high accuracy [41,48].

The majority of existing earthquake rapid response systems estimate building damages after an earthquake using fragility curves. Developing fragility curves requires a well-established building database and a large number of nonlinear dynamic analyses performed on the categorized building classes [49]. A limited number of countries and cities have well-developed building inventories [3]. Being dependent on a building inventory and computationally very intensive can be considered a limitation for deriving fragility functions for a specific region. Another advantage of the proposed methodology is its flexibility to be applied with or without a building inventory database. The proposed approach can be used for the regions where the fragility curves do not exist.

5. Conclusions

A response spectra-based post-earthquake structural damage estimation method aided by satellite-based remote sensing data was proposed in this study. A case study application was implemented in the Bayrakli region, the most affected area in Izmir, Turkey, by the Samos earthquake on 30 October 2020. According to the findings of this study, the following conclusions and recommendations can be made:

- The proposed approach appears to be efficient in diverting the site inspectors to the buildings expected to be heavily damaged. The accuracy assessment showed that the sensitivity (Recall) of the method is high (91%). The necessary time spent by the in situ damage assessment teams to detect the critical buildings that can cause casualties in a possible strong aftershock would have been significantly reduced (more than 50%) for the study area.

- The proposed approach uses remote sensing techniques to estimate the building heights from satellite images that can be provided anywhere on the earth, so the proposed methodology can be applied regardless of having an inventory database that includes building height information for the region.
- Integration with an existing building inventory in a GIS environment can provide more accurate results. Possible knowledge about the building capacity would be beneficial in addition to seismic demand provided from response spectra.
- The proposed approach can be used for the regions where the fragility curves do not exist.
- The critical seismic recording station selection is based on correlations between the selected region and the soil properties of the station locations and PGV, Housner, and Arias intensity values of the ground motions recorded at the surrounding stations. So, a dense seismic recording network is needed to conduct the proposed methodology, or empirical GMPM's or simulated datasets can be used.
- All peaks of the response spectra of ground motion at the selected critical station should be considered as crucial periods, especially if any of the peaks are above the design spectrum of the region.
- The empirical equation used to predict the elastic fundamental periods of the buildings should be region specific and represent the design spectra of the region and construction characteristics.
- Estimation of yield building periods can provide better damage predictions.
- The current state of the approach is limited to investigating RC buildings because of the used empirical period equation. Additionally, distinguishing masonry and RC buildings from each other using satellite images is a challenging task.
- The building height extraction methodology needs to be improved for the adjacent buildings and be automatized.

With the rapid development of remote sensing sensors and techniques, the results of the remote sensing classification can be improved in future studies as new data and new, more accurate classification techniques such as Mask RCNN [15], are becoming available every day. With the daily satellite imagery that can be easily obtained from various satellite sensors and updated drone images, post-earthquake data can be acquired in a short time, thus reducing the time required for in situ inspection. An automated method for height calculation from shadow information can be developed. Another option for this matter could be the use of ready-to-use data from urban digital twins if available. The proposed methodology would be of great importance in the case of automatization, and thus the results of the proposed methodology can be obtained within hours after the earthquake.

The results demonstrate that this approach can be used to speed up the damage assessment process, prioritizing the critical buildings based on their expected damage levels. The proposed prioritization aims to provide an optimum allocation of time and resources after a catastrophic event for the civil protection agencies and decision-makers responsible for managing the consequences of the disaster.

Author Contributions: The authors (O.K. and G.K.) contributed equally. All authors have read and agreed to the published version of the manuscript.

Funding: This study was supported by Eskisehir Technical University Scientific Research Projects Commission under grant No: 20ADP112, Project: "Building Inventory Information for Seismic Vulnerability Assessment Using Remote Sensing Techniques".

Informed Consent Statement: Not applicable.

Data Availability Statement: Not applicable.

Conflicts of Interest: The authors declare no conflict of interest.

References

1. Askan, A.; Gülerce, Z.; Roumelioti, Z.; Sotiriadis, D.; Melis, N.S.; Altindal, A.; Akbaş, B.; Sopaci, E.; Karimzadeh, S.; Kalogeras, I.; et al. The Samos Island (Aegean Sea) M7. 0 Earthquake: Analysis and engineering implications of strong motion data. *Bull. Earthq. Eng.* **2021**. [CrossRef]
2. Yakut, A.; Sucuoğlu, H.; Binici, B.; Canbay, E.; Donmez, C.; Ilki, A.; Caner, A.; Celik, O.C.; Ay, B. Performance of structures in İzmir after the Samos island earthquake. *Bull. Earthq. Eng.* **2021**. [CrossRef]
3. Erdik, M.; Şeşetyan, K.; Demircioğlu, M.; Hancılar, U.; Zülfikar, C. Rapid earthquake loss assessment after damaging earthquakes. *Soil Dyn. Earthq. Eng.* **2011**, *31*, 247–266. [CrossRef]
4. Tong, X.; Hong, Z.; Liu, S.; Zhang, X.; Xie, H.; Li, Z.; Yang, S.; Wang, W.; Bao, F. Building-damage detection using pre-and post-seismic high-resolution satellite stereo imagery: A case study of the May 2008 Wenchuan earthquake. *ISPRS J. Photogramm. Remote Sens.* **2012**, *68*, 13–27. [CrossRef]
5. Dong, L.; Shan, J. A comprehensive review of earthquake-induced building damage detection with remote sensing techniques. *ISPRS J. Photogramm. Remote Sens.* **2013**, *84*, 85–99. [CrossRef]
6. Karimzadeh, S.; Matsuoka, M. A Preliminary Damage Assessment Using Dual Path Synthetic Aperture Radar Analysis for the M 6.4 Petrinja Earthquake (2020), Croatia. *Remote Sens.* **2021**, *13*, 2267. [CrossRef]
7. Omarzadeh, D.; Karimzadeh, S.; Matsuoka, M.; Feizizadeh, B. Earthquake Aftermath from Very High-Resolution WorldView-2 Image and Semi-Automated Object-Based Image Analysis (Case Study: Kermanshah, Sarpol-e Zahab, Iran). *Remote Sens.* **2021**, *13*, 4272. [CrossRef]
8. Amini Amirkolaee, H.; Arefi, H. CNN-based estimation of pre-and post-earthquake height models from single optical images for identification of collapsed buildings. *Remote Sens. Lett.* **2019**, *10*, 679–688. [CrossRef]
9. Tang, R.; Liu, H.; Wei, J.; Tang, W. Supervised learning with convolutional neural networks for hyperspectral visualization. *Remote Sens. Lett.* **2020**, *11*, 363–372. [CrossRef]
10. Li, W.; Fu, H.; Yu, L.; Cracknell, A. Deep learning based oil palm tree detection and counting for high-resolution remote sensing images. *Remote Sens.* **2017**, *9*, 22. [CrossRef]
11. Chen, B.; Xiao, X.; Li, X.; Pan, L.; Doughty, R.; Ma, J.; Dong, J.; Qin, Y.; Zhao, B.; Wu, Z.; et al. A mangrove forest map of China in 2015: Analysis of time series Landsat 7/8 and Sentinel-1A imagery in Google Earth Engine cloud computing platform. *ISPRS J. Photogramm. Remote Sens.* **2017**, *131*, 104–120. [CrossRef]
12. Wang, Z.; Underwood, J.; Walsh, K.B. Machine vision assessment of mango orchard flowering. *Comput. Electron. Agric.* **2018**, *151*, 501–511. [CrossRef]
13. Mubin, N.A.; Nadarajoo, E.; Shafri, H.Z.M.; Hamedianfar, A. Young and mature oil palm tree detection and counting using convolutional neural network deep learning method. *Int. J. Remote Sens.* **2019**, *40*, 7500–7515. [CrossRef]
14. Timilsina, S.; Sharma, S.; Aryal, J. Mapping urban trees within cadastral parcels using an object-based convolutional neural network. *ISPRS Ann. Photogramm. Remote Sens. Spat. Inf. Sci.* **2019**, *4*, 111–117. [CrossRef]
15. Ocer, N.E.; Kaplan, G.; Erdem, F.; Kucuk Matci, D.; Avdan, U. Tree extraction from multi-scale UAV images using Mask R-CNN with FPN. *Remote Sens. Lett.* **2020**, *11*, 847–856. [CrossRef]
16. Zhang, Q.; Wang, Y.; Liu, Q.; Liu, X.; Wang, W. CNN based suburban building detection using monocular high resolution Google Earth images. In Proceedings of the 2016 IEEE International Geoscience and Remote Sensing Symposium (IGARSS), Beijing, China, 10–15 July 2016.
17. Hu, Y.; Guo, F. Building Extraction Using Mask Scoring R-CNN Network. In Proceedings of the 3rd International Conference on Computer Science and Application Engineering, Sanya, China, 22–24 October 2019.
18. Nie, S.; Jiang, Z.; Zhang, H.; Cai, B.; Yao, Y. Inshore ship detection based on mask R-CNN. In Proceedings of the IGARSS 2018–2018 IEEE International Geoscience and Remote Sensing Symposium, Valencia, Spain, 22–27 July 2018.
19. Nie, X.; Duan, M.; Ding, H.; Hu, B.; Wong, E.K. Attention Mask R-CNN for Ship Detection and Segmentation From Remote Sensing Images. *IEEE Access* **2020**, *8*, 9325–9334. [CrossRef]
20. Chopra, A.K. Elastic response spectrum: A historical note. *Earthq. Eng. Struct. Dyn.* **2007**, *36*, 3–12. [CrossRef]
21. Bommer, J.J.; Boore, D.M. *Engineering Geology/Seismology, Encyclopedia of Geology*; Elsevier Academic: Oxford, UK, 2005; pp. 499–514.
22. Pan, T.-C.; Goh, K.S.; Megawati, K. Empirical relationships between natural vibration period and height of buildings in Singapore. *Earthq. Eng. Struct. Dyn.* **2014**, *43*, 449–465. [CrossRef]
23. Hong, L.-L.; Hwang, W.-L. Empirical formula for fundamental vibration periods of reinforced concrete buildings in Taiwan. *Earthq. Eng. Struct. Dyn.* **2000**, *29*, 327–337. [CrossRef]
24. Kaplan, O.; Guney, Y.; Dogangun, A. A period-height relationship for newly constructed mid-rise reinforced concrete buildings in Turkey. *Eng. Struct.* **2021**, *232*, 111807. [CrossRef]
25. Akinci, A.; Cheloni, D.; Dindar, A. The 30 October 2020, M7. 0 Samos Island (Eastern Aegean Sea) Earthquake: Effects of source rupture, path and local-site conditions on the observed and simulated ground motions. *Bull. Earthq. Eng.* **2021**, *19*, 4745–4771. [CrossRef]
26. Erdik, M.; Demircioğlu, M.; Cüneyt, T. Forensic Analysis Reveals the Causes of Building Damage in İzmir in the Oct. 30 Aegean Sea Earthquake. 2020. Temblor Website. Available online: https://temblor.net/earthquake-insights/forensic-analysis-reveals-the-causes-of-building-damage-in-izmir-in-the-oct-30-aegean-sea-earthquake-12098/ (accessed on 25 October 2021).

27. Günay, S.; Mosalam, K.M.; Archbold, J.; Dilsiz, A.; Beyazit, A.Y.; Wilfrid, G.D.; Gupta, A.; Concern, O.; Hassan, W.; María, S.; et al. StEER Preliminary Virtual Reconnaissance Report (PVRR): Aegean Sea Earthquake (M7.0), 30 October 2020. Available online: https://www.designsafe-ci.org/data/browser/public/designsafe.storage.published/PRJ-2953/#details-47792654024623 34485-242ac116-0001-012 (accessed on 25 October 2021).
28. Vîrghileanu, M.; Săvulescu, I.; Mihai, B.-A.; Nistor, C.; Dobre, R. Nitrogen dioxide (NO_2) pollution monitoring with sentinel-5p satellite imagery over europe during the coronavirus pandemic outbreak. *Remote Sens.* **2020**, *12*, 3575. [CrossRef]
29. CNN. Powerful Earthquake Jolts Turkey and Greece, Killing at Least 27. 31 October 2020. Available online: https://edition.cnn.com/2020/10/30/europe/earthquake-greece-turkey-aegean-intl/index.html (accessed on 25 October 2021).
30. Csillik, O.; Cherbini, J.; Johnson, R.; Lyons, A.; Kelly, M. Identification of citrus trees from unmanned aerial vehicle imagery using convolutional neural networks. *Drones* **2018**, *2*, 39. [CrossRef]
31. Metropolitan Municipality of Izmir (MMI) Izmir Earthquake Masterplan. 2000. Available online: http://www.izmir.bel.tr/izmirdeprem/izmirrapor.htm (accessed on 25 October 2021). (In Turkish)
32. Turkish Accelerometric Database and Analysis System (TADAS). TADAS 2021. Available online: https://tadas.afad.gov.tr/login (accessed on 1 November 2021).
33. TBEC-2018. *Turkey Building Earthquake Code*; AFAD, Ministry of Interior, Disaster and Emergency Management Presidency: Ankara, Turkey, 2018.
34. *Turkish Earthquake Code (TEC-2007): Specifications for Buildings to Be Built in Seismic Areas*; Ministry of Public Works and Settlement: Ankara, Turkey, 2007. (In Turkish)
35. *Turkish Earthquake Resistant Design Code (TEC-1997): Specifications for Structures to Be Built in Disaster Areas*; Ministry of Public Works and Settlement: Ankara, Turkey, 1997. (In Turkish)
36. *Turkish Earthquake Code (TEC-1975): Specifications for Buildings to Be Built in Seismic Areas*; Ministry of Public Works and Settlement: Ankara, Turkey, 1975. (In Turkish)
37. Onder Cetin, K.; Mylonakis, G.; Sextos, A.; Stewart, J.P. Seismological and Engineering Effects of the M 7.0 Samos Island (Aegean Sea) Earthquake. Available online: http://learningfromearthquakes.org/images/earthquakes/2020_Samos_Greece_Izmir_Turkey/Samos_Island_Earthquake_Final_Report.pdf (accessed on 25 October 2021).
38. Stratoulias, D.; Nuthammachot, N. Air quality development during the COVID-19 pandemic over a medium-sized urban area in Thailand. *Sci. Total Environ.* **2020**, *746*, 141320. [CrossRef]
39. Ilki, A.; Halici, O.; Kupcu, E.; Comert, M.; Demir, C. Modifications on seismic damage assessment system of TCIP based on reparability. In Proceedings of the 17th World Conference on Earthquake Engineering 17WCEE, Sendai, Japan, 22 April 2020.
40. Boduroglu, H.; Ozdemir, P.; Binbir, E.; Ilki, A. Seismic damage assessment methodology developed for Turkish compulsory insurance system. In Proceedings of the 9th Annual International Conference of the International Institute for Infrastructure Renewal and Reconstruction, Brisbane, Australia, 8–10 July 2013.
41. Kaplan, G.; Kaplan, O. PlanetScope Imagery for Extracting Building Inventory Information. *Environ. Sci. Proc.* **2021**, *5*, 19. [CrossRef]
42. Boore, D.M.; Stewart, J.P.; Skarlatoudis, A.A.; Seyhan, E.; Margaris, B.; Theodoulidis, N.; Scordilis, E.; Kalogeras, I.; Klimis, N.; Melis, N.S. A ground-motion prediction model for shallow crustal earthquakes in Greece. *Bull. Seismol. Soc. Am.* **2021**, *111*, 857–874. [CrossRef]
43. Karimzadeh, S.; Askan, A.; Erberik, M.A.; Yakut, A. Seismic damage assessment based on regional synthetic ground motion dataset: A case study for Erzincan, Turkey. *Nat. Hazards* **2018**, *92*, 1371–1397. [CrossRef]
44. Bayhan, B.; Avcı, E.; Kayı, D.B. EGE DENİZİ SEFERİHİSAR AÇIKLIKLARI M6.6 DEPREMİ 30 EKİM 2020 YAPI ÖN DEĞERLENDİRME RAPORU; Bursa Teknik Üniversitesi, Deprem Mühendisliği Uygulama ve Araştırma Merkezi, Bursa, Turkey, 2020. (In Turkish). Available online: https://depo.btu.edu.tr/dosyalar/btu/Dosyalar/Rapor_Bursa.Teknik.Uni.Deprem.2020.10.30.pdf (accessed on 25 October 2021).
45. *National Earthquake Hazards Reduction Program: Recommended Provisions for the Development of Seismic Regulations for New Buildings*; Building Seismic Safety Council: Washington, DC, USA, 1994.
46. *National Research Council: The National Building Code (NBC)*; National Research Council: Ottawa, ON, Canada, 1995.
47. Ordaz, M.; Reinoso, E.; Jaimes, M.A.; Alcántara, L.; Pérez, C. High-resolution early earthquake damage assessment system for Mexico City based on a single-station. *Geofísica Int.* **2017**, *56*, 117–135. [CrossRef]
48. Comert, R.; Kaplan, O. Object Based Building Extraction And Building Period Estimation from Unmanned Aerial Vehicle Data. In Proceedings of the ISPRS Annals of the Photogrammetry, Remote Sensing and Spatial Information Sciences, Beijing, China, 7–10 May 2018.
49. Rossetto, T.; Elnashai, A. A new analytical procedure for the derivation of displacement-based vulnerability curves for populations of RC structures. *Eng. Struct.* **2005**, *27*, 397–409. [CrossRef]

 **buildings**

Article

Structural Analysis of Five Historical Minarets in Bitlis (Turkey)

Ercan Işık [1], Ehsan Harirchian [2,*], Enes Arkan [3], Fatih Avcil [1] and Mutlu Günay [1]

[1] Department of Civil Engineering, Bitlis Eren University, Bitlis 13100, Turkey; eisik@beu.edu.tr (E.I.); favcil@beu.edu.tr (F.A.); mutlugunay@gmail.com (M.G.)

[2] Institute of Structural Mechanics (ISM), Bauhaus-Universität Weimar, 99423 Weimar, Germany

[3] Department of Architecture, Bitlis Eren University, Bitlis 13100, Turkey; earkan@beu.edu.tr

* Correspondence: ehsan.harirchian@uni-weimar.de

Abstract: Bitlis has hosted many civilizations and is located in Turkey's significant strategic transit corridor. Many historical structures belong to different cultures in the city. The structural analysis of five minarets mentioned in folk songs and the brand value of Bitlis city in terms of historical buildings is the subject of this study. These minarets are precious because they witness important events in Bitlis city. Non-destructive test methods determined the material properties of the Bitlis stone used in constructing minarets. Within the scope of the study, detailed information about each minaret was given, and on-site measurements determined its dimensions and current structural conditions. For each minaret, its seismic behavior has been selected by using the vertical and horizontal design spectrum in the recent earthquake code of Turkey. Historical masonry minarets were modeled using the finite element method. In addition to stress distribution in the minarets under different loading conditions, period and displacement results are also investigated.

Keywords: historical heritage; finite element analysis; damage assessment; minaret

Citation: Işık, E.; Harirchian, E.; Arkan, E.; Avcil, F.; Günay, M. Structural Analysis of Five Historical Minarets in Bitlis (Turkey). *Buildings* **2022**, *12*, 159. https://doi.org/10.3390/buildings12020159

Academic Editor: Chiara Bedon

Received: 12 January 2022
Accepted: 28 January 2022
Published: 2 February 2022

Publisher's Note: MDPI stays neutral with regard to jurisdictional claims in published maps and institutional affiliations.

1. Introduction

Studies on historical artifacts have an important place in preserving their historical and cultural heritage and transferring it to the next generations. Such works contain information about the social life of the societies of the period and construction technologies. The fact that they have survived over time is an indication that such structures receive outstanding engineering services without any high-level technology [1–3]. The protection of cultural heritage and its safe transfer to the future are among essential engineering research and implementation subjects of the 21st century. Since this vital subject meets on common ground with fields of science such as engineering, architecture, art history, and archeology, it also attracts the attention of interdisciplinary working groups, which have gained importance in recent years [4–8].

Historical buildings are invaluable cultural assets that strongly connect the past and the future. Historical buildings are also an indicator of societies' engineering background, artistic understanding, and economic status. The Van Lake basin has hosted many civilizations in the historical process such as Hurrian, Urartian, Med, Persian, Sassanid, Seljuk, and Ottoman civilizations. Since the basin is a very old residential area, it has carried the historical structures and cultural values left behind by many civilizations until today. There are many historical buildings in the basin that were built in very old times and are still in use after restoration works. Bitlis is one of the centers in this basin that has cradled many civilizations. Lake Van basin is also a region that causes great loss of life and properties after destructive earthquakes.

Investigation of earthquake resistance of buildings, determining and examining earthquake safety, and the parameters affecting the safety of the buildings have increased its importance in recent years. For these reasons, determining earthquake behavior and the safety of structures is one of the most basic study fields of earthquake engineering. From the results of previous earthquakes, it can be observed that heavy damage and destruction in structures are relatively high depending on the level of development. However, it is crucial to distinguish and separately analyze the parameters that share in the formation of these damages and destructions. For this reason, when observing the behavior of structures under the effects of earthquakes, knowing the factors that will affect the earthquake resistance of structures attains particular importance. Studies and research on historical artifacts, which are a part of cultural heritage, both in our country and in different parts of the world are becoming more common day by day [9–12]. Some of these research studies include Bajrakli Mosque (Western Kosovo) [13], SS. Rosario Church Bell Tower [14], St. Mary of Carmel Church [15], Ben Ezra Synagogue [16], Athena Temple [17], Madre Santa Maria del Borgo Church [18], and Gazi Hasan Pasha mosque [19]. These studies are studies on monumental structures built in a masonry style. These studies can also be considered as case studies for the modelling and strengthening of monumental masonry structures with the finite element method, determining the material properties used, and determining their seismic behaviour using different analysis methods.

Minarets have an important place among historical monumental structures. Since these structures, which are symbols of faith, were built differently in different civilizations, they provide information about the construction and construction technologies of that period. Although minarets are built in different systems, historical minarets are commonly encountered as masonry structures. Generally, they were built by local cut stones and minaret masters of the region. There are many studies on minarets, which are an integral part of our historical assets. In studies on such minarets, the behavior under the influence of earthquakes has been examined in general. Çaktı et al. (2013) [20], by giving information about the damage to the minarets in the 2011 Van earthquake, provided the results of the study for forty-one new and historical minarets in Istanbul. In the study, the earthquake behavior of the Edirnekapı Mimrimah Sultan Mosque minaret has been specifically examined. In addition, information about earthquake recording and monitoring systems used in minarets is given. Işık and Antep (2018) [21] determined the seismic behavior of the minaret of the historical Kadı Mahmut Mosque in Ahlat district in terms of different load combinations, using the design spectra specified in the Turkish Seismic Design Code-2007 (TSDC-2007) [22]. In the study carried out by Kılıç et al. (2020) [23], the dynamic behavior of the Kırklareli Hızırbey Mosque Minaret was determined. In the study, analyses were investigated by the methods given in both TSDC-2007 and the Turkish Building Earthquake Code (TBEC-2018) [24]. Ural and Çelik (2018) [25] attempted to determine earthquake behavior and dynamic analysis of masonry minarets with a single balcony. In this context, seven different minarets in Aksaray district were chosen as examples. The finite element method was used in the analysis of minarets. Mutlu and Şahin (2016) [26], on the other hand, investigated the earthquake behavior of the historical Ulu Mosque minaret in Bursa by using different modeling techniques. Dynamic analyses were performed in the time history using acceleration time curves. Bayraktar et al. (2013) [27] determined the dynamic properties of the minaret of the historical Sundura Mosque, for which renovation was carried out, using environmental vibration test methods after restoration. Günaydın (2018) [28], in his study, determined the dynamic characteristics of the minaret of the Trabzon İskender Pasha Mosque after restoration processes. These processes were carried out experimentally. Çarhoğlu et al. (2013) [29] analyzed the seismic behavior of the minarets of the Hagia Sophia Mosque, one of the most important mosques of our country, in their work. The time-history analysis method was used in this study. Uğurlu et al. (2017) [30] carried out structural analyses by modeling the Four-Legged Minaret, one of the important historical buildings of Diyarbakır. Oğuzmert (2002) [31], in his master's thesis, information was given about the structural analysis of masonry minarets built differently and the methods used in masonry structures. Döven et al. (2018) [32], in the study, determined the dynamic behavior of the Green Minaret in the city of Kütahya in the case of closed and open balconies and compared them. Güneş et al. (2021) [33] focused on the seismic assessment of a reconstructed ruined mosque built between 1807 and 1820. Ertek and

Fahjan (2007) [34], in their study, provided information about the construction systems and technologies of the historical minarets of the Ottoman period, their classification, and how to model and analyze them. Çalık et al. (2017) [35] provided information on simplified natural frequency formulas for historical masonry minarets using experimental methods. Yetkin et al. in their study in 2021 [36] examined the damage to the minarets in Elazig city after the 6.8-magnitude earthquake that occurred in the Sivrice district of Elazig city on 24 January 2020. The sections where the damages occurred in the minarets examined were determined, and the reasons for the formation of these damages were evaluated.

Structural analyses of five minarets, which are the subject of the song Five Minarets in Bitlis written to describe the situation in the occupied city, were carried out using the horizontal and vertical design spectrum defined in the current Turkish Building Earthquake Code (TBEC-2018) and obtained using the Earthquake Hazard Maps Application of Turkey. The local name Bitlis stone was used in the construction of the minarets. This stone's material properties (modulus of elasticity and Poisson ratio) were obtained by using non-destructive test methods, and these values were used in the analyses. A survey study was conducted for each minaret, and finite element models of minarets were created. In this study, while giving detailed information about Five Minarets, their current situation is also stated based on observation. The safe transfer of these minarets to the next generations, which are of great value for Bitlis, is significant for preserving its historical and cultural heritage. The information obtained within the scope of this study will be an archive for such structures. For the first time, the modulus of elasticity and the Poison ratio were obtained for Bitlis stone. It is crucial to be the first and detailed study on Five Minarets, one of the brand structures of Bitlis city.

2. Materials and Methods

2.1. Five Minarets

This study is important in that it is the first study on the structural analysis of Five Minarets, which is one of the important cultural and historical advertising structures of Bitlis city. The most well known is the minaret of the Ulu Mosque of Five Minarets. This minaret was built after the mosque. The Ulu Mosque, to which the minaret belongs, is one of the oldest mosques in Bitlis. In addition to the Ulu Mosque in Bitlis, the Şerefiye Külliye, Dört Sandık Gökmeydan, Ayn'el Barit (Soğuk Pınar), Sultaniye, Meydan, Kızıl Mescit, Seyit İbrahim, Alemdar, Hacı Begiye, Kureyşi, and Memi Dede mosques are important historical places of worship of the city. The minarets, which are the subject of the song "Five Minarets in Bitlis", belong to the Ulu Mosque, Şerefiye Mosque, Meydan, and Gökmeydan mosques. It is estimated that the fifth minaret belonged to one of the Hatuniye, Kalealtı, or Kadiri mosques, and when it was destroyed, a new one was built in its place. The positions of Five Minarets are shown in Figure 1 and visuals are shown in Figure 2.

In general, minarets consist of the pulpit, the transition segment, the body, the balcony, the upper part of the body, the spire, and the end ornament from bottom to top. As an example, parts of the Ulu Mosque minaret are shown in Figure 3. The parts of the other four minarets are similar to the Ulu Mosque minaret. There is no difference in the parts of Five Minarets.

When the city of Bitlis is mentioned, the folk song "Five Minarets in Bitlis" comes to mind immediately. The story of this song is as follows. Bitlis, which was under Russian occupation for 5 months and 5 days and gained its freedom on 8 August 1916 took the appearance of a ruined city due to the fact that the city was destroyed during the occupation. A father and son, who fled from Bitlis during the war, returned to Bitlis after the withdrawal of the enemy and reached the foot of Dideban Mountain, where the city of Bitlis is visible. The father sends his son to the city to find out if somebody is alive in the city. After examining the city for a while, the son turns to his father and calls out from afar: "There is no trace of life in the city". When he said, "Only five minarets have survived", his father, hearing this, collapses and kneels down and calls his son to him with a lament: "Five Minarets in Bitlis, come back, son, come back, my heart is full of wounds, come back, son, come back". This lament, sung by the father, has survived to the present day as the subject of folk songs and poems. The folk song "Five Minarets in Bitlis" was composed in 1970 by Fatih Gündoğdu, who worked at Turkish Radio and Television (TRT)-Istanbul Radio. This song has become the advertisement words of Bitlis city over time [37–46].

Figure 1. (**a**) The location of the Bitlis city in Turkey. (**b**) The location of Five Minarets in Bitlis city. (**c**) The site plan of the Ulu Mosque and its minaret. (**d**) The site plan of the Şerefiye Külliye and its minaret. (**e**) The site plan of the Meydan Mosque and its minaret. (**f**) The site plan of the Kalealtı Mosque and its minaret. (**g**) Plan of Gökmeydan Mosque and Minaret.

Figure 2. Visuals of Five Minarets: (**1**) Ulu Mosque Minaret, (**2**) Şerefiye Mosque Minaret, (**3**) Meydan Mosque Minaret, (**4**) Kalealtı Mosque Minaret, and (**5**) Gökmeydan Mosque Minaret.

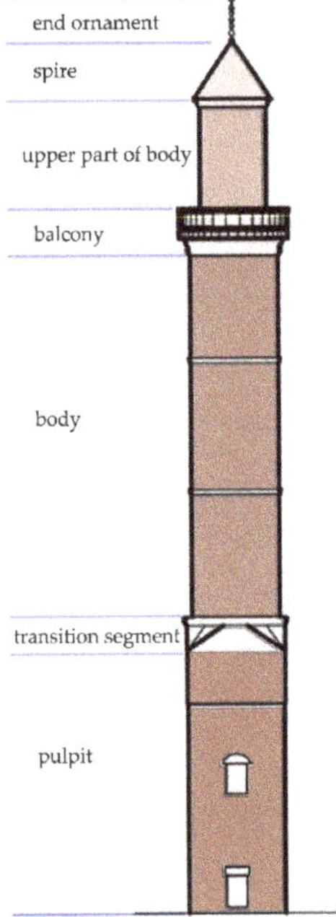

Figure 3. Parts of the minaret of the Ulu Mosque.

2.2. Ulu Mosque and Minaret

Although the exact date of construction of the Ulu Mosque in Gazibey district is not known, it is stated that the structure was one of the mosques destroyed by Byzantium in 928, as Şen reported from İbn'ül Esir in 2018. This means that the mosque was built in the city in the 7th or 8th century after the first Islamic conquests [44]. The mosque, together with the Bitlis Castle, is the oldest building in the city center of Bitlis.

There are three inscriptions on the construction/repair of the mosque, which are the most important historical documents among the immovable cultural assets. These are located on the upper part of the middle and western doors at the entrance to the sanctuary from the courtyard and the upper part of the minaret pulpit (boot). The inscriptions belong to the renovations and restorations of the mosque. The inscription dated 1150 on the middle door is an inscription belonging to the renovation of the building and corresponds to the reign of Dilmaçoğulları. The other renovation inscription dated 1651 corresponds to the period of the Serefhans under the rule of the Ottoman Empire. Ulu Mosque, which is one of the first examples of a rectangular horizontal plan mosque, also sets an example for the plan development of the Artuqid period [45–48].

The minaret is at the northwest part of the courtyard with the portico, which was unearthed as a result of the restoration works in 2012. There is an inscription on the construction of the minaret, dated 1492–1493, on the entrance door in the eastern direction [44,48]. Its construction date coincides with the period of Serefhans. The upper parts of the minaret, which has a square prism-shaped pulpit, were chamfered and a circular body was passed with a two-stage bracelet. The body is divided into three parts by two cornices. There is an inscription on the south face of the middle one of these partitions, which does not contain any ornamental elements. The transition to the balcony was made with a bracelet. The wall of the balcony is plain. The upper part and spire of the minaret were destroyed by a lightning strike, and these parts were repaired in an inaccurate manner. There is a door opening to the south in the plain upper part section of the minaret with a cylindrical body. The spire section was renovated in the form of an octagonal prism.

2.3. Gökmeydan Mosque and Minaret

It is written on the inscription on the east side of the Gökmeydan Mosque, which is located in the Taş District, that it was built in 1801. The southern part of the mosque, which was built on sloping topography, was designed as two floors. The building, in which the original function of the lower floor is used as a madrasa zawiya or khalwa and the upper floor as a place of worship, has a rectangular plan in the north–south direction.

The minaret was built to the west of the mosque, near the southwest corner. There is an inscription that also mentions Atatürk, dated 1924. It is the minaret with the most ornaments among the existing minarets. The square prism pulpit is entered through a door with a relieving arch in the southern part. The pulpit is divided into two with a cornice on the door. On the lower part of the south face, there is a circular inscription with floral motifs. Just above this, there is a relief decoration with lozenges. The part of the pulpit that connects to the bracelet on the body is built in the form of an inverted pendentive. The part of the pedestal that connects to the bracelet on the body is built in the form of an inverted pendentive. On the body part, an ornamental arrangement was preferred with motifs in the form of rosette-drop-rosette-triangle in four rows from bottom to top. In the lower part of the balcony, there is a row of ornaments made of white limestone in the form of lozenges, two cornices with ornaments in between, and a similar ornamentation series with a lozenge embossed from Bitlis stone. It has a metal railing balcony. The upper parts of the minaret section are octagonal prism, and there is a passage to the minaret on its four faces. There are four passages on four faces. The decorated eaves are covered with a stone dome [44,46].

2.4. Meydan(Çarşı) Mosque and Minaret

Meydan Mosque was destroyed in the 1915 Russian occupation, and its minaret stood alone in the bazaar square to the south of Bitlis Castle until the beginning of the 2000s. Reconstruction studies were carried out by the relevant institutions in 2003 [46,49]. There is an entrance door on the east face of the square prism pulpit of the minaret, which was built adjacent to the northwest part of the mosque. Passing from square prism to octagonal transition segment is provided with triangulations. There are eight blind arches in the transition segment. In the transition to the cylindrical body with blind arches, it is passed with a double cornice section. The body is divided into two with a double cornice. There is no ornament on the body, which has a window opening to the east in the lower part. The balcony wall was built with stone blocks. The upper parts of the minaret section with a door opening to the southwest are plain. The minaret, which is covered with an octagonal pyramid spire, has a crescent-tipped end ornament. This minaret resembles the minaret of the Şerefiye Mosque without ornaments.

2.5. Kalealtı (Aşağı Kale) Mosque and Minaret

Kalealtı (Aşağı kale) Mosque is located in the southwest of Bitlis Castle, next to the Kömüs stream. There is no inscription about its construction or repair in the mosque, which can also be classified as a neighborhood mosque. Arık (1971) [44] states that it is a 17,18th century structure typologically. The northwest corner and southeast part of this historical building, which has a square plan, were beveled due to the building it was built next to and the road. A lower section constructed at a different period than the main building was added to the entrance door of the sanctuary in the eastern part of the mosque [44]. The minaret of the mosque is not visible from time to time in old photographs. Today, in the southern part of the annex building, a minaret has a square prism-shaped pulpit built adjacent to the sanctuary wall. The minaret, which does not contain any ornamental elements, is transitioned from a cylindrical body to a stone-walled balcony with a five-stepped stone row. The cylindrical upper part of the minaret has a door opening to the south. A crescent-tipped end ornament is placed on the spire wrapped by lead sheets, which is uncommon in Bitlis [44,46].

2.6. Şerefiye Külliye and Minaret

The Şerefiye Külliye (complex of buildings) was built where the Kömüs (Hüsrev) and Rabat (Sapkor Suyu) Streams merged to form the Bitlis Stream. The mosque, minaret, imaret (the public soup kitchen), cupola, madrasah, and arasta (bazaar) were built on the west side of the streams, and the hammam was built on the east side. The connection between the Külliye is provided by the Şerefiye I Bridge on the Kömüs Stream and the Şerefiye II Bridge on the Rabat Stream. There is a mosque in the south of Külliye, the entrance gap, a cupola in the southeast, an imaret in the north, and a minaret in the northeast. In the eastern part of the külliye, there is a courtyard where the Kömüs stream is located. This section is entered through a portal opened from the south. There is an inscription on its portal states that it was built by Şeref Beg in 1528–1529. The door opens to a gap and there is the south window of the cupola. The left side opens to the portico of the mosque. Şerefiye Mosque is the only mosque in the city center with a portico place. The cupola, which was built by Emir Şemseddin in 1533, is the continuation of this section. There is a burial ground in the section between the cupola and the minaret. The imaret in the northern section was built at the same time as the mosque, and the northeast section was destroyed because of the overflow of the Kömüs stream. At the end of the imaret top cover of the square prism pulpit, there is an ornament element between the double cornices. In the south part, a panel with the inscription basmala in Kufic calligraphy was placed under the cornice. There is the entrance door of the minaret in the western part. Passing from square prism to octagonal transition segment is provided with triangulations. There are eight blind arches in the transition segment and, on the walls on the inner surface of these, panels with floral ornaments and the Kufic inscriptions are placed. In the transition to the cylindrical body with blind arches, it is passed with Kufic calligraphy inscribed ar the double cornice section. The body is divided into two by a plain double cornice. There are drop motif decorations in the lower part. The decorations in the section that passes to the balcony are made of limestone. The metal railing balcony was rebuilt with stone blocks during recent restoration works. The cylindrical upper part of the minaret section with a door opening to the southwest is plain. The minaret, which is covered with an octagonal pyramid spire, has an end ornament with a single-stage metal knob [44].

3. Results

3.1. The Observed Damages in the Minarets

Observational analyzes were made to reveal the current structural situation for each minaret. With these analyzes, information is given about the damages that occur in the minarets today. In short, before proceeding to the detailed structural analysis processes, structural analyses based on observation were made in order to provide information about the current status of Five Minarets. The purpose of this analysis is to reveal the current state of the building as well as to provide information about the damage and deformations that occur in the structure. Minarets have survived to the present day with protection and interventions by the relevant public institutions and organizations. However, partial damage occurred due to the high-temperature differences in the city and excessive precipitation. Bitlis Stone, with its local name, was used in all of the minarets. The stones obtained from the quarries in Bitlis were used in minarets after being subjected to the cutting process. One of the general features of Bitlis stone is discoloration and color change over time with the effects of natural conditions. Discoloration and color change were partially observed in almost all minarets studied. The Bitlis city, where the minarets are located, is one of the city centers with the highest snowfall in Turkey. Some of the minarets, which were examined due to excessive precipitation, have calcifications in places due to water effects. Some of the stones used in the building have occasional rupture and damage. It has been observed that there are almost no consolidation effects over time since the soil properties are good in the investigated structures. Although vegetative formations were noticeable in different parts of the minarets, they did not cause great damage to the minarets. It is seen that maintenance and restoration were conducted over time in all Five Minarets. Conducting such maintenance and restoration over time is an important step towards prolonging the life of the buildings and preserving the originality of the buildings. In addition, the bullet traces of the occupation forces during the years of occupation are also clearly visible. Neither random renovation nor restoration were made to the buildings. The images of the damages and deformations observed in Five Minarets are shown in Figure 4.

Figure 4. Damages observed in Five Minarets: a. high temperature difference; b. discoloration and color change; c. water effects; d. rupture damage; e. vegetation and f. bullet traces.

3.2. The Determination of the Bitlis Stone Properties Used in Minarets

It has been considered as a single type of the material, which is locally called Bitlis Stone, and is used in all the minarets considered in the study. Modulus of elasticity (E), Poisson ratio, and the weight per unit volume (γ) values for Bitlis stone was taken as a single value for all minarets. While determining these values, the unit volume and the specific weight values were taken directly from the study by Işık et al. (2020) [50]. For the other two properties, the results were obtained by using the nondestructive test method. The propagation changes of the ultrasonic pulse velocity (UPV) wave are analyzed and applied without causing any deterioration in the material in the UPV method. This method, which enables the investigation of material homogeneity, can be considered an important method in the evaluation of concrete or natural stone structures [51]. In this method, an idea of the strength of the specimen is obtained based on the propagation speed of ultrasonic sound waves at certain frequencies in the specimen. Sound waves give an idea of cracks in the sample. An ultrasonic pulse is applied to one side of the sample with an ultrasonic pulse velocity tester, and pressure waves (P waves) are generated and recorded from the other side of the specimen. The ultrasonic pulse velocity tester measures the time taken by the pulse to proceed through the specimen. UPV equipment consists of a receiver, a transmitter, and a digital display [52]. The propagation times of the waves read from the device display were divided by the size of the specimen, and the propagation rates were determined for each sample. UPV test was applied to cube specimens of 15 cm × 15 cm × 15 cm. The results obtained for the specimens are shown in Table 1.

Table 1. Test results of Bitlis stone.

Specimen Number	UPV (m/s)		V_s (m/s)		Poisson		Modulus of Elasticity	
	Direction 2	Direction 3	Direction 2	Direction 3	Direction 2	Direction 3	kN/m^2	
1	1574.5 ± 93.9	1804.5 ± 57.5	1032.9 ± 62.33	1185.557 ± 38.17	0.214295 ± 0.002	0.2085 ± 0.001	3755487 ± 4.77	4924239 ± 4.13
2	1751.0 ± 22.24	1805.5 ± 13.6	1150.4 ± 14.1	1184.95 ± 8.63	0.208497 ± 0.001	0.2125 ± 0.001	4604603 ± 1.21	4901648 ± 1.05
3	1543.25 ± 58.55	1687.5 ± 36.06	1009.3 ± 39.2	1105.313 ± 24	0.226122 ± 0.002	0.2202 ± 0.001	3570909 ± 2.82	4261941 ± 2.44
4	1340.0 ± 44.6	1449.25 ± 27.31	875.02 ± 29.72	947.8095 ± 18.19	0.232153 ± 0.002	0.2261 ± 0.001	2791537 ± 1.91	3259257 ± 1.65
5	1573.75 ± 41.84	1676.25 ± 25.62	1024.5 ± 27.92	1092.915 ± 17.1	0.244454 ± 0.002	0.2383 ± 0.001	3786705 ± 2.05	4287804 ± 1.77
6	1608.0 ± 6.94	1625.00 ± 4.25	1058.1 ± 5.23	1070.875 ± 3.2	0.202775 ± 0.002	0.1971 ± 0.001	3930450 ± 3.14	4007304 ± 2.72

According to the fracture test performed on cube specimens of 15 × 15 × 15 cm dimensions taken from natural stones known as Bitlis ignimbrites and used as building stones and the elastic property examinations obtained separately in two directions, the following results were obtained:

- Specimen 1 has the highest fracture load (179.5 kN). The weight per unit volume was calculated as 14.2 kN/m^3, average modulus of elasticity 4.34 × 10^6 (kN/m^2), average Poisson ratio 0.21, and average shear modulus 1.79 × 10^6 (kN/m^2);
- The fracture load of specimen 2 was obtained as 163.8 kN. The weight per unit volume was calculated as 14.12 kN/m^3, average modulus of elasticity 4.75 × 10^6 (kN/m^2), average Poisson ratio 0.21, and average shear modulus 1.96 × 10^6 (kN/ m^2);
- The fracture load of specimen 3 was obtained as 146.4 kN. The weight per unit volume was calculated as 14.02 kN/m^3, average modulus of elasticity 3.92 × 10^6 (kN/m^2), average Poisson ratio 0.22, and average shear modulus 1.60 × 10^6 (kN/m^2);
- The fracture load of specimen 4 was determined as 153.2 kN. The weight per unit volume was calculated as 14.51 kN/m^3, average modulus of elasticity 3.03 × 10^6 (kN/m^2), average Poisson ratio 0.23, and average shear modulus 1.23 × 10^6 (kN/m^2);
- The fracture load of specimen 5 was obtained as 153.2 kN. The weight per unit volume was calculated as 14.22 kN/m^3, average elastic modulus 4.04 × 10^6 (kN/m^2), average Poisson ratio 0.24, and average shear modulus 1.63 × 10^6 (kN/m^2);
- The fracture load of specimen 6 was obtained as 153.2 kN. The unit volume weight was calculated as 14.32 kN/m^3, average modulus of elasticity 3.97 × 10^6 (kN/m^2), average Poisson ratio 0.20, and average shear modulus 1.65 × 10^6 (kN/m^2).

Ignimbrites may be of different compositions due to their formation and may change over short distances when an evaluation is made about the specimens in general. The weight per unit volume of the six specimens examined is very close to each other and does not show great compositional differences. Porosity values are also in the range of 25%–26%, and it is understood that they do not have a significant difference in terms of their formation. However, strength properties show significant changes. Measurements of the specimens taken from both directions showed that there were different strength values in both directions. Although the specimens have similar physical properties, the main reason why they have different values in strength properties is considered to be meteorological conditions. Particularly, the small amount of cracking in the direction where the specimen is exposed to the natural environment may cause a decrease in material strength. However, in general terms, it has sufficient strength conditions as a building block. Using the average values of these results, the material properties that are the basis for structural analysis are shown in Table 2.

Table 2. Material properties considered for Bitlis stone used in minarets.

Material Type	Modulus of Elasticity (kN/m^2)	Specific Weight (kN/m^3)	Weight per Unit Volume (t/m^3)	Poisson Ratio
Bitlis Stone	4006824	20	1.46	0.22

Modelling of masonry walls is extremely important in the evaluation and design of historical and modern masonry structures. Masonry walls can be modelled using three different modelling techniques such as detailed micro-modelling, simplified micro-modelling, and macro-modelling. These models can be seen in Figure 5.

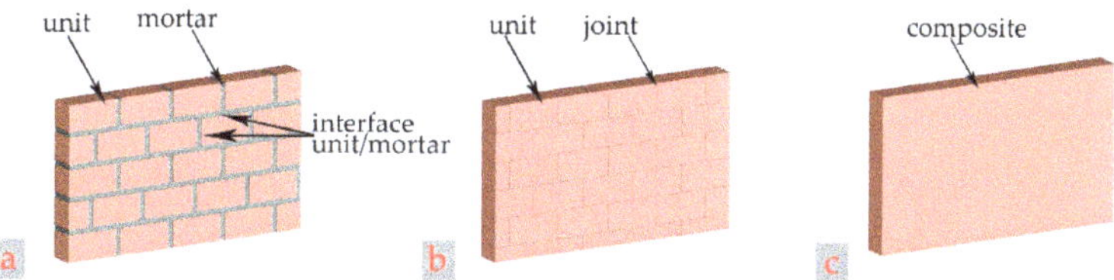

Figure 5. Modelling methods of masonry: (**a**) detailed micro modelling; (**b**) simplified micro modelling; and (**c**) Macro modelling.

In detailed micro-modelling, the mechanical properties of the masonry unit and the mortar forming the masonry wall are taken separately. In this approach, it is assumed that cracks will occur at the interfaces between the masonry unit and the mortar. In simplified micro-modelling, mass densification was made at each connection point consisting of a mortar and two masonry unit–mortar interfaces, and it was accepted that cracks that could occur in masonry could occur at the mean interface line, assuming the average interface. Findings differ slightly from detailed micro-modelling, as the Poisson ratio of the mortar is not taken into account here. However, this difference is so small that it can be neglected [53–55].

The macro modeling technique is one of the masonry structural modeling techniques and is widely used. While performing this type of modeling, analyses are carried out without making any distinction between the binding material (mortar, etc.) used in the building and the structural elements. In this modelling, the masonry unit and the properties of the mortar are homogenized and considered as a masonry composite material. The mechanical properties of this model are the values obtained as a result of the homogenization process. Macro modelling is more convenient in practice because it requires less memory and time. However, with macro modelling, stress distributions in masonry units and mortar can be obtained accurately [55–58] In this respect, structural masonry elements are considered composites, and an equivalent material model is used for all minarets models. Structural analyses were carried out for the Five Minarets in Bitlis using this macro-modeling technique. The mortar and Bitlis stone used in the minarets were considered as a single material. The sign criterion for the stress components of the elements used in the finite element model of the structure is shown in Figure 6, in accordance with the assumptions stipulated by the software [59] in which numerical modeling is made.

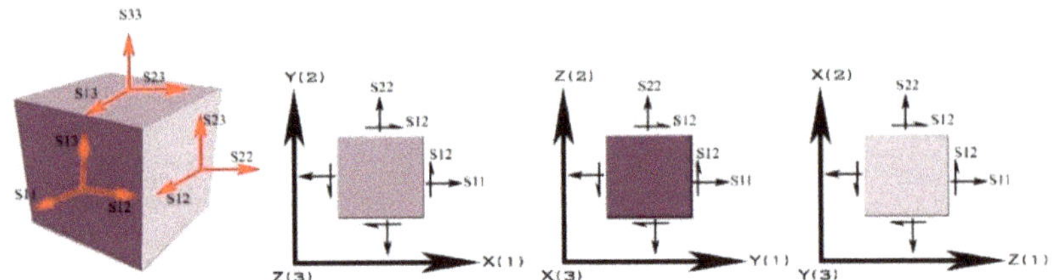

Figure 6. The sign criterion for the stress components.

As stated in Figure 6, S11 is vertical stress in (x) direction, S22 is vertical stress in (y) direction, S33 is vertical stress in (z) direction, and S12 = S21 constitute shear stresses in the x-y plane. With TBEC-2018, which was updated in 2018 and entered into force on 1 January 2019, the biggest change was the use of site-specific design spectra. Turkey Earthquake Maps Interactive Web Application has been developed to calculate design spectra and site-specific earthquake parameters. With the help of this application, horizontal and vertical design spectra can be obtained as well as earthquake parameters belonging to any desired geographical location. By using the coordinate values obtained for each minaret, design spectra and earthquake parameters were obtained with the help of this application. While obtaining these values, the design ground motion level DD-2 was chosen as the earthquake ground motion level. From the ground survey reports received from the relevant institutions, the ZB soil class was taken into account for all five minarets as the local soil class. As it can be seen from Table 3, the design spectra were obtained close to each other since the seismic parameters for the minarets are close to each other. The horizontal and vertical spectra obtained for the Ulu Mosque Minaret are shown in Figure 7 as an example. In these curves, the horizontal axis represents period values, while the vertical axes represent the horizontal and vertical elastic design spectral accelerations, respectively.

Table 3. The seismic parameters obtained for Five Minarets.

Parameter	Ulu Mosque	Gökmeydan Mosque	Meydan Mosque	Kalealtı Mosque	Şerefiye Mosque
Local soil classes	ZB	ZB	ZB	ZB	ZB
Short period map spectral acceleration coefficient (S_S)	0.614	0.614	0.614	0.614	0.614
Map spectral acceleration coefficient for a 1.0 s period (S_1)	0.172	0.172	0.172	0.172	0.172
Peak ground acceleration (PGA) (g)	0.260	0.260	0.260	0.260	0.260
Peak ground velocity (PGV) (cm/sn)	15.081	15.123	15.082	15.084	15.079
Local soil effect coefficient for the short period region (F_S)	0.900	0.900	0.900	0.900	0.900
Local soil effect coefficient for 1.0 s period (F_1)	0.800	0.800	0.800	0.800	0.800
Short period design spectral acceleration coefficient (unitless) (S_{DS})	0.553	0.553	0.553	0.553	0.553
Design spectral acceleration coefficient for a 1.0 s period (unitless) (S_{D1})	0.138	0.138	0.138	0.138	0.138
T_A	0.050	0.050	0.050	0.050	0.050
T_B	0.249	0.249	0.249	0.249	0.249
T_{AD}	0.017	0.017	0.017	0.017	0.017
T_{BD}	0.083	0.083	0.083	0.083	0.083

The earthquake parameters obtained for each minaret and used in structural analysis with the help of the Interactive Web Earthquake application are shown in Table 3.

While the dimensions of Five Minarets are shown in Figure 8, the three-dimensional models obtained from the software program are shown in Figure 9.

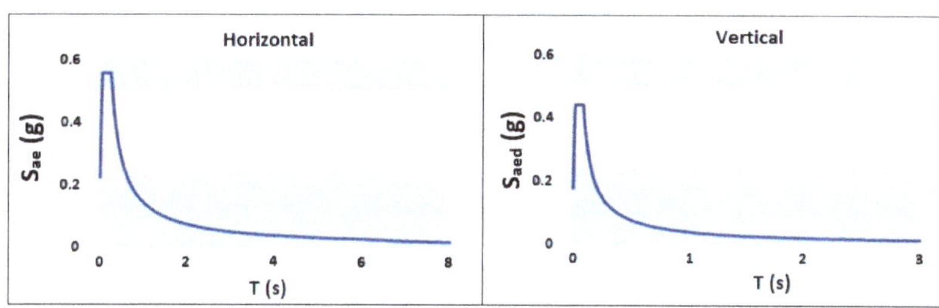

Figure 7. Horizontal and vertical design spectra obtained for the minaret of the Ulu Mosque.

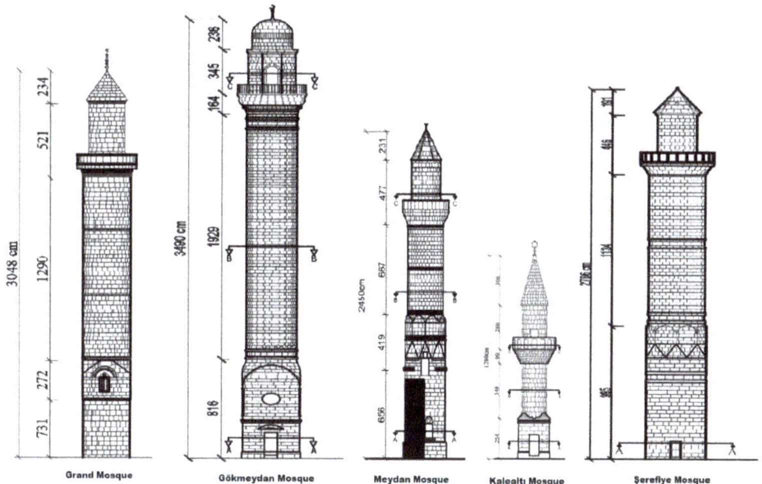

Figure 8. Dimensions of Five Minarets.

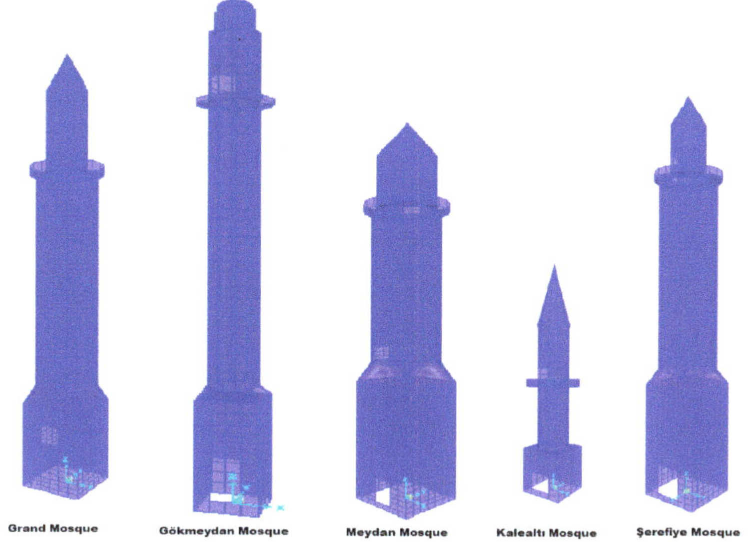

Figure 9. Three-dimensional models obtained from the software program of Five Minarets.

In the structural analysis of the minarets, the finite element method, where the cross-section and material properties can be easily defined, was used. In finite element analysis, the geometry of the structure or structural elements is determined by a finite number of nodal points. The general structural properties of the minarets, the number of nodes, and the number of shell elements modelled in the software are shown in Table 4.

Table 4. Characteristics and structural model properties of Five Minarets.

Minaret		Ulu Mosque	Şerefiye Mosque	Meydan Mosque	Kalealtı Mosque	Gökmeydan Mosque
Date of Construction		1492/93	1533	17th century	17–18th century	1924
Material	Stone	√	√	√	√	√
	Brick					
	Earth					
Balcony	Single	√	√	√	√	√
	Double					
Height (cm)		3048	2766	2450	1280	3490
Location		Northwest of courtyard (at outside)	Northeast of courtyard (at inside)	Northwest of mosque next to portico	East of mosque at annex	Southwest of mosque
Footing Dimensions (m)		3.10 × 3.10	3.0 × 3.0	3.75 × 3.75	2.15 × 2.15	2.90 × 2.90
Body Diameter (m)		3.1	2.7	3.1	1.8	2.6
Body wall thickness (m)		0.6	0.45	0.5	0.2	0.35
Height (m)		30.48	27.66	24.5	12.8	34.9
Number of nodes		6808	7096	7220	1558	2648
Number of Shell element		1710	1781	1812	393	666

Modal analysis is a dynamic analysis method that enables the determination of free vibration periods, frequency values, mass participation rates, and mode shapes of the structure. In order to determine the dynamic properties of the minarets, primarily modal analyzes were carried out. In TBEC-2018, it was requested to be determined according to the rule that the mass participation rates in the X and Y directions should not be less than 95%. In this case, modal analyses were carried out by considering the first 34 modes for the Ulu Mosque minaret; the first 24 modes for Gökmeydan mosque minaret; the first 44 modes for the Meydan Mosque minaret; the first 31 modes for the Kalealtı mosque minaret; and the first 43 modes for the Şerefiye Mosque minaret. The values related to the mass participation rates, natural vibration periods, and effective modes obtained by considering the first five modes of Five Minarets as a result of the modal analysis are shown in Table 5. Torsion in all minarets occurred in the fifth mode.

The dead-load, live-load, and earthquake-load are taken into account for stress calculations. The software program according to material properties made dead load calculations. Horizontal and vertical elastic design spectra obtained from Turkey Earthquake Hazard Maps Interactive Web Applications were used as design spectra. For the earthquake load, load definition was made in three directions as EQx, EQy, and EQz. Structural analyses were performed for different load combinations by using these values. The load combinations envisaged in the TBEC-2018, which is currently used in Turkey, have been taken into account. Load combinations have been selected in accordance with the definition under the title of Combining Earthquake Effect with other Effects in TBEC-2018. Load combinations are defined by the constant load effect, live load effect, earthquake effects defined in perpendicular directions, and the vertical earthquake effect together with the load coefficients. The stress diagrams for S11 (vertical stress in the x-direction) obtained from different load combinations are shown in Figure 10.

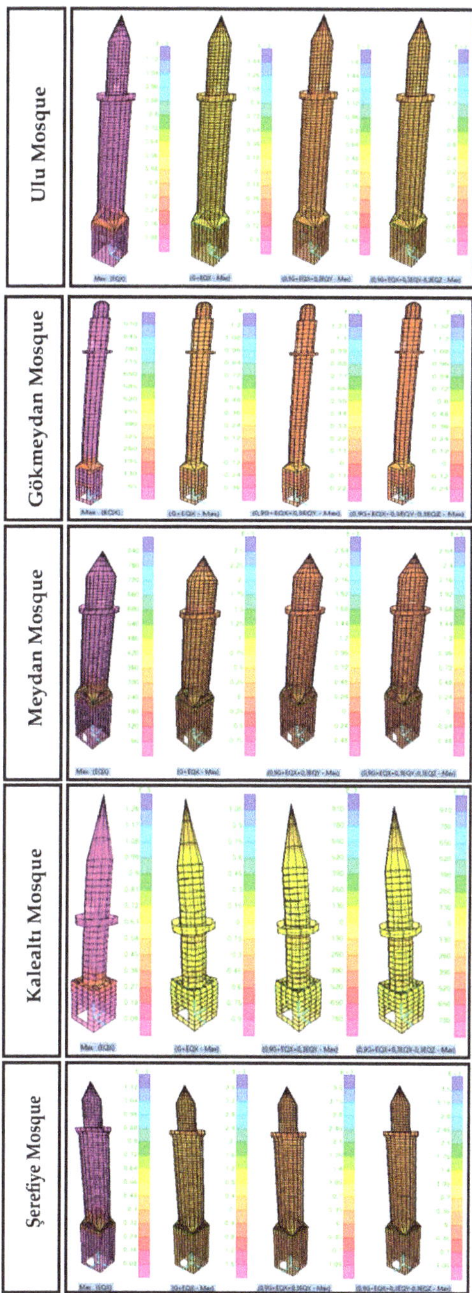

Figure 10. The maximum stress for Five Minarets (S11).

The maximum stress diagrams for the vertical stress in the y-direction (S22) obtained from different load combinations are shown in Figure 11.

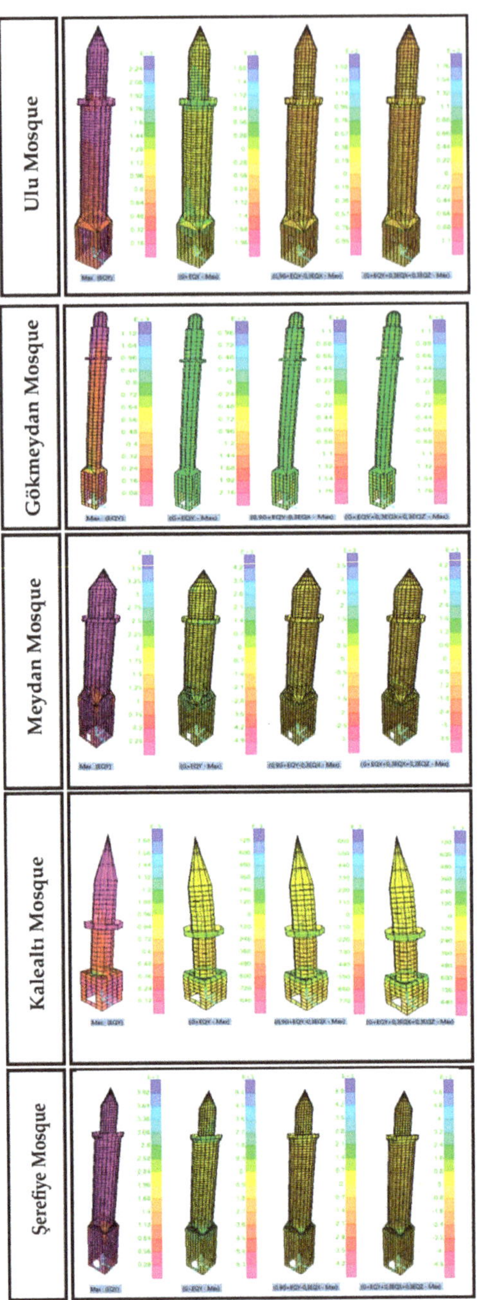

Figure 11. The maximum stress for Five Minarets (S22).

The diagrams of shear stress in the x-y directions (S12) for five minarets under different load combinations are shown in Figure 12.

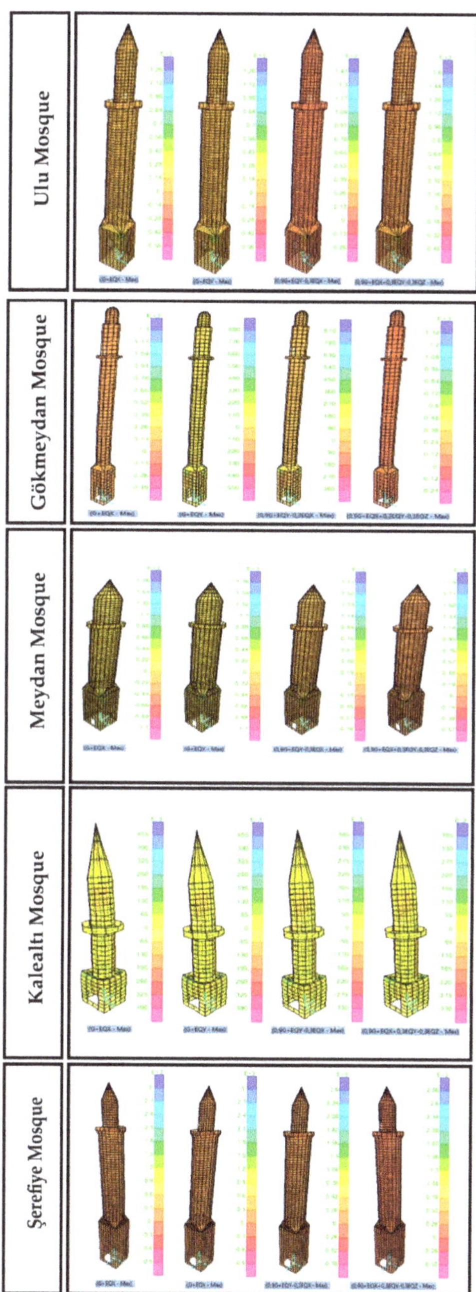

Figure 12. Maximum shear stress for Five Minarets (S12).

The comparison of the maximum tensile stresses for Five Minarets according to the results of the structural analysis is given in Table 6.

Table 5. Modal analysis results of Five Minarets.

Minaret	Mode	Period (s)	UX (m)	UY (m)	$\sum$UX (%)	$\sum$UY (%)	RX (m)	RY(m)	$\sum$RX (%)	$\sum$RY (%)
Ulu	1	0.7104	0.01394	0.55769	1.39	55.77	0.4139	0.01027	41.39	1.03
	2	0.7077	0.55858	0.01379	57.25	57.15	0.0104	0.41305	42.43	42.33
	3	0.1450	0.07872	0.09275	65.13	66.42	0.0648	0.05621	48.91	47.95
	4	0.1423	0.10062	0.07766	75.19	74.19	0.0555	0.07397	54.46	55.35
	5	0.0991	3.88×10^{-6}	1.82×10^{-5}	75.19	74.19	1.8×10^{-5}	2.86×10^{-7}	54.46	55.35
Gökmeydan	1	1.7760	0.5089	6.00×10^{-7}	50.89	0.00	4.7×10^{-7}	0.49011	0.00	49.01
	2	1.7410	4.89×10^{-7}	0.51229	50.89	51.23	0.4872	5.05×10^{-7}	48.72	49.01
	3	0.3126	0.00581	0.21057	51.47	72.29	0.0925	0.00258	57.97	49.27
	4	0.3103	0.22133	0.00575	73.60	72.86	0.0026	0.1001	58.23	59.28
	5	0.1486	0.04561	0.00032	78.16	72.89	0.0003	0.05404	58.25	64.68
Meydan	1	0.6206	0.01356	0.53314	1.36	53.31	0.4317	0.01086	43.17	1.09
	2	0.6135	0.53401	0.01338	54.76	54.65	0.011	0.43069	44.27	44.16
	3	0.1201	0.2624	0.00205	81.00	54.86	0.0014	0.17858	44.40	62.01
	4	0.1186	0.00214	0.2621	81.21	81.07	0.1711	0.00139	61.51	62.15
	5	0.1054	0.00829	3.85×10^{-5}	82.04	81.07	2.4×10^{-5}	0.00433	61.52	62.58
Kalealtı	1	0.2861	0.00246	0.49748	0.25	49.75	0.4953	0.00252	49.53	0.25
	2	0.2807	0.48622	0.00251	48.87	50.00	0.0025	0.49927	49.78	50.18
	3	0.0658	0.24789	6.76×10^{-6}	73.66	50.00	8.0×10^{-7}	0.08725	49.78	58.90
	4	0.0603	3.99×10^{-6}	0.22776	73.66	72.78	0.0926	1.37×10^{-7}	59.04	58.90
	5	0.0490	4.80×10^{-5}	0.00033	73.66	72.81	0.0003	0.00024	59.07	58.93
Şerefiye	1	0.7953	0.53912	0.01179	53.91	1.18	0.0096	0.43307	0.96	43.31
	2	0.7824	0.01191	0.53782	55.10	54.96	0.4332	0.0095	44.28	44.26
	3	0.1449	0.00015	0.26134	55.12	81.10	0.1848	9.16×10^{-5}	62.76	44.27
	4	0.1400	0.24586	0.0002	79.70	81.12	0.0002	0.16751	62.78	61.02
	5	0.1160	8.23×10^{-6}	0.00301	79.71	81.42	0.0007	6.39×10^{-6}	62.85	61.02

Table 6. Comparison of the maximum tensile stresses obtained for Five Minarets.

Minaret	Load Combination	S11 (MPa)	S22 (MPa)
Ulu Mosque	0.9G + EQX + 0.3EQY − 0.3EQZ	1.992	1.992
Gökmeydan Mosque	G + EQX	0.928	1.286
Meydan Mosque	0.9G + EQY − 0.3EQX	2.817	4.229
Kalealtı Mosque	1.4G	1.467	1.102
Şerefiye Mosque	G + EQY + 0.3EQX + 0.3EQZ	3.575	5.761

The comparison of the maximum compressive stresses according to the analysis results for Five Minarets is given in Table 7.

Table 7. Maximum compressive stresses obtained for Five Minarets.

Minaret	Load Combination	S11 (MPa)	S22 (MPa)
Ulu Mosque	0.9G + EQX + 0.3EQY − 0.3EQZ	2.953	8.859
Gökmeydan Mosque	G + EQX	2.268	4.901
Meydan Mosque	0.9G + EQY − 0.3EQX	2.517	12.099
Kalealtı Mosque	1.4G	0.967	1.280
Şerefiye Mosque	G + EQY + 0.3EQX + 0.3EQZ	4.062	14.460

The comparison of the maximum shear stress values obtained from the structural analyzes for Five Minarets is given in Table 8.

Table 8. Comparison of the maximum shear stresses obtained for Five Minarets.

Minaret	Load Combination	S12 (MPa)
Ulu Mosque	0.9G + EQX + 0.3EQY − 0.3EQZ	1.652
Gökmeydan Mosque	G + EQX	1.327
Meydan Mosque	G + EQY + 0.3EQX + 0.3EQZ	1.793
Kalealtı Mosque	1.4G	0.507
Şerefiye Mosque	G + EQY + 0.3EQX + 0.3EQZ	3.034

According to the analysis results of Five Minarets, the maximum displacement of connection elements in both negative and positive Ux (U_1) directions is given in Table 9.

Table 9. Maximum displacements of Five Minarets in Ux direction.

Minaret	Load Combination	Type	U_1 (mm)
Ulu Mosque	0.9G + EQX + 0.3EQY − 0.3EQZ	Negative	17.66536
Gökmeydan Mosque	G + EQX	Positive	44.45398
Meydan Mosque	0.9G + EQX + 0.3EQY − 0.3EQZ	Positive	14.28112
Kalealtı Mosque	G + EQX	Negative	0.13602
Şerefiye Mosque	0.9G + EQX + 0.3EQY − 0.3EQZ	Positive	19.88931

According to the analysis results of Five Minarets, the maximum displacement of connection elements in both negative and positive Uy (U_2) directions is given in Table 10.

Table 10. Maximum displacements of Five Minarets in U_y direction.

Minaret	Load Combination	Type	U_2 (mm)
Ulu Mosque	G + EQY + 0.3EQX + 0.3EQZ	Negative	−17.93548
Gökmeydan Mosque	G + EQY + 0.3EQX + 0.3EQZ	Negative	−40.81252
Meydan Mosque	G + EQY + 0.3EQX + 0.3EQZ	Negative	−14.92157
Kalealtı Mosque	G + EQY + 0.3EQX + 0.3EQZ	Negative	−0.23223
Şerefiye Mosque	G + EQY + 0.3EQX + 0.3EQZ	Positive	19.13056

According to the analysis results of Five Minarets, the maximum displacement of connection elements in both negative and positive Uz (U_3) directions is given in Table 11.

Table 11. Maximum displacements of Five Minarets in U_Z direction.

Minaret	Load Combination	Type	U_3 (mm)
Ulu Mosque	0.9G + EQX+0.3EQY − 0.3EQZ	Negative	−3.72910
Gökmeydan Mosque	G + EQY + 0.3EQX + 0.3EQZ	Negative	−6.17395
Meydan Mosque	G + EQY + 0.3EQX + 0.3EQZ	Negative	−4.44064
Kalealtı Mosque	G + EQY + 0.3EQX + 0.3EQZ	Negative	−0.39179
Şerefiye Mosque	G + EQY + 0.3EQX + 0.3EQZ	Negative	−5.48402

The highest period value was obtained for the Gökmeydan mosque minaret, which is the highest minaret, while the lowest period value was obtained for the Kalealtı mosque minaret. The variation of height directly affected the period values. The highest tensile stress values were obtained for the minaret of the Şerefiye Mosque, while the lowest values were obtained for the minaret of the Gökmeydan Mosque. The differences in the structural dimensions affected the tensile stresses. The highest values in terms of compressive stresses were obtained for the minaret of the Şerefiye Mosque, while the lowest values were obtained for the minaret of the Kalealtı Mosque. The highest shear stresses occurred in the minaret of the Şerefiye mosque, while the lowest shear stresses occurred in the minaret of the Kalealtı mosque. The largest displacements were obtained for the Gökmeydan mosque minaret, which is the highest minaret, while the smallest displacement values were obtained for the Kalealtı mosque minaret, which is the lowest one. All values obtained are considerably smaller than the minimum compression, tensile, and shear stress values in TS EN 1467 [60], which is used for natural stones and raw blocks in Turkey and includes natural stone properties. Accordingly, the minimum safe compressive stress is 34 MPa [60]. Therefore, the values found in all of the minarets were obtained below this value. In addition, the minimum tensile strength in bending for blasted stones can be taken as 8 MPa [61] For stone walls built in masonry, the safe shear stress (τ_s) can be calculated by Equation (1).

$$\tau_s = 0.10 + 0.5\,\sigma \qquad (1)$$

Here, σ indicates the compressive strength of the material. The compressive strength value, which was calculated as 34 MPa above, was substituted in Equation (1), and the safe shear stress value was calculated as 17 MPa. The maximum compressive, shear, and tensile stresses obtained from the analyses show thag these stresses can be safe to be carried by the structure. This result is in accordance with the fact that the structure survived in the process.

4. Conclusions

In the Bitlis stones, which formed the minarets under the influence of natural conditions, partial mass loss, rupture, and wear were observed over time. It has been surveyed that the vegetative cell formations formed on the minarets over time damaged the stones and the joining elements that formed the minaret. In some minarets, traces of moisture were observed, partly due to the harmful effects of water and the stone's characteristics. Many lead traces are still clearly observed on the minaret of the Ulu Mosque. Authorized institutions are proceeding to fix these damages and protect the structure from further damage in the future. Therefore, it was not possible to make a comparison between the damage conditions on the minarets and the results obtained.

In this study, structural analyses for Five Minarets, which are the significant cultural heritage of Bitlis province and have been the subject of songs, were carried out using observational and finite element methods. The fact that these minarets are exposed to many adverse effects over time makes these minarets' construction technologies and earthquake behavior more critical. Within the scope of this study, the seismic behavior of the minarets was determined by using the design spectrum given in the Turkish Building Earthquake Code (TBEC-2018), taking into account the different loading conditions for all five minarets. For this purpose, the mechanical properties of Bitlis stone used in minarets were determined for the first time within the scope of this thesis by using non-destructive test methods. Structural analysis of the minarets was carried out using the obtained values using the macro modeling method. S11, S22, and S12 stresses were founded for different loading cases of each minaret. Periods and mode shapes and data about minarets are given. The causes and results of the damage and destruction caused by the observational examinations made in the field are presented. Mainly, the maximum stress occurred in the transition zones of the minarets. It has been determined that the minaret can carry on these stresses. It expressed the engineering knowledge and experience when the minarets were built. In minarets, the effect of the cylindrical body and upper part of the minaret in the first mode is more significant than the pulpit. The common elements in the transition zones where cross-sectional changes occur in the minarets can be expressed as risky places. The fact that the elements that formed the pulpits have more rigidity indicates that the degree of damage will remain at lower values in this region. In this case, transition zones in minarets can be expressed as risky zones.

Bitlis stone, which is used in all of the minarets, is weak strength and has high porosity ratio, as well as the high-temperature difference in the city, the high and long-term snowfall causes fragmentation and partial rupture of these stones. In a city where the winter season is long, the freezing–thawing factor is one of the important reasons for decreasing the strength of the stone. Therefore, Bitlis stone, which is the main structural element of minarets, may lose its mechanical properties over time and affect its strength. In order to solve this problem, institutions/organizations related to minarets should observe the minarets structurally and ensure that the necessary engineering interventions are made in a timely. In this respect, the relevant public institutions and organizations have preserved the originality of the minarets by carrying out the necessary works and procedures. The continuity of such works and transactions is very important in order to transfer the minarets to the next generations.

The fact that the structural dimensions such as total height, diameter, wall thickness, and pulpit dimensions of the minaret had different values for Five Minarets caused the analysis results to be different from each other. Therefore, compressive, tensile, shear stresses, period, and displacement values differed.

Five Minarets are in Bitlis city center, they are close to each other, and the design spectra obtained are close to each other because the local soil class has the same values. In future studies, the effects of these variables on the behavior of masonry minaret structures will be examined. In this subject, this paper can be used as a source. In the study, only macro modeling was considered while performing structural analyses for each minaret. The material properties of the elements that make up the minarets will be determined in future studies. In addition, the analysis in the time history analyzes using micro modeling technique will also contribute.

It is vital that the Five Minarets are the most important historical structures of Bitlis city and be transferred to the next generations. For this purpose, these minarets should be observed according to the structural monitoring system, and when necessary, their maintenance work should be perfomed correctly. It is recommended that the necessary applications be made by the relevant institutions and organizations at the point of inclusion of Five Minarets in the UNESCO World heritage. This and similar studies will make important contributions to make this process happen faster and on a scientific basis.

Author Contributions: Conceptualization, E.I., F.A., and E.H.; methodology, M.G., E.I., E.A. and E.H.; software, M.G. and E.I.; validation, E.A., E.I., F.A. and E.H.; formal analysis, E.H.; investigation, E.I. and M.G.; resources, E.A., E.I., M.G. and E.H.; data curation, E.I. and M.G.; writing—original draft preparation, E.H. and E.A.; writing—review and editing, E.H., E.I. and F.A.; visualization, E.A.; supervision, E.H. and E.I.; project administration, E.I.; funding acquisition, E.H. All authors have read and agreed to the published version of the manuscript.

Funding: This research received no external funding.

Institutional Review Board Statement: Not applicable.

Informed Consent Statement: Not applicable.

Data Availability Statement: Most data are included in the manuscript.

Acknowledgments: This research was produced from master thesis of fifth author of the article. We acknowledge the support of the German Research Foundation (DFG) and the Bauhaus-Universität Weimar within the Open-Access Publishing Programme.

Conflicts of Interest: The authors declare no conflict of interest.

References

1. Hadzima-Nyarko, M.; Ademovic, N.; Pavic, G.; Sipos, T.K. Strengthening techniques for masonry structures of cultural heritage according to recent Croatian provisions. *Earthq. Struct.* **2018**, *15*, 473–485.
2. Isik, E.; Antep, B.; Buyuksarac, A.; Isik, M.F. Observation of behavior of the Ahlat Gravestones (TURKEY) at seismic risk and their recognition by QR code. *Struct. Eng. Mech.* **2019**, *72*, 643–652.
3. Pavić, G.; Hadzima-Nyarko, M.; Plaščak, I.; Pavić, S. Seismic vulnerability assessment of historical unreinforced masonry buildings in Osijek using capacity spectrum method. *Acta Phys. Pol. A* **2019**, *135*, 1138–1141. [CrossRef]
4. Bilgin, H. Typological classification of churches constructed during post-Byzantine period in Albania. *Gazi Univ. J. Sci. Part B Art Humanit. Des. Plan.* **2015**, *3*, 1–15.
5. Akan, A.E.; Başok, G.; Er, A.; Örmecioğlu, H.T.; Koçak, S.Z.; Cosgun, T.; Uzdil, O.; Sayin, B. Seismic evaluation of a renovated wooden hypostyle structure: A case study on a mosque designed with the combination of Asian and Byzantine styles in the Seljuk era (14th century AD). *J. Build. Eng.* **2021**, *43*, 103112. [CrossRef]
6. Karasin, I.B.; Isik, E. Protection of Ten-Eyed Bridge in Diyarbakır. *Budownictwo i Architektura* **2016**, *15*, 87–94. [CrossRef]
7. Hadzima-Nyarko, M.; Mišetić, V.; Morić, D. Seismic vulnerability assessment of an old historical masonry building in Osijek, Croatia, using Damage Index. *J. Cult. Herit.* **2017**, *28*, 140–150. [CrossRef]
8. Giordano, A.; Mele, E.; De Luca, A. Modelling of historical masonry structures: Comparison of different approaches through a case study. *Eng. Struct.* **2002**, *24*, 1057–1069. [CrossRef]
9. Işık, E.; Antep, B.; Büyüksaraç, A. Structural analysis and mapping of historical tombs in Ahlat District (Bitlis, Turkey). *Jcr-E-GFOS* **2019**, *10*, 22–35. [CrossRef]

10. De Backer, L.; Janssens, A.; Steeman, M.; De Paepe, M. Evaluation of display conditions of the Ghent altarpiece at St. Bavo Cathedral. *J. Cult. Herit.* **2018**, *29*, 168–172. [CrossRef]
11. Ortega, J.; Vasconcelos, G.; Rodrigues, H.; Correia, M.; Lourenço, P.B. Traditional earthquake resistant techniques for vernacular architecture and local seismic cultures: A literature review. *J. Cult. Herit.* **2017**, *27*, 181–196. [CrossRef]
12. Cosgun, T.; Sayin, B.; Gunes, B.; Osman Avşar, A.; Şengün, R.; Gümüşdağ, G. Rehabilitation of historical ruined castles based on field study and laboratory analyses: The case of Bigalı Castle in Turkey. *Rev. Constr.* **2020**, *19*, 52–67. [CrossRef]
13. Bilgin, H.; Ramadani, F. Numerical study to assess the structural behavior of the Bajrakli Mosque (Western Kosovo). *Adv. Civ. Eng.* **2021**, *2021*, 4620916. [CrossRef]
14. Formisano, A.; Milani, G. Seismic vulnerability analysis and retrofitting of the SS. Rosario church bell tower in Finale Emilia (Modena, Italy). *Front. Built. Environ.* **2019**, *5*, 70. [CrossRef]
15. Illampas, R.; Ioannou, I.; Lourenço, P.B. Seismic appraisal of heritage ruins: The case study of the St. Mary of Carmel church in Cyprus. *Eng. Struct.* **2020**, *224*, 111209. [CrossRef]
16. Hemeda, S. Geotechnical and geophysical investigation techniques in Ben Ezra Synagogue in Old Cairo area, Egypt. *Herit. Sci.* **2019**, *7*, 23. [CrossRef]
17. Carpinteri, A.; Lacidogna, G.; Manuello, A. The b-value analysis for the stability investigation of the ancient Athena Temple in Syracuse. *Strain* **2011**, *47*, e243–e253. [CrossRef]
18. Castellazzi, G.; Gentilini, C.; Nobile, L. Seismic vulnerability assessment of a historical church: Limit analysis and nonlinear finite element analysis. *Adv. Civ. Eng.* **2013**, *2013*, 517454. [CrossRef]
19. Karantoni, F.V.; Dimakopoulou, D. Displacement-based assessment of the Gazi Hasan Pasha mosque in Kos island (GR) under the 2017 M6. 6 earthquake and Eurocode 8, with proposals for upgrading. *Bull. Earthq. Eng.* **2021**, *19*, 1213–1230. [CrossRef]
20. Çaktı, E.; Saygılı, Ö.; Görk, S.; Zengin, E.; Oliveira, C.S.; Lemos, J.V. Earthquake behavior of the minaret of the Mihrimah Sultan mosque in Edirnekapı, İstanbul. *Restorasyon Yıllığı Dergisi* **2013**, *6*, 33–40.
21. Işık, E.; Antep, B. Structural analysis of historical masonry minaret in Ahlat. *BEU J. Sci.* **2018**, *7*, 46–56.
22. TSDC-2007. *Turkish Seismic Design Code*; T.C. Resmi Gazete: Ankara, Turkey, 2007.
23. Kılıç, İ.; Bozdoğan, K.B.; Aydın, S.; Gök, S.G.; Gündoğan, S. Determination of dynamic behaviour of tower type structures: The case of Kırklareli Hızırbey Mosque minaret. *J. Polytech.* **2020**, *23*, 19–26.
24. TBEC. *Turkish Building Earthquake Code*; T.C. Resmi Gazete: Ankara, Turkey, 2018.
25. Ural, A.; Çelik, T. Dynamic analyses and seismic behavior of masonry minarets with single balcony. *Aksaray J. Sci. Eng.* **2018**, *2*, 13–27. [CrossRef]
26. Mutlu, Ö.; Şahin, A. Investigating the effect of modeling approaches on earthquake behavior of historical masonry Minarets-Bursa Grand Mosque case study. *Sigma* **2016**, *7*, 123–136.
27. Bayraktar, A.; Çalık, İ.; Türker, T. Structural dynamic identification of a restored historical masonry Sundura mosque and minaret using ambient vibration. *Restorasyon Yıllığı* **2013**, *6*, 53–62.
28. Günaydın, M. Experimental determination of the dynamic characteristics of a historical masonry minaret after repairing. *GU J. Sci. Technol.* **2018**, *8*, 381–395.
29. Çarhoğlu, A.I.; Usta, P.; Korkmaz, K.A. Seismic behaviour investigation of historical minaret structures: Hagia Sophıa case. *SDU Int. Technol. Sci.* **2013**, *5*, 36–43.
30. Uğurlu, M.A.; Günaslan, S.E.; Karaşin, A. Modelling and structural analysis of the Four-legged minaret. *DU J. Eng.* **2017**, *8*, 413–422.
31. Oğuzmert, M. Dynamic Behavior of Masonry Minarets. Master's Thesis, Istanbul Technical University, İstanbul, Turkey, 2002.
32. Döven, M.S.; Serhatoğlu, C.; Kaplan, O.; Livaoğlu, R. Dynamic behaviour change of Kütahya Yeşil minaret with covered and open balcony architecture. *Eskişehir Tech. Univ. J. Sci. Technol. B-Theor. Sci.* **2018**, *6*, 192–203.
33. Gunes, B.; Cosgun, T.; Sayin, B.; Ceylan, O.; Mangir, A.; Gumusdag, G. Seismic assessment of a reconstructed historic masonry structure: A case study on the ruins of Bigali castle mosque built in the early 1800s. *J. Build. Eng.* **2021**, *39*, 102240. [CrossRef]
34. Ertek, E.; Fahjan, M.Y. Structural system of Ottoman minarets; classification, modelling and analysis. In Proceedings of the Sixth National Conference on Earthquake Engineering, İstanbul, Turkey, 16–20 October 2007; pp. 16–20.
35. Çalık, İ.; Bayraktar, A.; Türker, T. Simplified natural frequency formulas for historical masonry stone minarets based on experimental methods. In Proceedings of the Uluslararası Katılımlı 6. Tarihi Yapıların Korunması ve Güçlendirilmesi Sempozyumu, Trabzon, Turkey, 2–4 November 2007.
36. Yetkin, M.; Dedeoğlu, İ.Ö.; Calayır, Y. Investigation and assessment of damages in the minarets existing at Elazig after 24 January 2020 Sivrice earthquake. *Fırat Univ. J. Eng. Sci.* **2021**, *33*, 379–389.
37. Kültür Portalı. Türkiye Kültür Portalı. Available online: https://www.kulturportali.gov.tr/ (accessed on 25 December 2021).
38. Bitlis İl Kültür ve Turizm Müdürlüğü. Available online: https://bitlis.ktb.gov.tr/ (accessed on 20 December 2021).
39. Aslanapa, O. *Anadolu'da ilk Türk Mimarisi: Başlangıcı ve Gelişmesi*; Atatürk Kültür, Dil ve Tarih Yüks. Yayını: Ankara, Turkey, 1991.
40. Baş, G. *Bitlis'teki Mimari Yapılarda Süsleme*; Bitlis Valiliği Kültür Yayınları: Bitlis, Turkey, 2002.
41. Hasol, D. *Encyclopedic Architecture Dictionary*; Building Industry Center Publications: Istanbul, Turkey, 1998.
42. Güler, M.; Aktuğ, İ.K. 12th century Anatolian Turkish mosques. *İtüdergisi/a* **2006**, *5*, 83–90.
43. T.R. Directorate General of Foundations. Available online: https://www.vgm.gov.tr/home-page (accessed on 25 December 2021).
44. Arık, O. *Bitlis Yapılarında Selçuklu Rönesansı*; Selçuklu Tarih ve Medeniyeti Enstitüsü Yayını: Ankara, Turkye, 1971.

45. Şen, K. Two important inscriptions belong to the Bitlis grand mosque and Bitlis castle. *ASEAD* **2018**, *5*, 147–156.
46. Uluçam, A. *Ortaçağ ve Sonrasında Van Gölü Çevresi Mimarlığı-II-Bitlis*; Kültür Bakanlığı Yayınları: Ankara, Turkey, 2002.
47. Aslanapa, O. *Anadolu'da Türk sanatı Anadolu'da Büyük Selçuklulara Bağlanan Camiler*; Atatürk Kültür Merkezi Yayinlari: Ankara, Turkey, 2007.
48. Ülkü, C.; Yeğin, M. Courtyard in 11–12 century mosques/The newly discovered courtyard in Bitlis Ulu Mosque. *J. Soc. Sci.* **2017**, *17*, 17–37.
49. Öztürk, Ş. Bitlis Merkez Meydan Camii. *Vakıflar Dergisi* **2004**, *28*, 157–171.
50. Işık, E.; Büyüksaraç, A.; Avşar, E.; Kuluöztürk, M.F.; Günay, M. Characteristics and properties of Bitlis ignimbrites and their environmental implications. *Materiales de Construcción* **2020**, *70*, 214. [CrossRef]
51. Lorenzi, A.; Tisbierek, F.T.; Silva, L.C.P. Ultrasonic pulse velocity análysis in concrete specimens. In Proceedings of the IV Conferencia Panamericana de END, Buenos Aires, Argentina, 22–26 October 2007.
52. Sharma, P.K.; Khandelwal, M.; Singh, T.N. A correlation between schmidt hammer rebound numbers with impact strength index, slake durability index and P-wave velocity. *Int. J. Earth Sci.* **2011**, *100*, 189–195. [CrossRef]
53. Lourenço, P.B. *An Orthotropic Continuum Model for the Analysis of Masonry Structures*; Delft University of Technology: Delft, The Netherlands, 1995; pp. 3–21.
54. Lee, J.S.; Pande, G.N.; Middleton, J.; Kralj, B. Numerical modelling of brick masonry panels subject to lateral loadings. *Comput. Struct.* **1996**, *61*, 735–745. [CrossRef]
55. Pande, G.N.; Liang, J.X.; Middleton, J. Equivalent elastic modul for unit masonry. *Comput. Geotech.* **1989**, *8*, 243–265. [CrossRef]
56. Laurenco, P.B.; Rots, J.G.; Blaauwendraad, J. Two approaches for the analysis of masonry structures: Micro and macro-modeling. *Heron* **1995**, *40*, 1995.
57. Pande, G.N.; Middleton, J.; Kralj, B. Computer METHODS in structural Masonry 4. In Proceedings of the Fourth International Symposium on Computer Methods in Structural Masonry, Florence, Italy, 3–5 September 2017.
58. Lourenço, P.B. Computations on historic masonry structures. *Prog. Struct. Eng. Mater.* **2002**, *4*, 301–319. [CrossRef]
59. CSI Computers and Structures. *SAP2000: Integrated Software for Structural Analysis and Design Ver. 19*; CSI: Berkeley, CA, USA, 2017.
60. TS EN 1467. *Doğal Taş-Ham Bloklar-Özellikleri*; Türk Standardı: Ankara, Turkey, 2012.
61. Kuruşcu, A.O. Non-Lineer Modeling of Masonry Walls and Foundations. Ph.D. Thesis, YTU Graduate School of Science and Technology, İstanbul, Turkey, 2012.

 *buildings*

Article

Evaluation of Machine Learning and Web-Based Process for Damage Score Estimation of Existing Buildings

Vandana Kumari [1], Ehsan Harirchian [1,*], Tom Lahmer [1] and Shahla Rasulzade [2]

[1] Institute of Structural Mechanics (ISM), Bauhaus-Universität Weimar, 99423 Weimar, Germany; vandana.kumari@uni-weimar.de (V.K.); tom.lahmer@uni-weimar.de (T.L.)

[2] Research Group Theoretical Computer Science/Formal Methods, School of Electrical Engineering and Computer Science, Universität Kassel, Wilhelmshöher Allee 73, 34131 Kassel, Germany; shahla.rasulzade@uni-kassel.de

* Correspondence: ehsan.harirchian@uni-weimar.de

Abstract: The seismic vulnerability assessment of existing reinforced concrete (RC) buildings is a significant source of disaster mitigation plans and rescue services. Different countries evolved various Rapid Visual Screening (RVS) techniques and methodologies to deal with the devastating consequences of earthquakes on the structural characteristics of buildings and human casualties. Artificial intelligence (AI) methods, such as machine learning (ML) algorithm-based methods, are increasingly used in various scientific and technical applications. The investigation toward using these techniques in civil engineering applications has shown encouraging results and reduced human intervention, including uncertainties and biased judgment. In this study, several known non-parametric algorithms are investigated toward RVS using a dataset employing different earthquakes. Moreover, the methodology encourages the possibility of examining the buildings' vulnerability based on the factors related to the buildings' importance and exposure. In addition, a web-based application built on Django is introduced. The interface is designed with the idea to ease the seismic vulnerability investigation in real-time. The concept was validated using two case studies, and the achieved results showed the proposed approach's potential efficiency.

Keywords: rapid assessment; machine learning; seismic vulnerability; Django; damage classification

Citation: Kumari, V.; Harirchian, E.; Lahmer, T.; Rasulzade, S. Evaluation of Machine Learning and Web-Based Process for Damage Score Estimation of Existing Buildings. *Buildings* **2022**, *12*, 578. https://doi.org/10.3390/buildings12050578

Academic Editor: Fulvio Parisi

Received: 4 April 2022
Accepted: 26 April 2022
Published: 29 April 2022

Publisher's Note: MDPI stays neutral with regard to jurisdictional claims in published maps and institutional affiliations.

1. Introduction

Natural calamities can be unpredictable and may leave a cumbersome loss of life and economy. The infrequent seismic events (intermediate to large) eventually end up claiming lives and buildings or houses. The post-earthquake consequences are more evident in the metropolitan regions compared to rural areas. The complexities and unorganized distribution with inadequate design code implementation for mid-to-high-story buildings lead to unfortunate fatalities. In the past, significant earthquakes worldwide have provoked many human casualties and damages to structures. It reportedly killed 8519 persons and maligned 80,000 buildings in Nepal's territory [1]. City Bam, in southeastern Iran, was hit by an earthquake measuring 6.3–6.6 on the Richter scale early Friday morning on 26 December 2003. It claimed 26,271 lives and 51,500 fatalities [2]. On 15 August 2007, an earthquake of 7.9 Richter caused enormous devastation in Peru and claimed 514 lives and 1090 injuries. The dead-to-injured ratio was 32% [3]. On 12 May 2008, Sichuan, in China, faced one of the deadliest earthquakes in the past 30 years with an 8.0 magnitude. The authorities confirmed the death of 69,170 people, 17,426 missing, and 374,159 injuries [4]. Haiti was knocked by a giant earthquake with a magnitude of 7.2 on the Richter scale on 12 January 2010. The event aroused massive mortal and material casualties, with 212,000 dead, 512,000 wounded, and 1,000,000 homeless [5]. Many studies and records infer that notable earthquake-related fatalities occurred due to damages to buildings which caused approximately 75% of deaths as a result of earthquakes in the past decades [6].

It is not feasible to conduct a comprehensive analysis for every structure in an urban metropolis. Thus, a fast screening technique to filter out and prioritize the building stocks with high degrees of vulnerability, also known as Rapid Visual Screening or RVS, is felt to be highly essential. RVS, recognized for its fundamental calculations, provides a quick and reliable approach for assessing seismic risk. RVS was created in the 1980s by the Applied Technology Council (USA) [7] to assist a wide range of people, including building authorities, structural engineering consultants, inspectors, and property owners. This approach was created to be the preliminary screening phase in a multi-step process of assessing seismic risk. Ozcebe [8] provides a valuable overview of RVS. When screening out high-risk construction stocks, this technique is fast, dependable, and efficient. For executing RVS, a sheet with preset criteria is considered and a qualified assessor implements a preliminary walk-down assessment to determine the damage grade of buildings. The process is followed by the evaluator taking note of physical characteristics such as overhangs and floating columns. First, a Basic Score (BS) is calculated using a formula and then allocated to the building following several international codes.

Throughout the last 25 years, many research projects have used artificial intelligence (AI) capabilities in earthquake engineering, such as methods based on machine learning (ML) and neuro-fuzzy [9–11]. The intrinsic capacity of ML to embed and deploy solutions of issues with known input data in order to extract predictions for the solution of the same kind of problems with unknown input data instantaneously led to the idea of using them for the approximation of the seismic damage status of existing structures in real-time following an earthquake.

The current study examines different ML methods for the rapid classifying of building damage grades through the rapid screening of a building. With the input dataset, many supervised learning approaches were investigated. The final classifier was created using the highest-scoring algorithm. Furthermore, the study was focused on the critical factors that cause probable loss following an earthquake. Despite the capabilities of AI approaches, there is a significant gap in adopting the most efficient technique for evaluating the seismic vulnerability of structures for which some possible reasons could be the substantial knowledge, the required skills, and the proper place for the application of an AI-based method [12].

2. Background of the Study

Earthquakes are the most devastating natural calamities that humans have witnessed [13,14]. Every year earthquakes claim thousands of lives [15] and cause significant loss in the economy (direct or indirect), massive fatalities, and suspension in occupancy as well as in business, not only in developing but also in the developed and scientifically well-established (particularly in earthquake engineering) countries [16,17].

In general, an estimation of any damage involves the complications of defining, assessing, and modeling the associated factors along with controlling the uncertainty. Engineering communities have the challenge of predicting the structures' expected damages and face complexities, vagueness, and difficulties in their estimations [18], especially when it comes to dealing with a large building stock. Hence, restricting the damage to existing reinforced concrete (RC) buildings is emphasized, as they are more vulnerable than new buildings constructed according to the latest codes. The pre-earthquake assessment of existing structures has been introduced within the general framework of seismic risk management and aims to form a strategy for prioritizing buildings for further comprehensive analyses [19]. Although most new RC buildings fulfill the seismic safety criteria, many existing old RC buildings are not compliant with the principles of modern seismic codes, lacking ductility and sufficient seismic resistance, and hence their vulnerability analysis demands more attention. Lessons learned from the performance of buildings during previous earthquakes and research over the last three decades resulted in improved seismic design provisions and building resiliency [20,21]. The majority of the existing buildings are operational and require an assessment to analyze how safe they are to resist any seismic event in terms of life safety. A rapid and efficient risk-based seismic vulnerability assessment of existing buildings can help with estimating damage grades, prioritization for further analyses, saving time and costs, and assisting decision-makers with the provision of post-disaster crisis management plans [22].

2.1. Rapid Visual Screening

RVS estimates the seismic vulnerability of structures in a given territory. It depends on associations between the buildings' evaluated seismic performance and architectural typology (frame, shear wall, monolith, in-fill), material (steel, RC, masonry, wood, composite), design methods implementation, and several other factors [23]. Additional inputs to the analysis include the level of earthquake hazards. The predictions are based on expert opinions, pushover analyses, compelling response studies, and the performance of similar buildings in past seismic events. The RVS methods allow for quantifying structural vulnerabilities more easily than analytical approaches [24] by filtering the weak structures through faster implementation. There is no need for detailed calculations and multiple scenarios. A scale-based or score-based system quantifies the likelihood of a building collapse, while the scale or score represents the possible damages after an earthquake [25].

2.2. Machine Learning in Seismic Vulnerability Assessment

AI methods provide a device or computer with the ability to imitate human intelligence (cognitive process), procure from events, adapt to the newest information, and perform human-like activities. These techniques accomplish tasks intelligently and accurately while keeping adaptive and productive for the entire system. Abundant literature is available that attempted to integrate various branches of AI from several areas with an RVS method. There is a broad set of techniques that come in the domain of artificial intelligence, including Artificial Neural Networks (ANNs), ML, Fuzzy Logic (FL), and Multilayer Feedforward Perceptron Networks (MFPNs). A comprehensive review of different AI techniques developed in the literature for the application of RVS to existing buildings and damage classification can be found in [26]. Morfidis and Kostinakis [27] applied an ANN to explore the number and combination of seismic specifications and help forecast the seismic damage caused to the RC buildings. The desired combination of parameters was between five and fourteen, while the Housner Intensity stood the most important among all. In 2018, Morfidis and Kostinakis conducted another study [28] using a model based on an MFPN. Several algorithms trained the model and using the Maximum Inter-Story Drift Ratio, the damage index of the earthquake was examined. In another study published by Morfidis and Kostinakis [12], the seismic vulnerability was evaluated using an MFPN. The model was trained via datasets obtained from a number of RC buildings in Greece with heights of 3–30 m. The study aimed to exchange the knowledge base between a civil engineer and an ANN expert.

In 2010, Tesfamariam and Liu [29] conducted a comparative study on different classification methods of the Neural Network (NN), namely Naive Bayes (NB), K-Nearest-Neighbor (KNN), Fisher's Linear Discriminant Analysis (FLDA), Partial Least Squares Discriminant Analysis (PLS-DA), Multilayer Perceptron Neural Network (MLP-NN), Classification Tree (CT), Support Vector Machine (SVM), and Random Forest (RF). For different damage states including none (O), light (L), moderate (M), severe (S), and collapse (C), indicators 1, 2, 3, 4, and 5 were used, respectively. KNN and SVM performed optimally for segregating buildings with "Immediate Occupancy" (IO) and "Life Safety" (LS). A team of researchers (Zhang et al. [30]), in 2018, presented an analysis report on their approach based on ML. The selected mechanism was the Classification and Regression Tree (CART) and RF to sketch the response and damage patterns of buildings. They found an accuracy of 90% in predicting response patterns and 86% for both CART and RF. The traditional architecture of an ANN has each neuron connected to every other neuron, while a Convolutional Neural Network (CNN) has the last layer fully connected.

Harirchian et al. [21] utilized an optimized MLP-NN to study the earthquake perceptivity through six buildings' performance variables which can help in obtaining optimal foresight of the damage state of RC buildings using an ANN. The MLP network was trained and optimized, utilizing a database of 484 buildings damaged in the Düzce earthquake in Turkey. The effectiveness of this recommended NN design was displayed through the achieved results to build an introductory estimation procedure for the seismic risk prioritization of buildings.

Another study conducted by Allali et al. [31] focused on an approach based on FL for the post-earthquake damage evaluation of buildings. They proposed the creation of the global damage level of buildings with different lateral load resisting components. More than 27,000 buildings damaged in the destructive 2003 Boumerdes earthquake in Algeria (M_w = 6.8) were considered in the review process. The survey included the structural and non-structural components, and other factors such as soil conditions, etc., were overlooked. The damage levels of components and global damage levels were indicated using a scale system from "No damage" for D_1 up to "collapse" for D_5. The study applied a novel purpose of weighted fuzzy rules to associate the components' damage levels to the global damage levels. The outcomes showed an accuracy of 90%.

ML represents a branch of AI that uses the essential components of illustration, evaluation, and optimization to generate forecasts using computational algorithms that improve with time. From a natural hazard risk evaluation perspective, SVM approaches have been frequently used [32–35]. Few researchers compared an SVM performance with other soft computing techniques or used an integrated approach that included the characteristics of two different methodologies. Gilan [36] used an SVM combined with a fuzzy evolutionary function (EFF-SVR) to predict the concrete compressive strength and compared their findings with ANFIS, fuzzy function with least square estimation (FF-LSE), ANN, and enhanced FF-LSE. Simulation results and comparisons showed that the suggested EFF-SVR technique outperformed other methods in terms of performance. In their research investigations on compressive strength predictions for the no-slump concrete and concrete with a significant volume of fly ash additive, Sobhani [37] and Sun [38] utilized SVM and least square SVM (LS-SVM) models, respectively, and compared them with ANN models. SVM and LS-SVM models were high-speed in both tests and helped overcome the ANN model's disadvantages of over-training and poor generalization skills.

Harirchian [35] used a supervised ML methodology on the post-damage data of RC structures damaged by the 1999 Düzce earthquake in Turkey to develop a method applicable to risk assessment and crisis management strategies. They used an SVM optimization model to assess the efficacy of the suggested technique in damage prediction. The database contained 484 buildings, with 22 feature characteristics serving as inputs to the suggested method. When all 22 characteristics were used as inputs to the technique, their findings showed a 52% accuracy. The SVM technique has been used to estimate seismic risk for large-scale data. In fact, constructing a dataset with various parameters as inputs to an extensive computational model for many structures in an area rather than individual buildings would be a complicated task which would render the procedure computationally expensive. Zhang [39] fixed this problem by performing two case studies with 11 and 20 building features, respectively, utilizing an SVM as a data mining technique. For testing the performance and accuracy of the SVM in seismic vulnerability prediction in terms of spectral yield and ultimate points using capacity curves acquired by a pushover analysis [40–43] in individual as well as regional scales, samples were randomly selected from Taiwan's National Center for Research on Earthquake Engineering (NCREE) [44] database of 5400 school buildings in Taiwan. Their results showed an average accuracy of 64% on an individual scale, which could be raised to 74% by increasing the number of samples used in the dataset's training. A seismic risk assessment was completed on a regional scale when the SVM approach was combined with the seismic-demand curves based on building sites. Deep Neural Networks (DNN), also known as Deep Learning (DL), have been effectively employed in a range of applications [45], attracting the interest of academics since DL outperformed alternative ML approaches such as kernel machines [46–48].

Exposure models, or the inventory of people and facilities geographically separated at the regional scale impacted by earthquakes, are among essential variables that make up the fundamental inputs for a seismic risk assessment. Gonzalez [49] utilized a methodology that included deep learning and a CNN method, which was applied to Google Street View photos for image processing and categorizing individual buildings' structural typologies, which was then used for exposure models and a seismic risk assessment. A dataset from 9989 structures across eight typologies in Medellín, Colombia, was taken into consideration. The results indicated that non-ductile structures are most vulnerable in an earthquake, with an accuracy of 93% and a recall of 95%.

3. Data and Methodology

3.1. Input Data Source

Data are the observations of real-world phenomena, and they make the backbone of ML models. Data present a small window into a restricted viewpoint of reality. The accumulation of all of these observations provides a picture of the whole. However, it should be noted that even when one has a massive dataset, the adequate number of data points for some instances of interest might be pretty small.

The study presented in [50] incorporates cumulative data collected from RC buildings shaken under the past four seismic events in Ecuador, Haiti, Nepal, and South Korea. The data are archived in the open-access database of the DataCenterHub platform [51], simultaneously with the data obtained from the studies by different research groups on that platform [52–55]. Data from the individual earthquakes were not sufficient for predictive models to work efficiently. Therefore, to supplement the data and create a global method, the study consolidated all four datasets and merged them into one.

ML models require learning from data. ML algorithms acquire a mapping from input variables to a target variable on a predictive modeling project. In this study, the input dataset used for the predictive analysis consists of 526 data points corresponding to 8 independent building features and 1 dependent set of variables. The dependent variables in the dataset are the damage scales from 1 to 4 that have been assigned to the affected RC buildings after seismic events. The predictive models need independent and dependent variables in order to make predictions. In Table 1, the buildings' featured parameters, their relative units, and their category types have been presented. These features are the input variables for the given dataset, whereas the target or dependent variable contains the four damage scales, as shown in Table 2, for defining the damage and risk associated with the corresponding RC building.

Table 1. RC buildings' input feature collected for RVS study (based on the conducted study in [21]).

Features	Unit	Type
No. of Stories	$N\ (1, 2, \ldots)$	Integer
Total Floor Area	m^2	Integer
Column's Cross-Sectional Area	m^2	Integer
Concrete Wall Area (Y)	m^2	Integer
Concrete Wall Area (X)	m^2	Integer
Masonry Wall Area (Y)	m^2	Integer
Masonry Wall Area (X)	m^2	Integer
Captive Columns	$N\ (\text{exist} = \text{yes} = 1, \text{absent} = \text{no} = 0)$	Binomial

A feature depicts the numeric representation of raw data. The right features are essential to the task at hand and should be easy for the model to ingest. Feature engineering expresses the most appropriate features for the given data, model, and task. The number of features is also imperative. If there are not enough informative features, the model will not achieve the final goal. In case of many features or trivial features, the model will be more expensive and tricky to train. Something might go wrong in the training process that impacts the model's performance. Features and models sit between raw data and the desired insights. Thus, in an ML workflow, not only the model but also features are equally important.

Table 2. Damage scales associated with various risk factors.

Damage Scale	Risk Association
1	No visible damage to the structure. Safe to reoccupy.
2	Low damage. Hairline to wide cracks in the structural elements. Spalling of concrete may also be observed.
3	Significant loss. Failure of at least one element in the structure.
4	Severe damage. At least one floor slab or part of it loses its elevation.

3.2. Data Preprocessing

ML algorithms cannot fit solely on raw data. Instead, the data should be transformed by preprocessing to meet the requirements of individual ML algorithms. Figure 1 shows various steps followed in the study including merging four different datasets to form one, processing the consolidated raw dataset for improvising it, and making the data suitable for feeding into the predictive models. The study used various libraries available in *Python* programming language to preprocess the data and create ML predictive models. Data preprocessing is vast and can consist of numerous steps depending upon the data, features, and the model requirements. Below, sub-sections elaborate on the preprocessing steps carried on the raw dataset for implementation in the study.

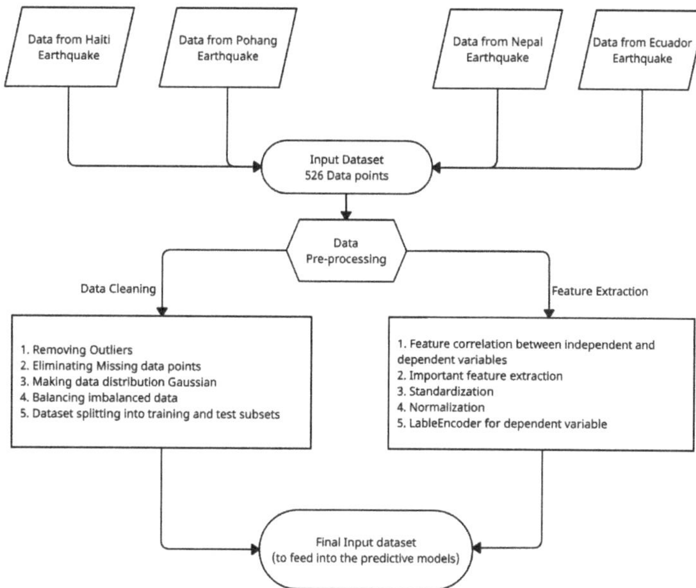

Figure 1. The workflow of data preprocessing for feeding data to predictive models.

3.2.1. Data Preparation

Data cleaning regards recognizing and fixing errors in the dataset that may negatively influence a predictive model. It is referred to all kinds of actions and activities which identify and adjust errors in the data [56]. One of the most common forms of preprocessing consists of a simple linear re-scaling of the input variables [57].

- Outlier detection—An outlier is a data point that is unlike the other data points. They are rare, discrete, or do not belong in some way. There is no definite technique to distinguish and recognize outliers as usual because of the specifics of each dataset. However, a domain expert can interpret the raw observations and verify if any given data are outliers or not. Identifying outliers can be tricky even after a thorough comprehension of the data. Proper attention should be taken not to eliminate or replace values rashly, specifically when the sample size is small [58].
  The input dataset for the study is medium size, so detecting the outliers on the basis of extreme value analysis was possible. There were few data points which were not in the range and distribution of attribute values. Those data points were eliminated to avoid creating any unforeseen circumstance to control the predicting model inaccurately.
- Missing data elimination—Major ML algorithms utilize numeric-type input data values arranged in rows and columns in a given dataset. Missing values in a dataset can create problems for the algorithms to function optimally. Therefore, it is a common practice to identify the missing values and substitute them with a corresponding numerical value. The method is known as missing data imputation or data imputing. A known approach for data imputation is replacing all missing values for that column with the respective column's mean or median value. *Scikit-learn* library provides a *SimpleImputer* with a mean or median strategy for missing value elimination. The input dataset for this study is moderately tiny with only 526 values; however, there were 3 NaN (Not a Number) data points raising errors while processing the algorithms. Therefore, using the conventional strategy, those NaN were replaced by the mean value of their respective columns.
- Gaussian data—ML models function better when the data have Gaussian distribution. The Gaussian is a standard distribution with the familiar bell shape. Data fitting techniques can modify each variable to make the distribution Gaussian, or if not Gaussian, then more Gaussian-like. These transforms are most efficient when the data population is nearly Gaussian, to begin with, and is skewed or affected by outliers. Figure 2 shows the histogram for each feature input variable. Floor number has Gaussian-like distribution of the data points, whereas most of the features are skewed toward the left. The captive column obtained a binomial value (0 or 1); therefore, the data distribution is discrete.
  Figure 3 illustrates the data distribution for each feature input after employing Power-Transformer class from *Scikit-learn* library. Total floor area and column area show a better Gaussian bell shape, whereas masonry wall area NS and masonry wall area EW have some skewness on the left side. Due to very few non-zero data points, the area of concrete wall area NS and concrete wall area EW did not show any significant improvement.
- Imbalanced Data—An imbalance in data distribution for each input feature creates objections for predictive modeling. In real-world cases, the dataset contains irregular data sharing several times, affecting the model prediction's performance. Figure 4 depicts that the input dataset contains an asymmetric distribution of feature data which resulted in imbalanced data in each output class. For example, damage class 4 has the highest number of samples, whereas damage class 1 has the least. This kind of data distribution prevents the model from performing optimally. Synthetic Minority Oversampling Technique (SMOTE) [59] generates synthetic data for the minority class, therefore producing symmetry for majority of classes. Table 3 illustrates that, using SMOTE, the imbalanced state of data distribution in the target variable is balanced.

- Dataset splitting—The suitable approach for performing data preparation with a train–test split evaluation is to fit the data preparation on the training set, then apply the transform to the train and test sets. Therefore, the input dataset was split into (80–20%) into the training set and testing set using function *train-test-split* from *model-selection* module in *Scikit-learn*.
- Cross-Validation—As a good practice, ML models should evaluate the dataset using k-fold cross-validation, particularly in small to medium-sized datasets. Cross-validation aims to examine the model's worth to predict unseen data, check issues such as model overfitting, or biasedness to give an insight into the model behavior against an independent dataset. The study implemented repeated stratified 10-fold cross-validation using the function *RepeatedStratifiedKFold* from module *ensemble* in *Scikit-learn*.

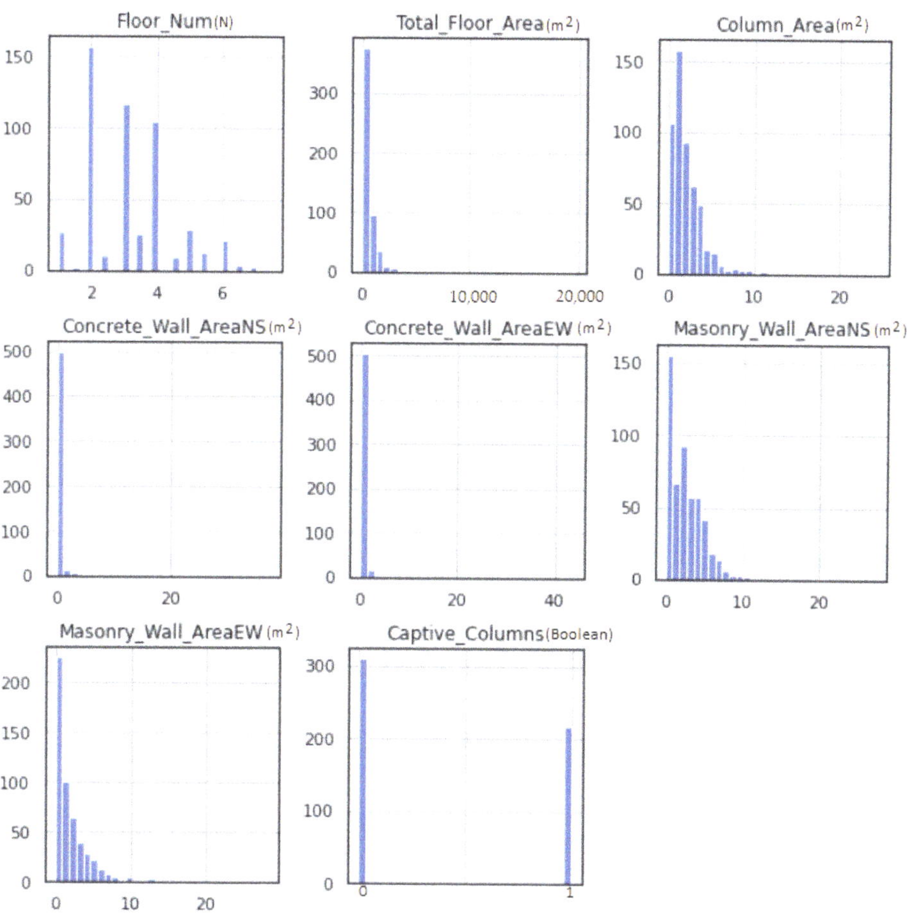

Figure 2. Distribution of data over each input variable.

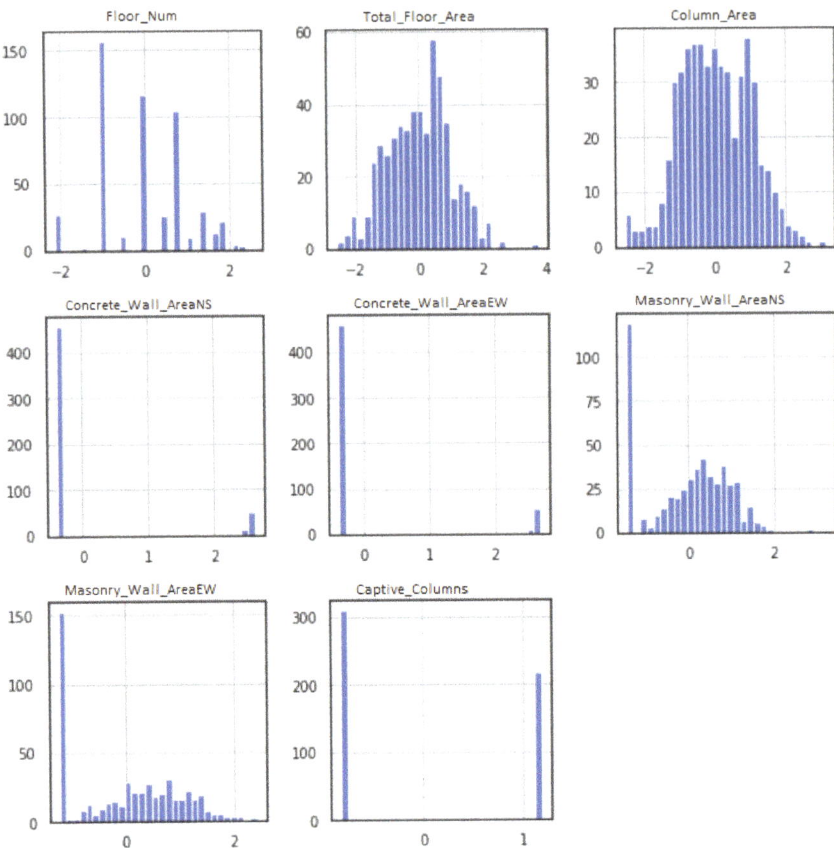

Figure 3. Distribution of data on each input variable after employing PowerTransformer.

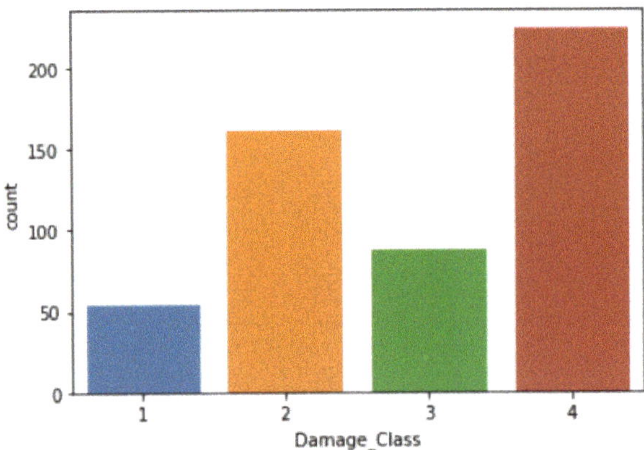

Figure 4. The damage class distribution of studied buildings.

Table 3. Balanced number of data in target variable with SMOTE [60].

Target Class	Imbalanced Data	Balanced Data
1	54	224
2	161	224
3	87	224
4	224	224

3.2.2. Feature Selection

Feature selection limits the number of independent variables in an input dataset while creating a predictive model. Reducing the number of input variables helps in terms of reducing the computational cost of modeling and also in order to increase the model's performance.

Figure 5 shows the application of statistical measures to draw a correlation between independent and dependent variables as the basis to filter out feature selection. For the available dataset, where the independent/input variable was numerical and dependent/output variable was categorical, ANOVA method was implemented to extract the essential features out of all given parameters.

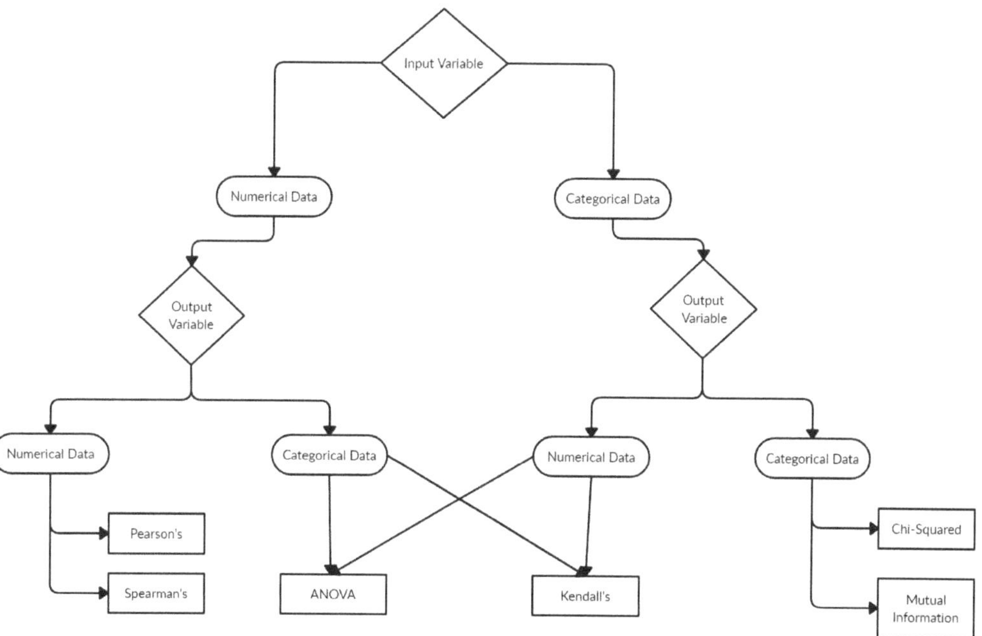

Figure 5. Types of statistical data.

3.2.3. Data Transformation

ML models are mapped from input variables to an output variable based on a suitable pre-defined algorithm. Variations in the scales over independent variables may raise the complexity of the problem to be modeled. Following approaches transform the data in a proportionate scale to achieve the optimum output.

- Standardization—Standardization of a dataset includes re-scaling the spread of values such that the mean of observed values is 0 and the standard deviation is 1. The method is offered by a function called *StandardScaler* available in the *preprocessing* module of

Scikit-learn library. Deducing the mean value from the given data is known as centering, and dividing the data by the standard deviation is known as scaling. The method is also referred to as center-scaling.

- Normalization—Data points in any dataset may scale differently from variable to variable. Often ML predictive models perform better if the variables are scaled in a standard range, for example, in the range between 0 and 1. The scaling of all variables in the range between 0 and 1 is known as Normalization. Class *MinMaxScaler* from *preprocessing* module in *Scikit-learn* library normalizes the input variable.
- Label Encoder—ML predictive models assume all the provided input and output variables to be numeric. Numerical data include data points that comprise numbers, such as integer or floating-point values. Categorical data involve label values instead of numerical. Categorical variables are frequently known as nominal. *Scikit-learn* library also has this requirement which implies that all categorical data must be transformed to numerical values.

For the dataset applied in this study, the target variable is in the categorical form: the different damage classes assigned to the earthquake-affected RC buildings. With the one-hot encoding method, the categorical output variable can be changed to an ordinal numerical form. The ordinal encoding transform is available in the *Scikit-learn* library via the *OrdinalEncoder* class.

3.3. Predictive Model Building

Designing an ML model involves several steps, as demonstrated in Figure 6. After preprocessing, the input dataset (including the five feature input parameters) is split into training and test subsets in a proportion of 80 and 20%, respectively. The training subsets underwent cross-validation, which further divided the training set into smaller training and validation subsets. In the study, there are ten subsets of the training data. There are six popular ML classifiers implemented: SVM, DT, RF, NB, KNN, and GB. While nine training subsets' data were training each classifier, the validation set evaluated the classifier's performance by reviewing their precision for each output class. Unlike binary classification [61], there is no positive or negative class in a multi-class classification problem, which means, in the case of binary classification, the focus is on a positive class to detect; however, in a multi-class classification problem, each sample needs to be categorized into 1 of N different classes. Nevertheless, factors such as true positive (*TP*), true negative (*TN*), false positive (*FP*), and false negative (*FN*) are available in multi-class classification help in calculating the *Precision*, *Recall*, and *F1-score*, which are helpful in determining the performance of the classifiers. Below, equation shows the details of each keyword:

- $Precision = TP/(TP + FN)$
- $Recall = TP/(TP + FN)$
- $F1\text{-}score = 2 \times ((Precision \times Recall)/(Precision + recall))$

To evaluate the classifier for general purposes, the model is exposed to the unseen test data. The accuracy was computed based on the model's predicting capability. The results could also be visualized using different plots such as a confusion matrix.

3.3.1. Support Vector Machine

SVM was introduced through a statistical research learning theory [48]. A generalized case study [62] published using SVM showed that data are not linearly separable. Gärtner et al. [63] introduced the technique of handling multi-instance learning problems using regular single-instance learners to summarize bag-level data combined with SVM.

The study implemented a GridSearchCV function from model_selection module in the *Scikit-learn* library for implementing the SVM algorithm with all the possible hyperparameters. The optimal accuracy achieved by the grid was 52% for the training data prediction and 57 % for the test data prediction using the hyperparameter values of C = 10, gamma = 1, and kernel = RBF.

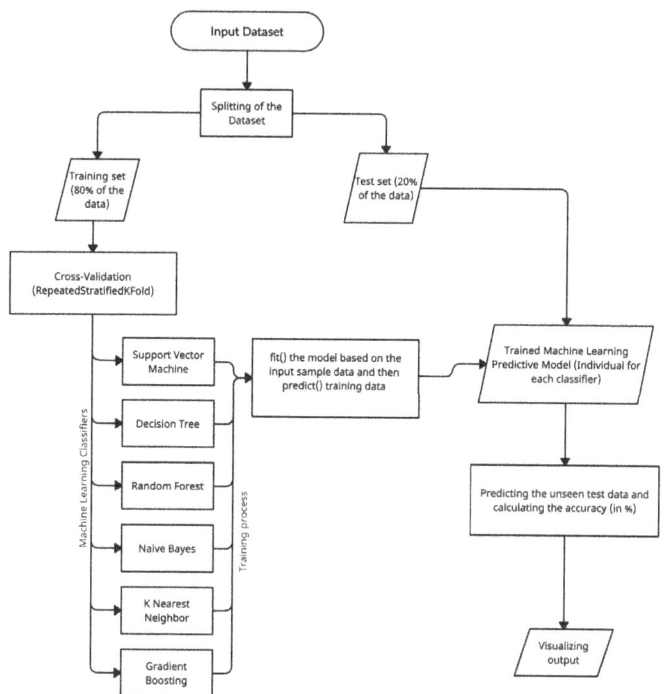

Figure 6. Machine-learning-model building schematic flowchart.

3.3.2. Decision Tree Classifier

Developing a tree to classify samples from the training dataset is the goal of the decision tree algorithm, which is also known as Classification and Regression Tree (CART) [64]. The tree can be considered as a division of the training dataset, with samples progressing down the tree's decision points until they reach the tree's leaves, where they are assigned a class label. It follows the approach of "divide-and-conquer" to a set of independent instances.

This research analyzed the given input dataset with the DT algorithm while using *entropy* as purity criteria and made the tree with a maximum depth of three. The trained model achieved 47% accuracy for the training set, while it evaluated the test data with an accuracy of 42%.

3.3.3. Naive Bayes

Determining the conditional probability of a class label that is provided with a data sample can be regarded as the problem of classification predictive modeling. The Bayes Theorem [65] provides a systematic method for estimating this conditional probability, but it is computationally expensive and requires many samples (a huge dataset). In the current study, Gaussian NB fits well as the attributes contain continuous values, and it was assumed that the data associated with each class have a Gaussian distribution. As a result, the model achieved 43% accuracy for the training set and 42% accuracy for the test set.

3.3.4. K-Nearest Neighbor

Proposed by Evelyn Fix [66] in 1951 and further modified by Thomas Cover [67], KNN is a simple non-parametric procedure in ML that is used for classification and regression problems. It is assumed that the same class data will remain close to each other, and the data will be assigned to a specific class with the most votes from their neighbors. Because the classification is locally approximated, it is also known as lazy learning. For the present dataset, KNN performed fairly good where it scored 64% accuracy for the training data, whereas for the test data, the accuracy was 45%.

3.3.5. Random Forest

RF is an ensemble classifier that originates from a significant advancement in DT variance. Each tree uses bagging and feature randomization to construct an uncorrelated forest of trees whose committee prediction is more accurate than that of any individual tree. Leo Breiman [68] proposed the idea, which combined concepts from bagging [69] and random feature selection by Ho [70,71] and Amit and Geman [72,73]. RF classifier attained almost equal accuracy for the training and test subsets for the given dataset, i.e., 45 and 37%, respectively.

3.3.6. Gradient Boost

GB follows ensemble learning by sequentially building simpler (weak) prediction models, with each model attempting to identify the error issued over by the previous model. Interestingly, because the loss function is optimized using a gradient descent, it is referred to as GB. GB employs a short, simpler DT, which makes it GBDT. It combines many weak learners to form one strong learner. Trees are linked in a series, and each tree attempts to minimize the error of the previous tree. GBDT has efficiency, accuracy, and interpretability features [74], and it is a popular ML ensemble algorithm. In addition, GBDT delivers high performance in a variety of tasks, including multi-class classification [75,76], click prediction [77], and learning to rank [78].

For the input dataset in this research, GBDT scored 78% on the training dataset and 51% on the test dataset. The outcome of the prediction is shown is Figure 7. The confusion matrix represents the classifier's outcome on test data. Overall, GBDT classifier evaluated moderate numbers of true positives for class 2, class 3, and class 4, whereas, for class 1, it achieved the highest numbers of correct predictions.

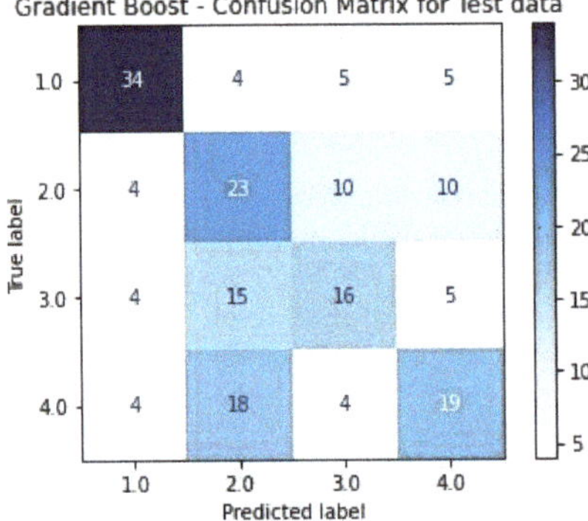

Figure 7. Gradient Boost Decision Tree—Confusion Matrix for Test Data. Class 1 obtained the majority of correct predictions, whereas the rest of the classes have a modest number of accurate data predictions.

3.4. Selection of ML Classifier

The models' accuracies were recorded to evaluate their performance on datasets before and after the preprocessing steps. Before preprocessing, the dataset contained eight parameters, including floor number, column's cross-sectional area, total floor area, existence of captive columns, masonry wall area in east–west, masonry wall area in north–south, concrete wall area in east–west, and concrete wall area in north–south, respectively.

Table 4 represents the accuracy (in %) scored by each of the trained classifiers on the set of training and test data before and after preprocessing of the dataset. GBDT performed good in comparison to all other classifiers.

Table 4. Accuracy scored by candidate classifiers with training and test subsets before and after postprocessing of raw dataset.

ML Classifier	Dataset (before Preprocessing)		Dataset (after Preprocessing)	
	Training Set (in %)	Test Set (in %)	Training Set (in %)	Test Set (in %)
Decision Tree	48	41	47	42
Gradient Boost Decision Tree	77	49	78	51
K-Nearest Neighbor	64	46	64	45
Naive Bayes	46	35	43	42
Random Forest	38	33	45	37
Support Vector Machine	57	45	52	47

Based on the performance of each candidate classifier on the test set, GBDT fits best for further creation of the browser-based application for rapid screening of RC buildings. Compared to other non-parametric models, ensemble models are robust and accurate because of their architecture, as shown by the performance of GBDT. However, a single DT is likely to overfit. Therefore, the classifier with a combination of DT models is preferred to reduce the chances of overfitting. These combined models are more efficient in terms of accuracy, as it can be seen for RF. To create an ensemble, RF employs a technique called *bagging* to merge multiple DTs. Combining in parallel a number of DTs is called bagging, while processing them in a series is referred to as boosting. Boosting is the process of successively connecting weak learners to produce a strong learner. The weak learners in the GBDT model are DTs. Each tree tries to decrease previous mistakes and boost the accuracy and efficiency of the method. Figure 8 shows the working principle behind the GBDT classifier, as stated above.

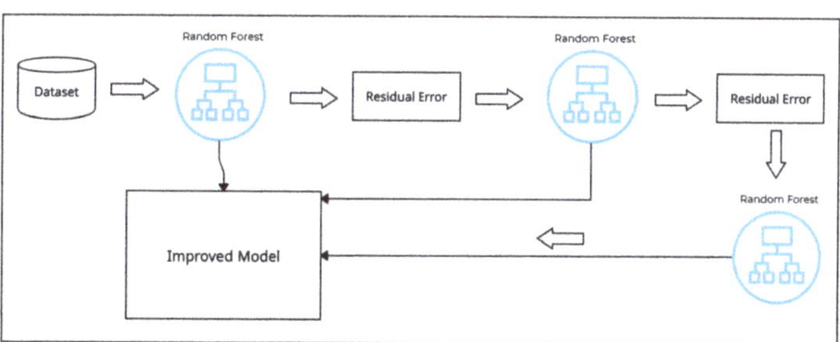

Figure 8. Schematic diagram of Gradient Boost Decision Tree (GBDT) algorithm.

3.5. Web-Development Using ML Model

One of the aims of this study is to incorporate the best-performing ML model in a browser-based UI. The purpose is to evolve from the traditional RVS methods to develop a fast, easily accessible, accurate, and scalable technique that can efficiently minimize the seismic risk assessment and reduce any imprecision that comes with human interference. Django is a free web framework based on Python language [79] that allows creating tested, scalable [80], and reliable online applications. Adrian Holovaty and Simon Willison created Django in 2003 for the *Lawrence Journal-World* newspaper, and it was freely distributed under the BSD license in 2005.

3.5.1. Design Approach

This section will introduce the design and implementation of the web application based on Django as defined in the previous section. First, the models require to be trained by the given input data. The data would be labeled in the case of supervised learning models [81] and unlabeled for unsupervised learning models [82]. Nevertheless, most of these models demand different amounts of data, depending on the algorithm, to reach a desired level of accuracy. After the model has been trained and is ready to predict, it will be given new and previously unseen data as input to make predictions about the target or independent variable in the second phase. Once the model is trained, the next concern is to utilize it in the web application to make predictions based on the unseen data that the user will feed. The trained models have acquired the state of a set of attributes that need to be kept in the application to predict the unseen data.

The GBDT is trained outside of the web application by the training dataset. Dataset is cleaned before training phase by preprocessing, and the model's performance has been recorded in terms of accuracy by the test data. Figure 9 represents the correlation coefficient of each attribute in the dataset for each given target class. Based on the outcome, the following five attributes have the highest predictive ability to predict the seismic vulnerability of any RC building.

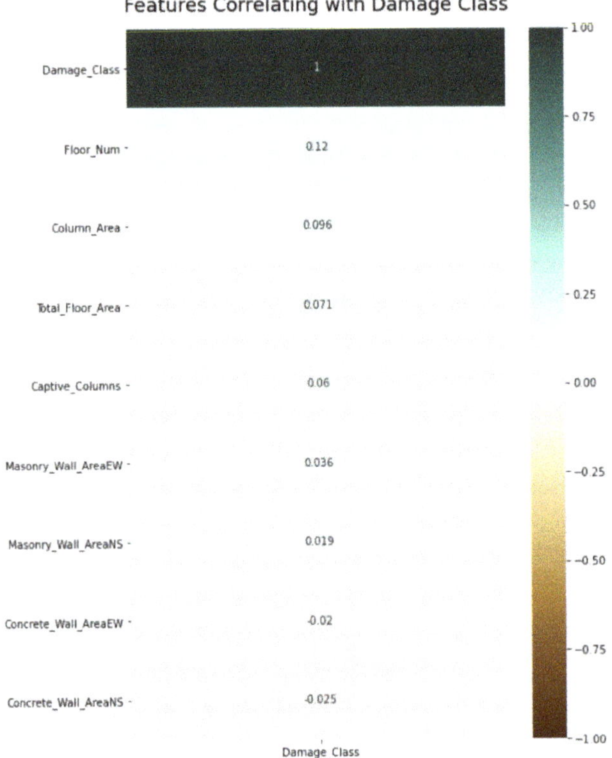

Figure 9. Heatmap illustrating correlation coefficient between independent and dependent variables.

Figure 10 describes the flow of the process in Django architecture. The Django Schema Model (or Model) defined the fields for the input attributes to save in the database. The right side has the pre-trained ML model feed to the logic layer; controller or View. Based on the provided rules from the ML model, the controller will perform and pass the data to the appropriate Template. If there is no input data from the user, the Template will not show any prediction, but the data will be executed if there are valid data. The predicted damage class of the building will be displayed in the Template.

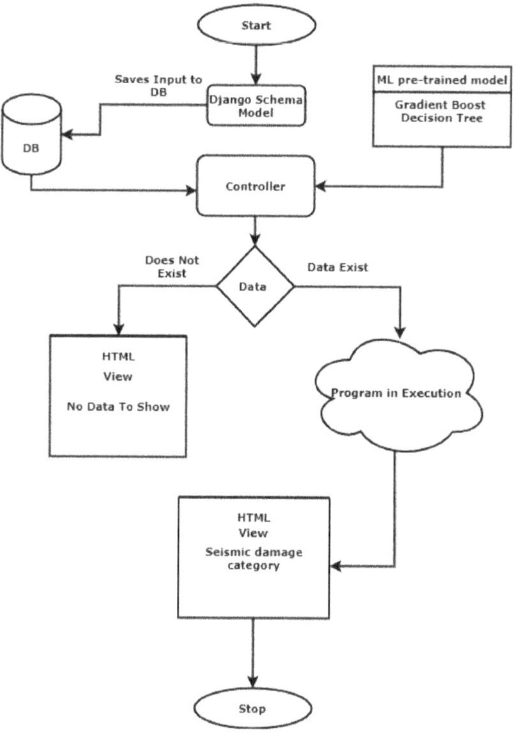

Figure 10. The flow of the web application based on the Django framework.

3.5.2. Project Structure

The project is divided into two main applications to separate concerns for different responsibilities across the project. The web-interface created is not hosted online; however, Github links are present to take a reference of the code and further contribute to it. The link for developing the classifiers is available at https://github.com/vandana1145/MasterThesisProjectCode/tree/feature/branch (accessed on 3 April 2022), and the Django code for the web-interface is linked at https://github.com/vandana1145/MasterThesisProjectWeb/tree/master (accessed on 3 April 2022). Figure 11 shows the sample index page of the web application. The web-based application developed for this research has the below main sections:

1. Home Page: This application provides the basic home view of the project. The user can see the blank input fields to fill the values of required attributes.
2. Predictor: The Submit button will have the functionality to use the ML model and to generate predictions.
3. Database (DB): The database or DB button on the home page will connect to another page where all the user-inserted inputs with their predicted values will be saved. In case of no data, the database table will show a blank page.

Modern websites come with a web-based administration backend as standard. Trusted site administrators can utilize the administrative interface or admin to create, update, and publish content, manage site users, and conduct other administrative activities.

Django includes a built-in administration interface. Django's admin allows authenticating users automatically, showing and managing forms, and validating data. Django also has a user-friendly interface for managing model data. Figure 12 shows the screenshot for the admin page. As an example, one admin is registered for the site.

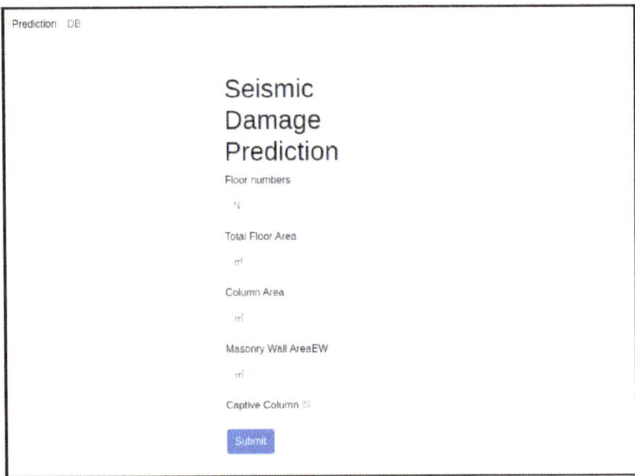

Figure 11. The home page with input fields and the submit button.

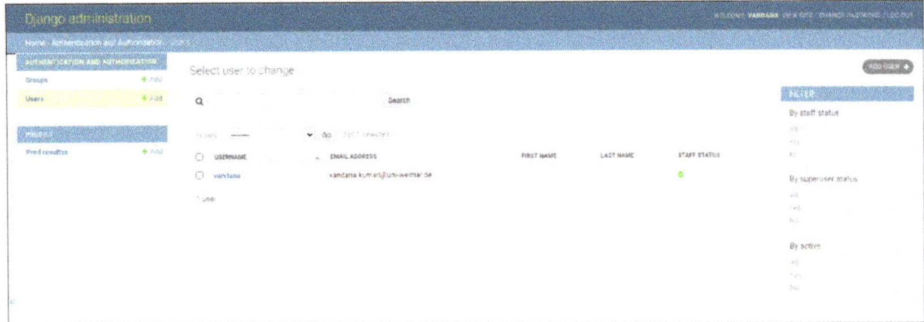

Figure 12. Django admin page.

Once the model in Django is defined, the admin application interface can populate the GBDT model table in the database. The admin application is accessible by reaching this URL: "localhost: 8000/admin". Figure 13 shows the "Pred results" in the admin panel.

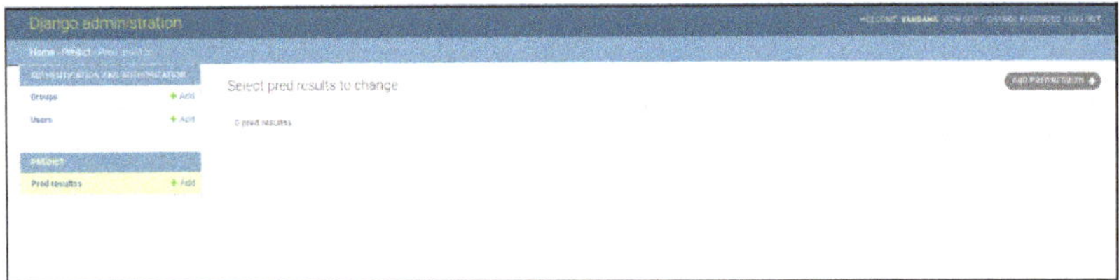

Figure 13. The table for predicted results available in the admin panel.

4. Results and Discussion

In past years, there have been substantial improvements in the design codes and safety measures of modern buildings. However, existing RC buildings which do not satisfy code-based design criteria are in danger of severe damages under future seismic events. Consequently, an efficient, rapid, and reliable approach is of crucial importance to assess the seismic vulnerability and risk levels of a large stock of such buildings in a reasonable timeframe which can directly enhance post-disaster crisis management plans.

An ML model for estimating the post-earthquake structural safety assessment is created by integrating ML methods with an input dataset comprising 526 data points and four damage classification scales. The response and damage patterns are mapped to their categorized structural safety states using ML techniques such as DT, GBDT, RF, NB, KNN, and SVM. The predictor space is iteratively partitioned to capture the underlying connection. The predictive models reported in this study, trained using ML techniques, can make moderately accurate, sensitive, and specific predictions. When applied to response patterns with unknown safety states, the highest prediction accuracy is 57% for GBDT. A trained GBDT classifier was incorporated into a web tool based on the Django framework.

The proposed approach is applied to two case studies. Because there was no separate exclusive dataset to examine the performance of the classifier, the test was performed on several data points from available datasets of the Nepal [52] and Pohang [53] earthquakes. The tool was examined, and the obtained predictions were recorded as shown in Tables 5 and 6 for the seismic data of the Nepal and Pohang earthquakes, respectively.

Table 5. Tested Predictions based on Nepal Earthquake Data.

Target Class	Number of Data Predicted	Correctly Predicted Data	Accuracy (in %)
1	7	2	28.57
2	7	4	57.14
3	7	2	28.57
4	7	5	71.40

Table 6. Tested Predictions based on Pohang Earthquake Data.

Target Class	Number of Data Predicted	Correctly Predicted Data	Accuracy (in %)
1	7	3	42.85
2	7	4	57.14
3	7	1	14.28
4	7	4	57.14

There were seven data points selected from each damage class for this study. In the case of the Nepal earthquake, the tool predicted class 4 with a higher accuracy than other damage classes. The only two true positives are for class 1 and class 3. Similarly, for the Pohang earthquake, there were seven input points for each target class. Here, the classifier predicted four true positives for class 2 and five true positives for class 4. Only one correct prediction occurred for class 3.

Building damage assessment is time-consuming and challenging due to the geographical dispersion of effects in an earthquake-affected area. The proposed technology might allow for quick evaluations of seismic damages. However, depending on the technique employed for the assessment, forecasting the spatial distribution of building damage with fair accuracy may vary.

The technique's performance was estimated based on the case study results accomplished in the previous section. Therefore, one can interpret that the tool is fairly robust. However, although the realized model predicts with more than 50% accuracy, the performance has to be validated for utilization of the tool on a more extensive scale and in real-time.

5. Conclusions

There are existing RC structures which do not satisfy the code-based design and strength criteria and hence are not code-compliant. These structures are vulnerable to significant damages in the case of the next moderate to severe seismic events. Several factors contribute to seismic vulnerabilities of such buildings, including the application of obsolete building codes, poor design procedures, and non-standard construction techniques. The majority of these old RC buildings are still in use; therefore, the application of a rapid, efficient, and reliable seismic vulnerability assessment is of vital importance. This study has tried to integrate some of the most modern soft computing techniques with the traditional RVS methods to achieve a more enhanced rapid risk-based analysis methodology. The literature has demonstrated promising capabilities of soft computing techniques in achieving high levels of performance for solving a variety of intricate problems. The application of these methodologies to RVS techniques for the rapid vulnerability assessment of existing buildings can substantially minimize the subjectivity of evaluators, uncertainties, and vagueness and increase accuracy. Following an earthquake, the proposed framework in this research may be used to make a quick probabilistic evaluation of whether a damaged structure is safe to reoccupy. In addition, the model created by the ML algorithms may be used to prioritize field inspections after an earthquake. Furthermore, the probabilistic safety state forecasts might be employed in community resilience assessments to examine and improve the life-cycle performance of individual buildings.

A web development framework was necessary in order to construct the web application for the purpose of this study. Django was an appropriate choice because Python has widely been applied to analyze and process the datasets for constructing ML models. There are open-source and robust web development frameworks based on Python. A GBDT-trained model was implemented into a web browser-based simple application using the Django framework. The obtained results confirmed that the proposed approach for a rapid and inexpensive vulnerability assessment can be conducted. Information on seismic zone hazards is essential to extend the developed tool to other regions. Meanwhile, the more data are available, the more satisfactorily the ML technique converges statistically.

Nevertheless, the results emphasize that ML could be an exciting alternative in making a support tool for evaluating the vulnerability of buildings. Indeed, further investigations by increasing and diversifying the type of construction in the learning state could strengthen this last conclusion. Despite these limitations, this work has demonstrated the enormous potential of AI methods for the identification of vulnerable structures, decreasing future seismic hazards, and assisting decision-makers in post-disaster crisis management plans with direct outputs of saving time and costs.

Future Recommendations

This study covered a literature review of existing RVS methods for the vulnerability assessment of buildings and examined the integration of ML techniques with the traditional RVS to enhance its capabilities. Moreover, a web-based application was designed using the Django library of Python for the purpose of this research. Future investigations are necessary to validate the application of the proposed web-based tool to extend its utilization to other regions of the world, taking into account the seismicity characteristics and available seismic data in a given zone. Several recommendations for future research are given below:

- Future studies should consider additional data taking into account the structural system, scale, and damage classifications. In addition, the overall accuracy and robustness of prediction models can be enhanced in future research by adding more extensive datasets (e.g., numerous incidents) and additional site- (e.g., soil conditions) and building-specific predictor variables.
- Concept drift [83] is an ML phenomenon that focuses on data changes, resulting in the ML model's testing performance to deteriorate over time. Finally, in the case of RVS, a model's incorrect forecast might be dangerous as, over time, its performance may decrease. However, this effect can be checked by constantly updating the model and periodically re-fitting the model with new data. Therefore, in the long term, the effect of the concept drift needs to be addressed in any ML-based methodology.
- The web-based application is built in a test-driven development environment. The application is based on the Django framework and uses an internal WSGI gateway and an SQLite3 database. The server and database must be uploaded to the proper

server and databases expressly developed for production environments. For heavy-load traffic settings, open-source servers such as Apache and Nginx can be utilized. Other open-source databases, such as MySQL, offer greater security and a broader range of features in a production setting. These ideas will need further research and experimentation, which will be left to future projects.

Author Contributions: Conceptualization, E.H. and V.K.; methodology, E.H. and V.K.; validation, T.L. and E.H.; formal analysis, E.H., V.K. and S.R.; investigation, E.H. and V.K.; resources, E.H.; data curation, V.K. and E.H.; writing—original draft preparation, E.H. and V.K.; writing—review and editing, E.H., S.R. and V.K.; visualization, E.H. and V.K.; supervision, E.H. and T.L.; project administration, E.H. All authors have read and agreed to the published version of the manuscript.

Funding: This research received no external funding.

Institutional Review Board Statement: Not applicable.

Informed Consent Statement: Not applicable.

Data Availability Statement: Not applicable.

Acknowledgments: We acknowledge the support of the German Research Foundation (DFG) and the Bauhaus-Universität Weimar within the Open-Access Publishing Programme.

Conflicts of Interest: The authors declare no conflict of interest.

Abbreviations

The following abbreviations are used in this manuscript:

ANOVA	Analysis of Variance
AI	Artificial Intelligence
ANN	Artificial Neural Network
CART	Classification and Regression Tree
CT	Classification Tree
CNN	Convolution Neural Network
DB	Database
DT	Decision Tree
FEMA	Federal Emergency Management Agency
FLDA	Fisher's Linear Discriminant Analysis
FL	Fuzzy Logic
GB	Gradient Boost
GBDT	Gradient Boost Decision Tree
K	Kernel
KNN	K-Nearest Neighbor
ML	Machine Learning
MISDR	Maximum Inter-Story Drift Ratio
MFPN	Multilayer Feedforward Perceptron Networks
MLP-NN	Multilayer Perceptron Neural Network
MVC	Model-View-Controller
MVT	Model-View-Template
NB	Naive Bayes
NCREE	National Center for Research on Earthquake Engineering
NN	Neural Network
PLS-DA	Partial Least Squares Discriminant
RBF	Radial basis function
RF	Random Forest
RVS	Rapid Visual Screening
RC	Reinforced Concrete
SVM	Support Vector Machine
SMOTE	Synthetic Minority Oversampling Technique
UI	User Interface
WSGI	Web Server Gateway Interface

References

1. Dixit, A.; Shrestha, S.; Parajuli, Y.; Thapa, M. Preparing for a Major Earthquake in Nepal: Achievements and Lessons. In Proceedings of the 12th World Conference on Earthquake Engineering, Lisbon, Portugal, 24–28 September 2012.
2. Emami, M.J.; Tavakoli, A.R.; Alemzadeh, H.; Abdinejad, F.; Shahcheraghi, G.; Erfani, M.A.; Mozafarian, K.; Solooki, S.; Rezazadeh, S.; Ensafdaran, A.; et al. Strategies in evaluation and management of Bam earthquake victims. *Prehospital Disaster Med.* **2005**, *20*, 327–330. [CrossRef] [PubMed]
3. Doocy, S.; Daniels, A.; Aspilcueta, D.; Team, I.J.C.S.; others. Mortality and injury following the 2007 Ica earthquake in Peru. *Am. J. Disaster Med.* **2009**, *4*, 15–22. [CrossRef] [PubMed]
4. Nie, H.; Tang, S.Y.; Lau, W.B.; Zhang, J.C.; Jiang, Y.W.; Lopez, B.L.; Ma, X.L.; Cao, Y.; Christopher, T.A. Triage during the week of the Sichuan earthquake: A review of utilized patient triage, care, and disposition procedures. *Injury* **2011**, *42*, 515–520. [CrossRef] [PubMed]
5. Gamulin, A.; Villiger, Y.; Hagon, O. Disaster medicine: Mission Haiti. *Rev. Med. Suisse* **2010**, *6*, 973–977.
6. Stein, S.; Geller, R.J.; Liu, M. Why earthquake hazard maps often fail and what to do about it. *Tectonophysics* **2012**, *562*, 1–25. [CrossRef]
7. Federal Emergency Management Agency. *Rapid Visual Screening of Buildings for Potential Seismic Hazards: Supporting Documentation*; Government Printing Office: Washington, DC, USA, 2015.
8. Ozcebe, G.; Yucemen, M.S.; Aydogan, V. Statistical seismic vulnerability assessment of existing reinforced concrete buildings in Turkey on a regional scale. *J. Earthq. Eng.* **2004**, *8*, 749–773. [CrossRef]
9. Alvanitopoulos, P.; Andreadis, I.; Elenas, A. Neuro-fuzzy techniques for the classification of earthquake damages in buildings. *Measurement* **2010**, *43*, 797–809. [CrossRef]
10. Xie, Y.; Ebad Sichani, M.; Padgett, J.E.; DesRoches, R. The promise of implementing machine learning in earthquake engineering: A state-of-the-art review. *Earthq. Spectra* **2020**, *36*, 1769–1801. [CrossRef]
11. Lazaridis, P.C.; Kavvadias, I.E.; Demertzis, K.; Iliadis, L.; Vasiliadis, L.K. Structural Damage Prediction of a Reinforced Concrete Frame under Single and Multiple Seismic Events Using Machine Learning Algorithms. *Appl. Sci.* **2022**, *12*, 3845. [CrossRef]
12. Morfidis, K.; Kostinakis, K. Use of Artificial Neural Networks in the R/C Buildings' Seismic Vulnerabilty Assessment: The Practical Point of View. In Proceedings of the 7th ECCOMAS Thematic Conference on Computational Methods in Structural Dynamics and Earthquake Engineering, Crete, Greece, 24–26 June 2019; pp. 24–26.
13. Yong, C.; Ling, C.; Güendel, F.; Kulhánek, O.; Juan, L. Seismic hazard and loss estimation for Central America. *Nat. Hazards* **2002**, *25*, 161–175. [CrossRef]
14. Bilgin, H.; Shkodrani, N.; Hysenlliu, M.; Ozmen, H.B.; Isik, E.; Harirchian, E. Damage and performance evaluation of masonry buildings constructed in 1970s during the 2019 Albania earthquakes. *Eng. Fail. Anal.* **2022**, *131*, 105824. [CrossRef]
15. Anbarci, N.; Escaleras, M.; Register, C.A. Earthquake fatalities: The interaction of nature and political economy. *J. Public Econ.* **2005**, *89*, 1907–1933. [CrossRef]
16. Rashid, M.; Ahmad, N. Economic losses due to earthquake-induced structural damages in RC SMRF structures. *Cogent Eng.* **2017**, *4*, 1296529. [CrossRef]
17. Işık, E.; Harirchian, E.; Büyüksaraç, A.; Ekinci, Y.L. Seismic and structural analyses of the eastern anatolian region (Turkey) using different probabilities of exceedance. *Appl. Syst. Innov.* **2021**, *4*, 89. [CrossRef]
18. Mandas, A.; Dritsos, S. Vulnerability assessment of RC structures using fuzzy logic. *WIT Trans. Ecol. Environ.* **2004**, *77*, 1–10. [CrossRef]
19. Demartinos, K.; Dritsos, S. First-level pre-earthquake assessment of buildings using fuzzy logic. *Earthq. Spectra* **2006**, *22*, 865–885. [CrossRef]
20. Tesfamariam, S.; Saatcioglu, M. Risk-based seismic evaluation of reinforced concrete buildings. *Earthq. Spectra* **2008**, *24*, 795–821. [CrossRef]
21. Harirchian, E.; Lahmer, T. Improved Rapid Assessment of Earthquake Hazard Safety of Structures via Artificial Neural Networks. In *IOP Conference Series: Materials Science and Engineering*; IOP Publishing: Bristol, UK, 2020; Volume 897, p. 012014.
22. Riedel, I.; Guéguen, P.; Dalla Mura, M.; Pathier, E.; Leduc, T.; Chanussot, J. Seismic vulnerability assessment of urban environments in moderate-to-low seismic hazard regions using association rule learning and support vector machine methods. *Nat. Hazards* **2015**, *76*, 1111–1141. [CrossRef]
23. Shah, M.F.; Ahmed, A.; Kegyes-B, O.K. A Case Study Using Rapid Visual Screening Method to Determine the Vulnerability of Buildings in two Districts of Jeddah, Saudi Arabia. In Proceedings of the 15th International Symposium on New Technologies for Urban Safety of Mega Cities in Asia, Tacloban, Philippines, 7–9 November 2016.
24. Calvi, G.M. A displacement-based approach for vulnerability evaluation of classes of buildings. *J. Earthq. Eng.* **1999**, *3*, 411–438. [CrossRef]
25. Fäh, D.; Kind, F.; Lang, K.; Giardini, D. Earthquake scenarios for the city of Basel. *Soil Dyn. Earthq. Eng.* **2001**, *21*, 405–413. [CrossRef]
26. Harirchian, E.; Hosseini, S.E.A.; Jadhav, K.; Kumari, V.; Rasulzade, S.; Işık, E.; Wasif, M.; Lahmer, T. A review on application of soft computing techniques for the rapid visual safety evaluation and damage classification of existing buildings. *J. Build. Eng.* **2021**, *43*, 102536. [CrossRef]

27. Morfidis, K.; Kostinakis, K. Seismic parameters' combinations for the optimum prediction of the damage state of R/C buildings using neural networks. *Adv. Eng. Softw.* **2017**, *106*, 1–16. [CrossRef]

28. Morfidis, K.; Kostinakis, K. Approaches to the rapid seismic damage prediction of r/c buildings using artificial neural networks. *Eng. Struct.* **2018**, *165*, 120–141. [CrossRef]

29. Tesfamariam, S.; Liu, Z. Earthquake induced damage classification for reinforced concrete buildings. *Struct. Saf.* **2010**, *32*, 154–164. [CrossRef]

30. Zhang, Y.; Burton, H.V.; Sun, H.; Shokrabadi, M. A machine learning framework for assessing post-earthquake structural safety. *Struct. Saf.* **2018**, *72*, 1–16. [CrossRef]

31. Allali, S.A.; Abed, M.; Mebarki, A. Post-earthquake assessment of buildings damage using fuzzy logic. *Eng. Struct.* **2018**, *166*, 117–127. [CrossRef]

32. Yao, X.; Tham, L.; Dai, F. Landslide susceptibility mapping based on support vector machine: A case study on natural slopes of Hong Kong, China. *Geomorphology* **2008**, *101*, 572–582. [CrossRef]

33. Esteban, M.; Valenzuela, V.P.; Yun, N.Y.; Mikami, T.; Shibayama, T.; Matsumaru, R.; Takagi, H.; Thao, N.D.; De Leon, M.; Oyama, T.; et al. Typhoon Haiyan 2013 evacuation preparations and awareness. *Int. J. Sustain. Future Hum. Secur.* **2015**, *3*, 37–45. [CrossRef]

34. Tripathi, S.; Srinivas, V.; Nanjundiah, R.S. Downscaling of precipitation for climate change scenarios: A support vector machine approach. *J. Hydrol.* **2006**, *330*, 621–640. [CrossRef]

35. Harirchian, E.; Lahmer, T.; Kumari, V.; Jadhav, K. Application of Support Vector Machine Modeling for the Rapid Seismic Hazard Safety Evaluation of Existing Buildings. *Energies* **2020**, *13*, 3340. [CrossRef]

36. Gilan, S.S.; Ali, A.M.; Ramezanianpour, A.A. Evolutionary fuzzy function with support vector regression for the prediction of concrete compressive strength. In Proceedings of the 2011 UKSim 5th European Symposium on Computer Modeling and Simulation, Madrid, Spain, 16–18 November 2011; pp. 263–268.

37. Sobhani, J.; Khanzadi, M.; Movahedian, A. Support vector machine for prediction of the compressive strength of no-slump concrete. *Comput. Concr.* **2013**, *11*, 337–350. [CrossRef]

38. Sun, J.; Li, H.; Adeli, H. Concept drift-oriented adaptive and dynamic support vector machine ensemble with time window in corporate financial risk prediction. *IEEE Trans. Syst. Man Cybern. Syst.* **2013**, *43*, 801–813. [CrossRef]

39. Zhang, Z.; Hsu, T.Y.; Wei, H.H.; Chen, J.H. Development of a data-mining technique for regional-scale evaluation of building seismic vulnerability. *Appl. Sci.* **2019**, *9*, 1502. [CrossRef]

40. Cannizzaro, F.; Pantò, B.; Lepidi, M.; Caddemi, S.; Caliò, I. Multi-directional seismic assessment of historical masonry buildings by means of macro-element modelling: Application to a building damaged during the L'Aquila earthquake (Italy). *Buildings* **2017**, *7*, 106. [CrossRef]

41. Fagundes, C.; Bento, R.; Cattari, S. On the seismic response of buildings in aggregate: Analysis of a typical masonry building from Azores. *Structures* **2017**, *10*, 184–196. [CrossRef]

42. Casapulla, C.; Argiento, L.U.; Maione, A. Seismic safety assessment of a masonry building according to Italian Guidelines on Cultural Heritage: Simplified mechanical-based approach and pushover analysis. *Bull. Earthq. Eng.* **2018**, *16*, 2809–2837. [CrossRef]

43. Greco, A.; Lombardo, G.; Pantò, B.; Famà, A. Seismic vulnerability of historical masonry aggregate buildings in oriental Sicily. *Int. J. Archit. Herit.* **2020**, *14*, 517–540. [CrossRef]

44. Lin, P.C.; Tsai, K.C.; Wang, K.J.; Yu, Y.J.; Wei, C.Y.; Wu, A.C.; Tsai, C.Y.; Lin, C.H.; Chen, J.C.; Schellenberg, A.H.; et al. Seismic design and hybrid tests of a full-scale three-story buckling-restrained braced frame using welded end connections and thin profile. *Earthq. Eng. Struct. Dyn.* **2012**, *41*, 1001–1020. [CrossRef]

45. Bengio, Y.; Courville, A.; Vincent, P. Representation learning: A review and new perspectives. *IEEE Trans. Pattern Anal. Mach. Intell.* **2013**, *35*, 1798–1828. [CrossRef]

46. Schmidhuber, J. Deep learning. *Scholarpedia* **2015**, *10*, 32832. [CrossRef]

47. Singh, R.; Qi, Y. Character based string kernels for bio-entity relation detection. In Proceedings of the 15th Workshop on Biomedical Natural Language Processing, Berlin, Germany, 12 August 2016; pp. 66–71.

48. Vapnik, V.N. An overview of statistical learning theory. *IEEE Trans. Neural Netw.* **1999**, *10*, 988–999. [CrossRef]

49. Gonzalez, D.; Rueda-Plata, D.; Acevedo, A.B.; Duque, J.C.; Ramos-Pollan, R.; Betancourt, A.; Garcia, S. Automatic detection of building typology using deep learning methods on street level images. *Build. Environ.* **2020**, *177*, 106805. [CrossRef]

50. Harirchian, E.; Kumari, V.; Jadhav, K.; Rasulzade, S.; Lahmer, T.; Raj Das, R. A Synthesized Study Based on Machine Learning Approaches for Rapid Classifying Earthquake Damage Grades to RC Buildings. *Appl. Sci.* **2021**, *11*, 7540. [CrossRef]

51. Catlin, A.C.; HewaNadungodage, C.; Pujol, S.; Laughery, L.; Sim, C.; Puranam, A.; Bejarano, A. A cyberplatform for sharing scientific research data at DataCenterHub. *Comput. Sci. Eng.* **2018**, *20*, 49–70. [CrossRef]

52. Shah, P.; Pujol, S.; Puranam, A.; Laughery, L. *Database on Performance of Low-Rise Reinforced Concrete Buildings in the 2015 Nepal Earthquake*; DEEDS, Purdue University Research Repository: Lafayette, IN, USA, 2015.

53. Sim, C.; Laughery, L.; Chiou, T.; Weng, P.W. *2017 Pohang Earthquake: Reinforced Concrete Building Damage Survey*; DEEDS, Purdue University Research Repository: Lafayette, IN, USA, 2018.

54. Sim, C.; Villalobos, E.; Smith, J.P.; Rojas, P.; Pujol, S.; Puranam, A.Y.; Laughery, L.A. *Performance of Low-Rise Reinforced Concrete Buildings in the 2016 Ecuador Earthquake*; Purdue University Research Repository: Purdue, IN, USA, 2018. [CrossRef]

55. Patton, J. Earthquake Report: 2010 Haiti M 7.0. Avaliable online: http://earthjay.com/?p=9178 (accessed on 2 January 2022).
56. Ilyas, I.F.; Chu, X. *Data Cleaning*; ACM: New York, NY, USA, 2019.
57. Bishop, C *Neural Networks for Pattern Recognition*; Oxford University Press: Oxford, UK, 1995.
58. Kuhn, M.; Johnson, K. *Applied Predictive Modeling*; Springer: New York, NY, USA, 2013; Volume 26.
59. Chawla, N.V.; Bowyer, K.W.; Hall, L.O.; Kegelmeyer, W.P. SMOTE: Synthetic minority over-sampling technique. *J. Artif. Intell. Res.* **2002**, *16*, 321–357. [CrossRef]
60. Villalobos, E.; Sim, C.; Smith-Pardo, J.P.; Rojas, P.; Pujol, S.; Kreger, M.E. The 16 April 2016 Ecuador earthquake damage assessment survey. *Earthq. Spectra* **2018**, *34*, 1201–1217. [CrossRef]
61. Bouazizi, M.; Ohtsuki, T. Sentiment analysis: From binary to multi-class classification: A pattern-based approach for multi-class sentiment analysis in Twitter. In Proceedings of the 2016 IEEE International Conference on Communications (ICC), Kuala Lumpur, Malaysia, 22–27 May 2016; pp. 1–6.
62. Cortes, C.; Vapnik, V. Support-vector networks. *Mach. Learn.* **1995**, *20*, 273–297. [CrossRef]
63. Gärtner, T.; Flach, P.A.; Kowalczyk, A.; Smola, A.J. Multi-instance kernels. In Proceedings of the Nineteenth International Conference on Machine Learning, San Francisco, CA, USA, 8–12 July 2002; Volume 2, p. 7.
64. Timofeev, R. Classification and Regression Trees (CART) Theory and Applications. Master's Thesis, Humboldt University, Berlin, Germany, 2004; pp. 1–40.
65. Lindley, D.V. Fiducial distributions and Bayes' theorem. *J. R. Stat. Soc. Ser. B (Methodol.)* **1958**, *20*, 102–107. [CrossRef]
66. Fix, E. *Discriminatory Analysis: Nonparametric Discrimination, Consistency Properties*; USAF School of Aviation Medicine, Randolph field, TX, USA : 1985; Volume 1.
67. Cover, T.; Hart, P. Nearest neighbor pattern classification. *IEEE Trans. Inf. Theory* **1967**, *13*, 21–27. [CrossRef]
68. Breiman, L. Random forests. *Mach. Learn.* **2001**, *45*, 5–32. [CrossRef]
69. Breiman, L. Bagging predictors. *Mach. Learn.* **1996**, *24*, 123–140. [CrossRef]
70. Ho, T.K. Random decision forests. In Proceedings of the 3rd International Conference on Document Analysis and Recognition, Montreal, QC, Canada, 14–16 August 1995; Volume 1, pp. 278–282.
71. Ho, T.K. The random subspace method for constructing decision forests. *IEEE Trans. Pattern Anal. Mach. Intell.* **1998**, *20*, 832–844.
72. Amit, Y.; Geman, D. Shape quantization and recognition with randomized trees. *Neural Comput.* **1997**, *9*, 1545–1588. [CrossRef]
73. Fawagreh, K.; Gaber, M.M.; Elyan, E. Random forests: From early developments to recent advancements. *Syst. Sci. Control Eng. Open Access J.* **2014**, *2*, 602–609. [CrossRef]
74. Friedman, J.H. Greedy function approximation: A gradient boosting machine. *Ann. Stat.* **2001**, *29*, 1189–1232. [CrossRef]
75. Li, P. Robust logitboost and adaptive base class (abc) logitboost. *arXiv* **2012**, arXiv:1203.3491.
76. Richardson, M.; Dominowska, E.; Ragno, R. Predicting clicks: Estimating the click-through rate for new ads. In Proceedings of the 16th International Conference on World Wide Web, Banff, AB, Canada, 8–12 May 2007; pp. 521–530.
77. Burges, C.J. From ranknet to lambdarank to lambdamart: An overview. *Learning* **2010**, *11*, 81.
78. Friedman, J.; Hastie, T.; Tibshirani, R.; others. Additive logistic regression: A statistical view of boosting (with discussion and a rejoinder by the authors). *Ann. Stat.* **2000**, *28*, 337–407. [CrossRef]
79. Izquierdo, J.L.C.; Cabot, J. The role of foundations in open source projects. In Proceedings of the 40th International Conference on Software Engineering: Software Engineering in Society, Gothenburg, Sweden, 27 May–3 June 2018; pp. 3–12.
80. Robenolt, M. Scaling Django to 8 Billion Page Views. 2013. Available online: https://blog.disqus.com/scaling-django-to-8-billion-page-views (accessed on 2 January 2022).
81. Sathya, R.; Abraham, A. Comparison of supervised and unsupervised learning algorithms for pattern classification. *Int. J. Adv. Res. Artif. Intell.* **2013**, *2*, 34–38. [CrossRef]
82. Ghahramani, Z. Unsupervised learning. In *Summer School on Machine Learning*; Springer: New York, NY, USA, 2003; pp. 72–112.
83. Tsymbal, A. *The Problem of Concept Drift: Definitions and Related Work*; Computer Science Department, Trinity College Dublin: Dublin, Ireland, 2004; Volume 106, p. 58.

 *buildings*

Article

Seismic Fragility Analysis of Mega-Frame with Vibration Control Substructure Based on Dual Surrogate Model and Active Learning

Yanjie Xiao, Xun'an Zhang, Feng Yue, Muhammad Moman Shahzad, Xinwei Wang and Buqiao Fan *

School of Mechanics, Civil Engineering and Architecture, Northwestern Polytechnical University, Xi'an 710072, China; yjxiaoyj@163.com (Y.X.); jiaoping@nwpu.edu.cn (X.Z.); yuefeng@mail.nwpu.edu.cn (F.Y.); m.mominshahzad@mail.nwpu.edu.cn (M.M.S.); wangxw@mail.nwpu.edu.cn (X.W.)
* Correspondence: buqiao@mail.nwpu.edu.cn

Abstract: Seismic fragility analysis of a mega-frame with vibration control substructure (MFVCS) considering structural uncertainties is computationally expensive. Dual surrogate model (DSM) can be used to improve computational efficiency, whereas the proper selection of design of experiments (DoE) is a difficult work in the DSM-based seismic fragility analysis (DSM-SFA) method. To efficiently assess the seismic fragility with sufficient accuracy, this paper proposes an improved DSM-SFA method based on active learning (AL). In this method, the Kriging model is employed for surrogate modeling to obtain the predicted error of approximation. An AL sampling strategy is presented to update the DoE adaptively, and the refinement of the surrogate models can reduce the error of the probability result computed by the Monte Carlo (MC) simulation. A numerical example was studied to verify the effectiveness and feasibility of the improved procedure. This method was applied to the fragility analysis of an MFVCS and a mega-frame structure (MFS). The finite element models were established using OpenSeesPy and SAP2000 software, respectively, and the correctness of the MFVCS model was verified. The results show that MFVCS is less vulnerable than MFS and has better seismic performance.

Keywords: seismic fragility analysis; dual surrogate model; Kriging model; active learning; mega-frame with vibration control substructure

Citation: Xiao, Y.; Zhang, X.; Yue, F.; Shahzad, M.M.; Wang, X.; Fan, B. Seismic Fragility Analysis of Mega-Frame with Vibration Control Substructure Based on Dual Surrogate Model and Active Learning. *Buildings* **2022**, *12*, 752. https://doi.org/10.3390/buildings12060752

Academic Editors: Tom Lahmer, Ehsan Harirchian and Viviana Novelli

Received: 14 April 2022
Accepted: 30 May 2022
Published: 1 June 2022

Publisher's Note: MDPI stays neutral with regard to jurisdictional claims in published maps and institutional affiliations.

1. Introduction

With the growth of the urban population, the demand for high-rise buildings is increasing. Mega-frame structure (MFS) is a type of structure that appeared in response to this demand. The megastructure of MFS consists of mega beams and mega columns. Solid web tubes or trusses are adopted as mega columns, and large hollow rectangular members or trusses are used as mega beams. The substructure is composed of ordinary beams and columns. Vertical loads on the substructure are transferred to the mega beams. The lateral stiffness of MFS depends on the stiffness of mega beams and columns. The substructures can be designed in various forms [1]. To reduce the dynamic response, Feng et al. [2] proposed an improved MFS called mega-frame with a vibration control substructure (MFVCS) with reference to the vibration control principle of tuned mass damper (TMD). MFVCS is also known as a mega-sub controlled structure. In this new structure, dampers or isolators were installed between the substructures and megastructures (Figure 1), and the function of the substructure is similar to that of the additional mass in TMD [3]. Many experiments and analyses showed that MFVCS has better response control ability than MFS [4–7]. Recent investigations have been devoted to further expanding and improving the design theory of MFVCS. Kalehsar et al. [8] compared the along-wind and crosswind responses of MFVCS and TMD, and investigated the influence of the mass ratio and stiffness ratio of the megastructure to the substructure on wind-induced vibration control of MFVCS.

Fan et al. [9] calculated the failure path of MFVCS under random earthquake excitation by using the weighted rank–sum ratio method, and analyzed the weak members of the structure. Abdulhadi et al. [10] explored the appropriate stiffness ratio and mass ratio to reduce the seismic response of MFVCS. Shahzad et al. [11] proposed a modified MFVCS structure with isolators installed on mega columns.

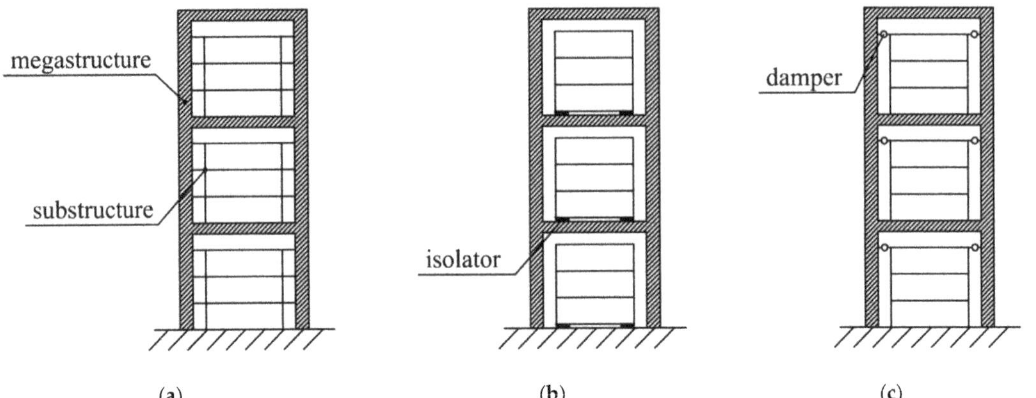

Figure 1. Configurations of MFVCS. (**a**) MFS. (**b**) MFVCS with isolators. (**c**) MFVCS with dampers.

Since seismic design is an essential part of the design of MFVCS, seismic risk assessment is an issue worthy of attention. Seismic fragility analysis is an effective tool for this assessment, which provides an engineering indicator for decision-making in the design of new structures and the retrofitting of existing structures [12]. Seismic fragility analysis plays an important role in performance-based seismic engineering. There are three main types of fragility analysis methods: the expert opinion method, seismic damage investigation method, and analytical method [13]. The expert opinion method obtains the analysis results by collecting the evaluation opinions of experts, which is influenced by the expert experience. The seismic damage investigation method evaluates the fragility according to the actual seismic damage data, which is not suitable for areas lacking survey data. The analytical method uses numerical simulation to analyze the seismic risk, which is more commonly used than the first two methods. The fragility curves obtained from the fragility analysis can be used to assess the post-earthquake loss, which provides a basis for the seismic design of structures. In the past study [14], the incremental dynamic analysis (IDA) method was used to analyze the seismic fragility of MFVCS without considering the influence of structural uncertainties on seismic risk. In fact, although the record-to-record (RTR) variability is the main factor causing the variation of structural response [15,16], structural uncertainties may also affect the fragility curves [17]. In addition, the vibration control effect of MFVCS is related to the structural stiffness and damping device properties, and there are certain uncertainties in these parameters. Therefore, it is more reasonable to take the randomness of structural parameters into account in the seismic fragility analysis of MFVCS. Monte Carlo (MC) simulation can directly incorporate structural uncertainties into seismic fragility analysis, but needs a great number of calls to nonlinear time history analysis (NLTHA). Due to the complexity of MFVCS, repeated finite element analysis (FEA) will cause a huge computational burden.

Surrogate model approaches can give the approximations of structural responses and are very popular in reducing the computational cost. The dual surrogate model (DSM) framework is often used to describe the variability of optimization objectives in robust optimization, which employs two surrogate models to predict the mean and standard deviation of structural performance, respectively [18–21]. Towashiraporn [22] proposed the DSM-based seismic fragility analysis (DSM-SFA) method, in which DSM was introduced into seismic fragility analysis to tackle the RTR variability. DSM-SFA has been widely used in studies of structures subjected to seismic excitation, including bridges [23], base-isolated nuclear power plants [24], liquid storage tanks [25], and frame structures [26]. In addition, DSM has also been used to deal with problems considering other loads with RTR variabilities, such as wind [27], wave [28,29], and blast [30]. However, DSM-SFA has high requirements for the selection of the DoE used to build the surrogate models. The need for the number and distribution of experiment points is different for different structures, which is difficult to know before surrogate modeling. The one-shot sampling commonly used in the DSM framework may not obtain a suitable DoE, which will lead to the inaccuracy of fragility curves and waste of computing time.

In recent years, active learning (AL) sampling has been introduced into reliability analysis to improve the DoE [31,32]. The AL-based reliability analysis (AL-RA) methods usually adopt the Kriging model as the surrogate model. Taking advantage of the model's ability to provide probability distributions of unknown values, the AL-RA methods select new samples sequentially from the candidate set to obtain the DoE, instead of choosing experiments by one-shot sampling. These methods can significantly reduce the calculation burden of MC simulation without much compromise on the accuracy of results, and have advantages in reliability analysis of large complex structures. However, the AL sampling strategies in AL-RA generally cannot be used in DSM-SFA. On the one hand, the prediction of each performance function in AL-RA is evaluated by one surrogate model, while two surrogate models are needed in DSM-SFA to estimate the seismic demand. On the other hand, reliability analysis is to calculate the failure probability of the structure, while seismic fragility analysis aims at obtaining the fragility curves which can describe the conditional failure probability corresponding to a given intensity measure (IM).

To accurately assess the seismic fragility of MFVCS in an efficient way, an improved DSM-SFA method based on AL is proposed. In the paper, the DSM framework for fragility analysis and the Kriging model are first reviewed. Then, the proposed method is described, and its effectiveness is validated by using an example. Finally, the finite element model of MFVCS is established and verified in OpenSeesPy, and the fragility analysis results of MFVCS and MFS considering structural uncertainties are compared.

2. DSM for Seismic Fragility Analysis

2.1. DSM-SFA

In DSM-SFA, both structural uncertainties and RTR variability are taken into account. Given the structural parameters $\mathbf{s}$ and IM value im, the engineering demand parameter (EDP) D of the structure under seismic load is assumed to be a random variable [24]. The mean and standard deviation of EDP vary with the structural parameters and IM. Usually, the lognormal EDP assumption [14,33] is adopted in seismic fragility analysis. To consider the RTR variability, multiple seismic records are required in structural analysis.

The surrogate model is constructed based on a DoE $\mathbf{X} = \left[\mathbf{x}^{(1)}, \mathbf{x}^{(2)}, \ldots, \mathbf{x}^{(m)}\right]$ and the outputs $\mathbf{y} = \left[y^{(1)}, y^{(2)}, \ldots, y^{(m)}\right]^{\mathrm{T}}$, in which $\mathbf{x}^{(i)}$ denotes an n-dimensional vector of input variables and $y^{(i)}$ denotes the function output of $\mathbf{X}^{(i)}$. Surrogate model approaches commonly used in DSM-SFA include the Kriging model and polynomial response surface.

The input variable vector $\mathbf{x}$ of the surrogate models in DSM-SFA consists of the structure parameters $\mathbf{s}$ and the IM variable im, i.e., $\mathbf{x}^{\mathrm{T}} = [\mathbf{s}^{\mathrm{T}}, im]$. To establish the model, a group of experiments is first sampled in the space of $\mathbf{x}$. At each experiment point, the mean $\mu_{\ln D}$ and standard deviation $\sigma_{\ln D}$ of the logarithms of EDP results can be computed by carrying out NLTHA for all selected earthquakes. Then, the surrogate models of $\mu_{\ln D}$ and $\sigma_{\ln D}$ can be constructed, respectively. The approximation of the unknown EDP is obtained [26] by:

$$\ln \hat{D} = \hat{\mu}_{\ln D}(\mathbf{x}) + N\left(0, \hat{\sigma}_{\ln D}^2(\mathbf{x})\right) \tag{1}$$

where $\hat{\mu}_{\ln D}(\mathbf{x})$ and $\hat{\sigma}_{\ln D}(\mathbf{x})$ are the predictions of $\mu_{\ln D}$ and $\sigma_{\ln D}$ given by the surrogate models, respectively, and $N\left(0, \hat{\sigma}^2_{\ln D}(\mathbf{x})\right)$ denotes a normally distributed variable whose mean is 0 and variance is $\hat{\sigma}^2_{\ln D}(\mathbf{x})$.

Once the surrogate models of the standard deviation and mean are built, the failure probability for a given IM can be calculated by MC simulation, in which $\hat{D}$ is used as a substitute for the true response to avoid performing NLTHA. Generally, a set of performance levels are considered in seismic fragility analysis, and each level corresponds to a limit state. The fragility curve corresponding to a performance level can be obtained by computing the failure probabilities for a series of IM values.

DSM-SFA requires much less computational time compared with the fragility analysis approach based on MC simulation, because the surrogate models of mean and standard deviation avoid repeated calls to FEA. However, since the DoE is usually obtained by one-shot sampling, the number of the selected experiments may not be appropriate. In addition, the generation of test samples used to verify the accuracy of the surrogate models also incurs additional computational costs.

2.2. Kriging Model

The Kriging model is an interpolation model composed of a parametric part and a nonparametric part [34], which is expressed as:

$$y(\mathbf{x}) = \mathbf{f}^{\mathrm{T}}(\mathbf{x})\boldsymbol{\beta} + z(\mathbf{x}) = \sum_{l=1}^{p} \beta_l f_l(\mathbf{x}) + z(\mathbf{x}) \tag{2}$$

where $y(\mathbf{x})$ denotes the unknown function to be fitted, $\mathbf{f}(\mathbf{x}) = [f_1(\mathbf{x}), f_2(\mathbf{x}), \ldots, f_p(\mathbf{x})]^{\mathrm{T}}$ is the vector of given basis functions, $\boldsymbol{\beta} = [\beta_1, \beta_2, \ldots, \beta_p]^{\mathrm{T}}$ is the vector of the regression coefficients, and $z(\mathbf{x})$ is a stationary Gaussian stochastic process with zero mean. The covariance of $z(\mathbf{x})$ is:

$$cov(z(\mathbf{x}^{(i)}), z(\mathbf{x}^{(j)})) = \sigma_z^2 R(\mathbf{x}^{(i)}, \mathbf{x}^{(j)}) \tag{3}$$

where σ_z^2 represents the process variance, and $R(\mathbf{x}^{(i)}, \mathbf{x}^{(j)})$ is the correlation function between the experiment points $\mathbf{x}^{(i)}$ and $\mathbf{x}^{(j)}$. The anisotropic squared-exponential correlation function is usually adopted, which is described as:

$$R(\mathbf{x}^{(i)}, \mathbf{x}^{(j)}) = \prod_{k=1}^{n} \exp[-\theta_k (x_k^{(i)} - x_k^{(j)})^2] \tag{4}$$

where θ_k is the k-th component of the undetermined parameters $\boldsymbol{\theta}$, and $x_k^{(i)}$ is the k-th component of $\mathbf{x}^{(i)}$.

At point $\mathbf{x}$, the prediction $\hat{y}(\mathbf{x})$ and the predicted variance $\sigma_{\hat{y}}^2(\mathbf{x})$ are given by:

$$\hat{y}(\mathbf{x}) = \mathbf{f}^{\mathrm{T}}(\mathbf{x})\hat{\boldsymbol{\beta}} + \mathbf{r}^{\mathrm{T}}(\mathbf{x})\mathbf{R}^{-1}(\mathbf{y} - \mathbf{F}\hat{\boldsymbol{\beta}}) \tag{5}$$

and

$$\sigma_{\hat{y}}^2(\mathbf{x}) = \hat{\sigma}_z^2[1 - \mathbf{r}^{\mathrm{T}}(\mathbf{x})\mathbf{R}^{-1}\mathbf{r}(\mathbf{x}) + \mathbf{u}^{\mathrm{T}}(\mathbf{x})(\mathbf{F}^{\mathrm{T}}\mathbf{R}^{-1}\mathbf{F})^{-1}\mathbf{u}(\mathbf{x})] \tag{6}$$

where $\mathbf{R}$ represents the $m \times m$ correlation matrix with component $R_{ij} = R(\mathbf{x}^{(i)}, \mathbf{x}^{(j)})$, $\mathbf{r} = [R(\mathbf{x}^{(1)}, \mathbf{x}), R(\mathbf{x}^{(2)}, \mathbf{x}), \ldots, R(\mathbf{x}^{(m)}, \mathbf{x})]^{\mathrm{T}}$ is the correlation vector composed of the correlation function values between $\mathbf{x}$ and the experiment points, $\mathbf{F}$ denotes the $m \times p$ matrix of basis function values with component $F_{il} = f_l(\mathbf{x}^{(i)})$, $\mathbf{u}(\mathbf{x}) = \mathbf{F}^{\mathrm{T}}\mathbf{R}^{-1}\mathbf{r}(\mathbf{x}) - \mathbf{f}(\mathbf{x})$, and $\hat{\boldsymbol{\beta}}$ and $\hat{\sigma}_z^2$ are the estimations of $\boldsymbol{\beta}$ and σ_z^2.

$\hat{\boldsymbol{\beta}}$ and $\hat{\sigma}_z^2$ are calculated by:

$$\hat{\boldsymbol{\beta}} = (\mathbf{F}^{\mathrm{T}}\mathbf{R}^{-1}\mathbf{F})^{-1}\mathbf{F}^{\mathrm{T}}\mathbf{R}^{-1}\mathbf{y} \tag{7}$$

and

$$\hat{\sigma}_z^2 = \frac{1}{m}(\mathbf{y} - \mathbf{F}\hat{\boldsymbol{\beta}})^{\mathrm{T}}\mathbf{R}^{-1}(\mathbf{y} - \mathbf{F}\hat{\boldsymbol{\beta}}) \tag{8}$$

The parameters $\boldsymbol{\theta}$ can be determined by maximizing the log-likelihood function expressed as:

$$lnL(\boldsymbol{\theta}) = -\frac{1}{2}\left(m\ln(\hat{\sigma}_z^2) + \ln|\mathbf{R}|\right) \tag{9}$$

Kriging model can be classed among the realm of Gaussian process methods [35]. The unknown function value in the Kriging model is considered to follow a normal distribution as:

$$y(\mathbf{x}) \sim N\left(\hat{y}(\mathbf{x}), \sigma_{\hat{y}}^2(\mathbf{x})\right) \tag{10}$$

In this work, the ordinary Kriging model was adopted, where $\mathbf{f}^{\mathrm{T}}(\mathbf{x})\boldsymbol{\beta} = \beta_1$.

3. Proposed Seismic Fragility Analysis Method

3.1. Surrogate Models for Estimating the EDP

To overcome the difficulty of controlling the accuracy of surrogate models in DSM-SFA, the Kriging model is employed for surrogate modeling in the improved method. Although the Kriging model has been applied in existing literature on fragility analysis [23,36], its ability to analyze errors has not been utilized.

To be able to evaluate the error of the EDP approximation, Equation (1) is transformed into:

$$\ln \hat{D} = \hat{\mu}_{\ln D}(\mathbf{x}) + v \cdot \hat{\sigma}_{\ln D}(\mathbf{x}) \tag{11}$$

where v follows the standard normal distribution. It can be seen from Equation (10) that $\mu_{\ln D} \sim N\left(\hat{\mu}_{\ln D}(\mathbf{x}), \sigma_{\hat{\mu}_{\ln D}}^2(\mathbf{x})\right)$ and $\sigma_{\ln D} \sim N\left(\hat{\sigma}_{\ln D}(\mathbf{x}), \sigma_{\hat{\sigma}_{\ln D}}^2(\mathbf{x})\right)$, where $\sigma_{\hat{\mu}_{\ln D}}(\mathbf{x})$ and $\sigma_{\hat{\sigma}_{\ln D}}(\mathbf{x})$ are the predicted standard deviations of $\mu_{\ln D}$ and $\sigma_{\ln D}$ provided by the surrogate models. Then, when $\mathbf{s}$, im, and v are fixed, the unknown value of logarithmic EDP follows the normal distribution and its predicted standard deviation can be obtained as:

$$\sigma_{\ln \hat{D}} = \sqrt{\sigma_{\hat{\mu}_{\ln D}}^2(\mathbf{x}) + v^2 \cdot \sigma_{\hat{\sigma}_{\ln D}}^2(\mathbf{x})} \tag{12}$$

Based on a group of MC samples $\mathbf{X}^{\mathrm{MC}} = [\mathbf{x}_1^{\mathrm{MC}}, \mathbf{x}_2^{\mathrm{MC}}, \ldots, \mathbf{x}_M^{\mathrm{MC}}]$, the probability that the EDP exceeds the h-th limit state for a given IM value im can be obtained by:

$$P[D \geq z_h | IM = im] \approx \frac{\sum\limits_{j=1}^{M} I[z_h - \hat{D}|_{\mathbf{X}_j^{\mathrm{MC}}}]}{M} \tag{13}$$

where z_h is the threshold of EDP corresponding to the h-th limit state $D = z_h$, and $I(\cdot)$ is an indicator function used to count the number of negative values. If the input is negative, the output of $I(\cdot)$ is 1; otherwise, it is equal to 0. Each sample in $\mathbf{X}^{\mathrm{MC}}$ consists of the given im and randomly generated $\mathbf{s}$ and v.

3.2. AL Sampling

To refine the surrogate models adaptively, an AL sampling strategy was presented.

The initial experiments were uniformly selected using the Latin hypercube sampling (LHS) method [32] from the space of $\mathbf{x}$ between the lower and upper bounds of each variable. The lower and upper bounds of the i-th element X_i in $\mathbf{x}$ can be $F_i^{-1}[\Phi(\pm 4)]$, where $F_i^{-1}(\cdot)$ denotes the inverse function of the probability distribution function of X_i, and $\Phi(\cdot)$ represents the probability distribution function of the standard normal distribution. The bounds of im can be the IM range concerned in the problem.

In the improvement of DoE, a candidate sample set $\mathbf{X}^{\mathrm{c}} = [\mathbf{x}_1^{\mathrm{c}}, \mathbf{x}_2^{\mathrm{c}}, \ldots, \mathbf{x}_r^{\mathrm{c}}]$ needs to be first generated in the space of variables containing v, $\mathbf{s}$, and im. It is worth noting that there is no need to perform a time history analysis on the candidate points. Then, the point $\mathbf{x}_{new}$ which is most favorable for reducing the error of the result is selected from the candidate samples in terms of the learning function value of each point, and the values of $\mu_{\ln D}$ and $\sigma_{\ln D}$ at this point are calculated. By adding $\mathbf{x}_{new}$ to the DoE as a new test point, the surrogate models are updated once. The surrogate models can be gradually refined by choosing new experiments sequentially in this way.

It can be seen that the prediction of the sign of $z_h - D$ in Equation (13) depends on the accuracy of $\hat{D}$ in the vicinity of the threshold z_h. Thus, more attention should be paid to the accuracy of the predictions related to the limit states. The AL sampling strategy should be able to improve this accuracy. U function [32,37] is a widely used learning function in AL sampling. For the approximation of logarithmic EDP, the U function is described as:

$$U(\mathbf{x}|z_h) = \left| \frac{\ln z_h - \ln \hat{D}|_{\mathbf{x}}}{\sigma_{\ln \hat{D}}|_{\mathbf{x}}} \right| \tag{14}$$

Considering that a point with an approximate seismic demand close to a threshold and a large predicted error has a small U function value, the U function is used to identify the points that can reduce the error of $\hat{D}$ near z_h. In seismic fragility analysis, the accuracy of EDP approximation at multiple limit states needs to be considered. Consequently, the minimum U (MU) value at a sample for all limit states is treated as the learning function value of this sample, which is denoted as:

$$MU(\mathbf{x}) = \min_{h=1}^{Q}(U(\mathbf{x}|z_h)) \tag{15}$$

where Q is the number of performance levels. The point $\mathbf{x}_{new}$ with the minimum MU value in the candidate set is taken as the new experiment, which is described as:

$$\mathbf{x}_{new} = \arg\min_{\mathbf{x}\in X^c}(MU(\mathbf{x})) \tag{16}$$

In the process of adding samples to DoE, the new experiment points will switch in the areas near these limit states as the accuracy of the surrogate models changes, so that the accuracy of EDP approximation in the vicinities of all thresholds will be improved together.

In AL-RA, if the U function value of a point is greater than 2, the probability of a wrong prediction of the safety state for this point is less than 2.3%. It is usually taken as the stopping condition of the AL-RA methods that the U function values of all candidate points are greater than 2 [37]. With reference to this condition, the proportion of candidate points with an MU value less than 2 is used in the proposed method as the indicator for the decision to stop sampling, which is expressed as:

$$p_{MU<2} = \frac{N_{MU<2}}{r} = \frac{\sum\limits_{j=1}^{r} I[MU(\mathbf{x}_j^c) - 2]}{r} \tag{17}$$

where $N_{MU<2}$ denotes the number of points whose MU value is less than 2 in the candidate sample set. When $p_{MU<2}$ is less than a certain tolerance ζ, the sampling of new experiments can be stopped. ζ is taken as 0.02 in this research.

The selection of candidate points is performed according to the probability distribution of variables, similar to the selection of MC samples. In the generation of the candidate set, *im* is regarded as a uniformly distributed variable. LHS is employed for sampling to avoid excessive aggregation of candidate points. Based on a group of points $\mathbf{H} = [h_{ij}]_{(n+1)\times r}$ uniformly sampled in the space of $(0,1)^{n+1}$, the candidate set can be obtained by the transformation [17] as:

$$\mathbf{X}^c = [F_i^{-1}(h_{ij})]_{(n+1)\times r} \tag{18}$$

3.3. Computational Process of Seismic Fragility Analysis

The proposed seismic fragility analysis method can be implemented as the following process:

Step 1: Build a finite element model of the structure and choose a group of earthquake records.

Step 2: Select a DoE containing several experiments in the space of $\mathbf{x} = [\mathbf{s}^{\mathrm{T}}, im]^{\mathrm{T}}$ using LHS, and carry out NLTHA to compute the mean $\mu_{\ln D}$ and standard deviation $\sigma_{\ln D}$ of the logarithmic EDP for each experiment.

Step 3: Generate a candidate set consisting of r samples ($r = 1000$ in this work) in the space of $\bar{\mathbf{x}} = [\mathbf{s}^{\mathrm{T}}, im, v]^{\mathrm{T}}$ using LHS according to Equation (18).

Step 4: Based on the DoE, establish the surrogate models of the mean $\mu_{\ln D}$ and standard deviation $\sigma_{\ln D}$ using Kriging model.

Step 5: Find the point $\mathbf{x}_{\mathrm{new}}$ with the minimum MU value from the candidate samples in terms of Equation (16). If the stopping condition $p_{MU<2} < \zeta$ (ζ is taken as 0.02) is unsatisfied, calculate the values of $\mu_{\ln D}$ and $\sigma_{\ln D}$ at the point $\mathbf{x}_{\mathrm{new}}$, add this point to the DoE, and return to step 4; otherwise, go to the next step.

Step 6: Compute the failure probabilities of the structure for a series of IM values using MC simulation with Equation (13), and plot the results as fragility curves.

The above process is drawn as a flowchart in Figure 2.

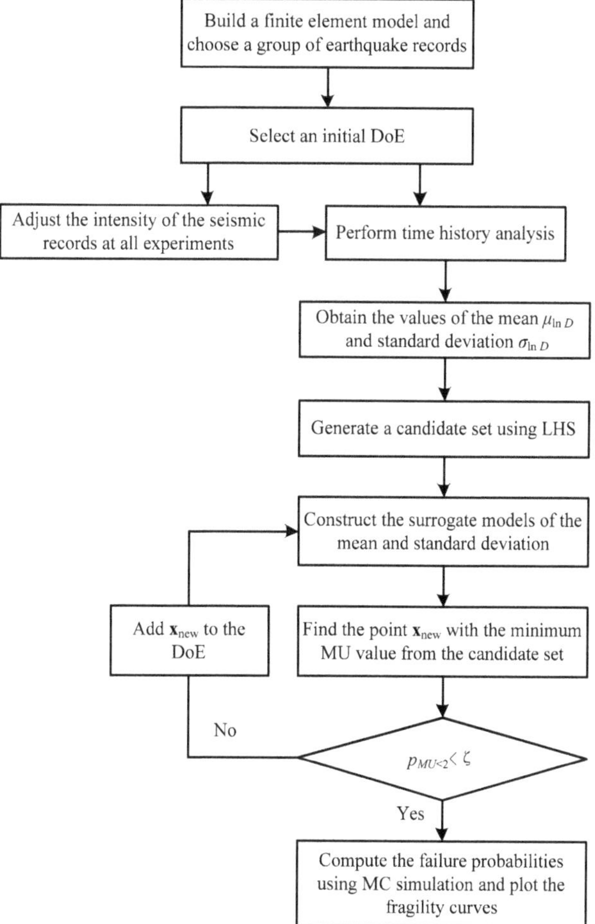

Figure 2. Flowchart of seismic fragility analysis.

3.4. Validation Example

To demonstrate the effectiveness and feasibility of the proposed method, a three-dimensional five-storey steel frame was studied, and the results were compared with those of the MC simulation and DSM-SFA methods. As the calculation time is mainly spent on NLTHA, the calculation efficiency was evaluated by N_c, the number of calls to FEA, which is equal to the number of experiments in DoE.

It was assumed that the site classification was II, and the seismic fortification intensity was 8. According to the requirements of seismic records for time history analysis in the current Chinese code for seismic design of buildings (GB50011-2010) [38], 12 records corresponding to the site condition were selected, which are listed in Table 1. The design spectrum from the seismic code and the response spectra of the selected earthquakes are shown in Figure 3. The maximum inter-storey drift angle is a widely accepted parameter for evaluating structural damage, and it was employed as EDP. The peak ground acceleration (PGA) is a commonly used index to measure the ground motion intensity, and it was adopted as IM in this study. Three performance levels, namely, immediate occupancy, life safety, and collapse prevention, were considered. With reference to the control target of the maximum inter-storey drift angle of steel structures corresponding to different failure states in GB50011-2010, the thresholds of these three performance levels were taken as $z_1 = 1/200$, $z_2 = 1/100$, and $z_3 = 1/50$, respectively.

Table 1. Seismic records.

No.	Event	Year	Station Name	Arias Intensity (m/s)	Magnitude
1	Parkfield	1966	Cholame—Shandon Array #12	0.1	6.19
2	San Fernando	1971	Whittier Narrows Dam	0.2	6.61
3	Tabas, Iran	1978	Boshrooyeh	0.3	7.35
4	Imperial Valley-06	1979	El Centro Array #13	0.3	6.53
5	Imperial Valley-06	1979	Niland Fire Station	0.2	6.53
6	Loma Prieta	1989	Salinas—John & Work	0.2	6.93
7	Landers	1992	Barstow	0.2	7.28
8	Northridge-01	1994	El Monte—Fairview Av	0.3	6.69
9	Northridge-01	1994	LA—Pico and Sentous	0.3	6.69
10	Iwate, Japan	2008	Oomagari Hanazono-cho, Daisen	0.2	6.9
11	Iwate, Japan	2008	Takanashi Daisen	0.4	6.9
12	Darfield, New Zealand	2010	MAYC	0.1	7

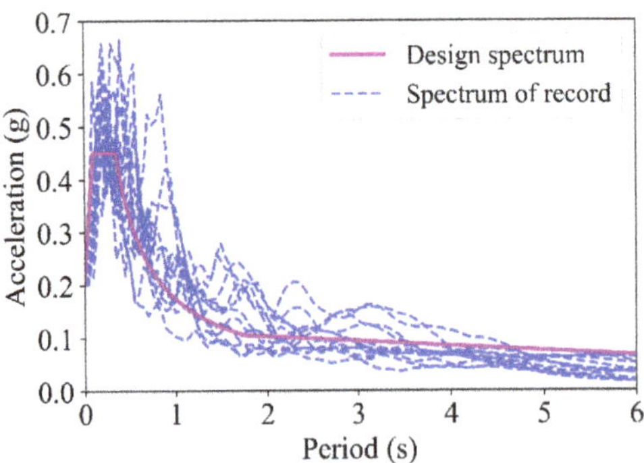

Figure 3. Spectra of the earthquakes.

The finite element model of the structure was constructed in OpenSeesPy (Figure 4a). The storey height was 3.8 m, and the span of each beam was 5 m. The density of steel was 7850 kg/m^3. The cross-section information of beams and columns is shown in Figure 4b.

The displacement-based beam–column element was adopted to model the beams and columns. The stress-strain behavior of steel was simulated by the Steel02 material (uniaxial Giuffre–Menegotto–Pinto material with isotropic strain hardening), where the strain-hardening ratio B was 0.02 and the three parameters controlling the transition of steel from elastic stage to hardening stage were 18, 0.925 and 0.15, respectively [39]. The material model is shown in Figure 5, where f_s and E_s represent the yield strength and initial elastic modulus of steel. The vertical load of 7 kN/m^2 was uniformly distributed on each floor. The slabs were modeled through a diaphragm constraint at each floor [40]. Rayleigh damping model was used in dynamic analysis.

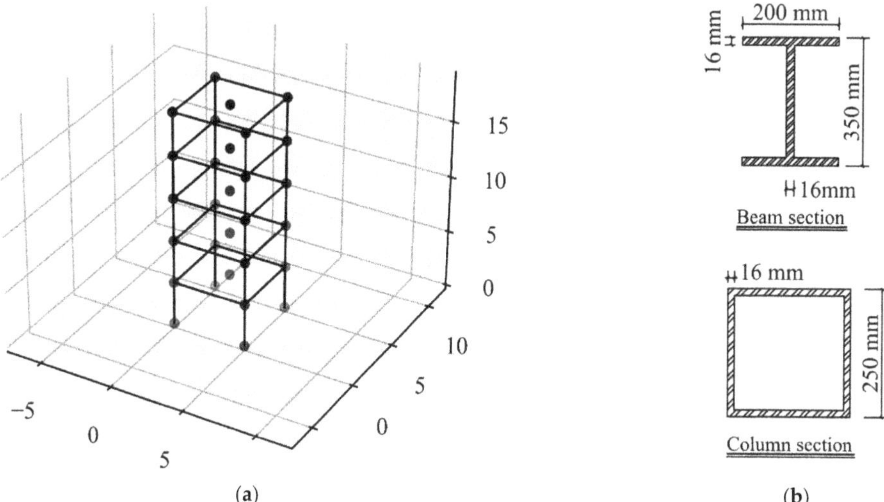

Figure 4. Five-storey steel frame. (**a**) Finite element model. (**b**) Cross-sections.

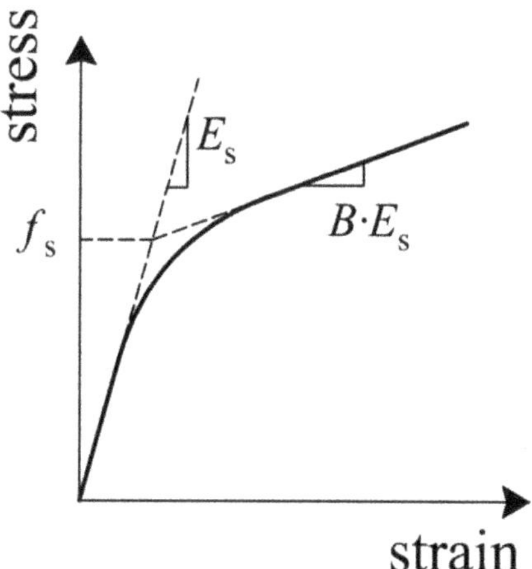

Figure 5. Stress-strain behavior of steel.

In the aspect of structural uncertainties, the randomness of structural damping ratio ξ [41,42], yield strength f_s, and initial elastic modulus E_s were considered. The probability distribution, mean, and coefficient of variation (CoV) of the structural parameters are listed in Table 2.

Table 2. Statistical properties of the structural parameters.

Variable	Distribution	Mean	CoV
f_s (Pa)	Lognormal	3×10^8	0.1
E_s (Pa)	Lognormal	2×10^{11}	0.05
ζ	Lognormal	0.05	0.4

The seismic fragility of the structure was analyzed using the proposed method, in which 6 initial experiments were selected and 42 samples were added to the DoE by AL sampling. In the process of AL sampling, the maximum inter-storey drift angle φ and the approximation $\hat{\varphi}$ at the new samples are shown in Figure 6. It can be observed that the EDP approximations of the new points are mainly located in the vicinities of the thresholds of limit states. As the number of experiments in DoE increased, $p_{MU<2}$ decreased gradually. To validate the accuracy and efficiency of the developed method, MC simulation and DSM-SFA were used to calculate the fragility curves. The curves obtained by these three approaches are shown in Figure 7.

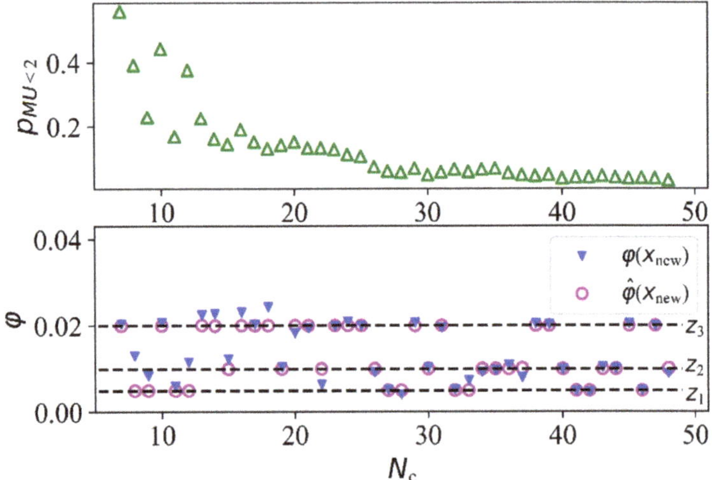

Figure 6. The seismic demand of the added samples.

In the calculation of MC simulation, the MC samples were selected by LHS, and the number of samples was 1000. The failure probabilities of the structure were calculated for PGAs of 0.02, 0.1, 0.2, ... , 1.0 g, respectively, and the number of calls to FEA to obtain the fragility curves was 1000×11. The fragility curves plotted by the proposed procedure had little difference from those by MC simulation, and the N_c value of the former was only 0.4% of that of the latter. Since MC simulation is a very accurate reliability method, the closeness of the two results verifies the accuracy of the proposed method.

In the calculation of the DSM-SFA method, the number of the experiments was made equal to the N_c value in the calculation of the proposed method, and the Kriging model was used for surrogate modeling. Although the numbers of calls to FEA in the analyses of the two methods are the same, the fragility curves obtained by the DSM-SFA method had obvious errors. This indicates that the proposed method needs less computation than the DSM-SFA method to obtain sufficiently accurate results, which verifies its efficiency. Compared with the traditional DSM framework for fragility analysis, this method gives full play to the ability of the Kriging model to predict errors and avoids overreliance on one-shot sampling.

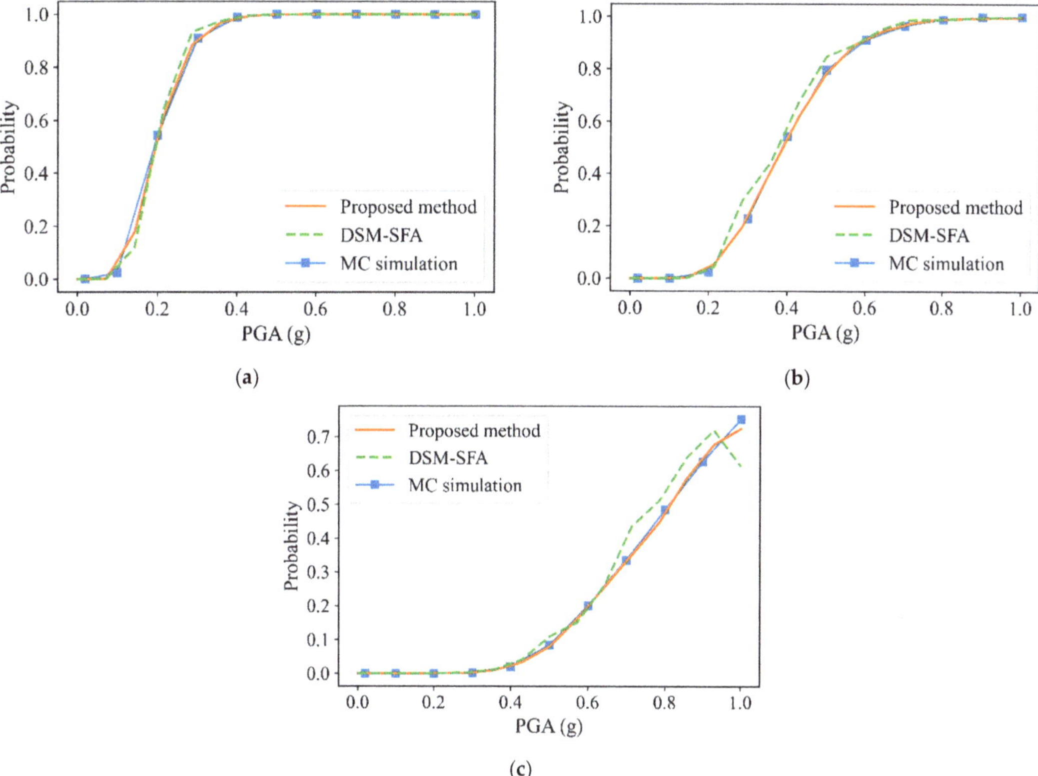

Figure 7. Fragility curves obtained using different approaches. (**a**) Immediate occupancy. (**b**) Life safety. (**c**) Collapse prevention.

4. Seismic Fragility Analysis of MFVCS

In this section, the developed method was applied to MFVCS. The influence of structural uncertainties on the fragility curves of MFVCS was studied, and the fragility of MFVCS and MFS was compared.

4.1. Establishment of Finite Element Model

The components of MFVCS in this research were arranged with reference to the experimental model in previous studies [43,44] and the configuration of the structure is shown in Figure 8a. The structure has four mega storeys. The beam span and storey height of the substructures were 5 and 4 m, respectively. The cross-section information of the members is listed in Table 3. The yield strength and the elastic modulus of steel were 300 MPa and 200 GPa. The dead load (including the self-weight of the slab) and live load on the slab surface were 4 and 2 kN/m^2. Viscous dampers connected to the mega columns were installed on the substructures of the second, third, and fourth mega storeys. In this work, the damper constitutive law [45,46] is described in the form:

$$F_{\mathrm{d}}(\dot{u}) = c_{\mathrm{d}} \cdot |\dot{u}|_\alpha \cdot \mathrm{sgn}(\dot{u})$$ (19)

where F_{d} represents the damper resisting force, c_{d} denotes the damping coefficient, α represents the velocity exponent, $\dot{u}$ denotes the velocity between the damper's end, and $\mathrm{sgn}(\cdot)$ is the sign extractor operator.

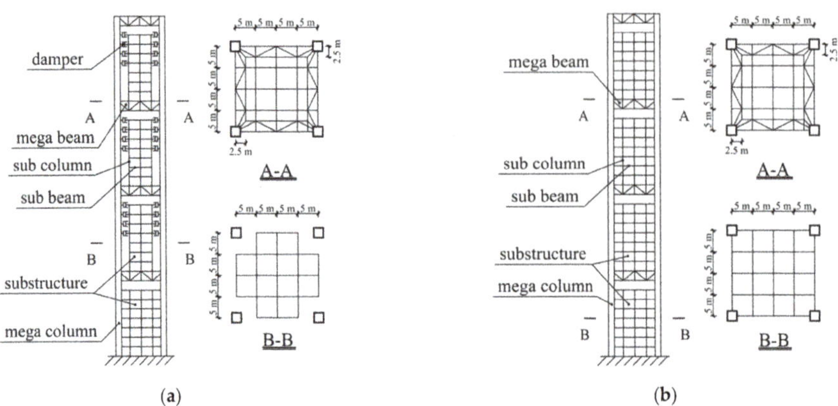

Figure 8. Schematics of MFS and MFVCS. (**a**) MFVCS. (**b**) MFS.

Table 3. Member section information.

Member	Cross-Section (m)
Sub beam	H0.5 × 0.25 × 0.016 × 0.016
Sub column	□0.5 × 0.5 × 0.02 × 0.02
Chord of mega beam	H0.5 × 0.25 × 0.016 × 0.016
Vertical bar of mega beam	□0.5 × 0.5 × 0.02 × 0.02
Diagonal brace of mega beam	□0.2 × 0.2 × 0.01 × 0.01
Mega column	□2.5 × 2.5 × 0.03 × 0.03

The finite element model of MFVCS was built in OpenSeesPy. The material model of steel and the element used to model beams and columns were the same as those in the previous section. The hysteretic response of the damper was simulated using the Maxwell material model.

To verify the finite element model, another model was established in SAP2000 (Figure 9). The modal analysis of two finite element models was carried out, and the information of the first six natural vibration periods is listed in Table 4. The calculation results of the two software were very close. Therefore, the finite element model in this study could be considered correct and could be used for subsequent research.

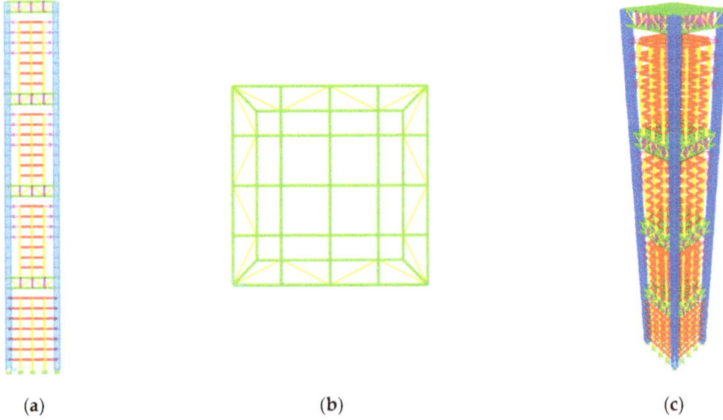

Figure 9. Finite element model of MFVCS. (**a**) Elevation View. (**b**) Plan View. (**c**) 3D View.

Table 4. Modal analysis results.

Order	Natural Vibration Period (s)	
	OpenSeesPy	**SAP2000**
1	4.35587	4.357356
2	4.34556	4.357356
3	2.90373	2.904145
4	1.52061	1.486444
5	1.51575	1.486444
6	1.2149	1.1628

4.2. Comparison of Results

In the fragility analysis of MFVCS, the uncertainties of material properties and damper parameters [45] were taken into account. The statistical properties of random variables are shown in Table 5. The same seismic records as those in the validation example were used for NLTHA.

Table 5. Statistical properties of random variables.

Variable	Distribution	Mean	CoV
$f_{s,\,CM}$ (Pa), yield strength of steel of mega column	Lognormal	3×108	0.1
$E_{s,\,CM}$ (Pa), initial elastic modulus of steel of mega column	Lognormal	2×10^{11}	0.05
$f_{s,\,CS}$ (Pa), yield strength of steel of sub column	Lognormal	3×10^8	0.1
$E_{s,\,CS}$ (Pa), initial elastic modulus of steel of sub column	Lognormal	2×10^{11}	0.05
α, velocity exponent	Lognormal	0.3	0.2
c_d (N·s/m), damping coefficient	Lognormal	2.2×10^4	0.2
ξ, damping ratio	Lognormal	0.05	0.4

The seismic fragility of MFVCS was calculated using the proposed procedure in three different cases, in which the numbers of initial experiments N_I were 6, 9, and 12, respectively. The values of φ, $\hat{\varphi}$, and $p_{MU<2}$ of the new samples in the sampling process are plotted in Figure 10. The fragility curves are plotted in Figure 11. The results of the three analyses were in good agreement, which indicates that the proposed method can obtain the fragility curves stably.

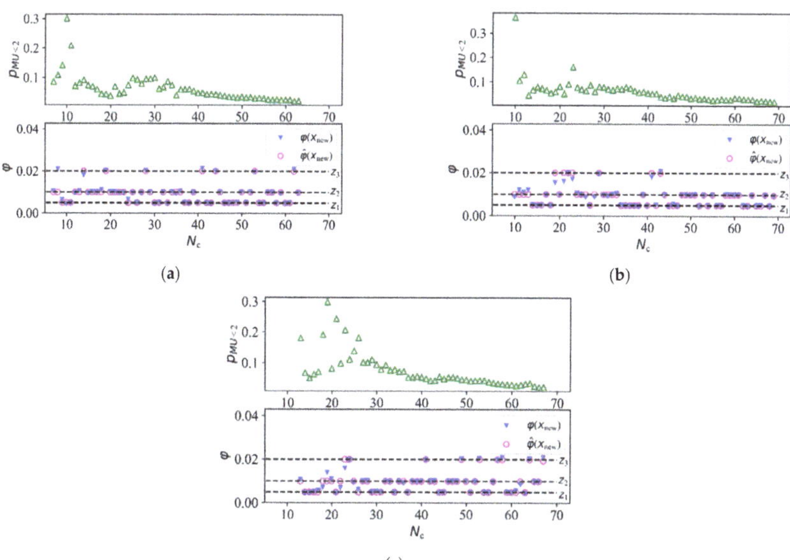

Figure 10. The seismic demand of the added samples of MFVCS. (**a**) N_I = 6. (**b**) N_I = 9. (**c**) N_I = 12.

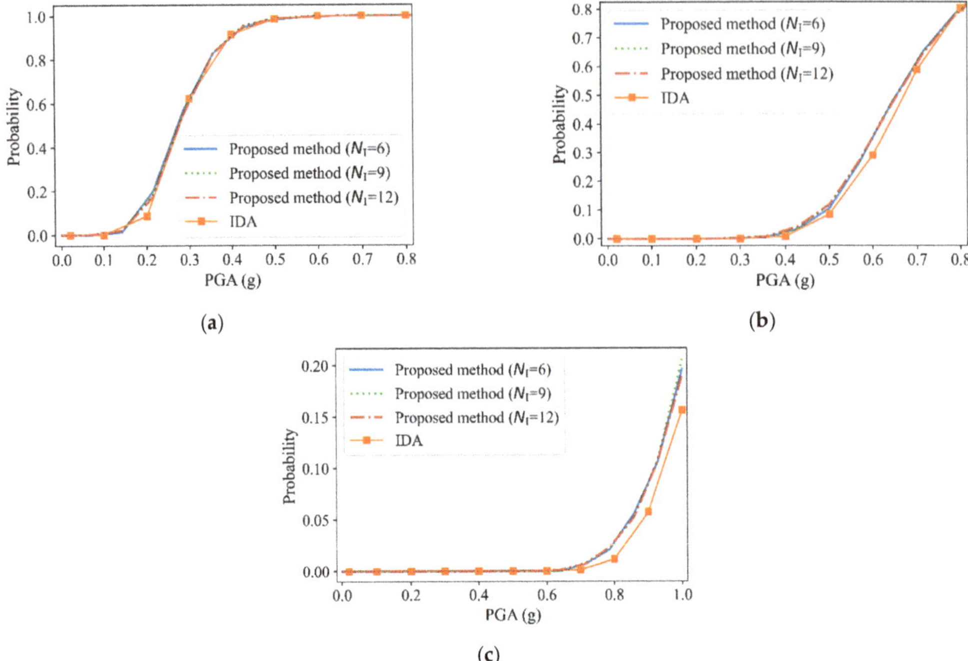

Figure 11. Fragility curves of MFVCS obtained using different approaches. (**a**) Immediate occupancy. (**b**) Life safety. (**c**) Collapse prevention.

To study the effect of structural randomness on the seismic fragility of MFVCS, the fragility curves were calculated by the IDA method neglecting model uncertainty. The CoV of each parameter was treated as 0. The probability of EDP exceeding z_h for a given IM value *im* calculated by IDA [14] is:

$$P[D \geq z_h | IM = im] = 1 - \Phi \left(\frac{\ln z_h - \mu_{\ln D}|_{im}}{\sigma_{\ln D}|_{im}} \right) \tag{20}$$

The results are also drawn in Figure 11. For the immediate occupancy level, there is not much difference between the results corresponding to considering and ignoring the model uncertainty. For the collapse prevention level, the difference between the two results is obvious. It shows that structural uncertainties can cause significant variation of the fragility curve corresponding to the collapse prevention level of MFVCS.

The fragility of MFS was also analyzed by the proposed method. The component arrangement of MFS is shown in Figure 8b. The member cross-sections were the same as those of MFVCS. Due to the absence of dampers, the parameters of the damping device were not considered in the fragility analysis. The statistical properties of other random parameters were the same as those of MFVCS. The number of calls to FEA in the calculation was 38. The fragility curves are shown in Figure 12. By comparing the curves of the two structures, it can be found that the failure probability of MFS is obviously higher than that of MFVCS under the same PGA, indicating that MFVCS is less vulnerable than MFS. Therefore, compared with MFS, MFVCS has better seismic performance. Compared with the previous studies on MFVCS based on IDA [14,47], this method can incorporate the randomness of structural parameters into fragility analysis, which provides more accurate reliability information for the seismic design of MFVCS.

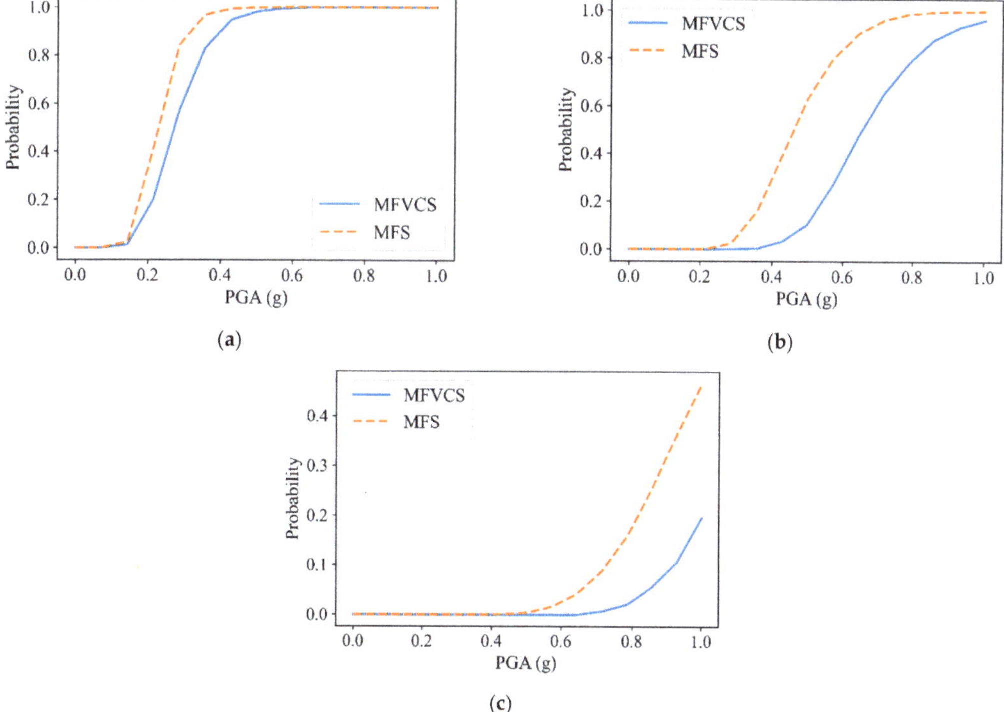

Figure 12. Fragility curves of MFS and MFVCS. (**a**) Immediate occupancy. (**b**) Life safety. (**c**) Collapse prevention.

5. Conclusions

To save the computational time of the seismic fragility analysis of MFVCS considering structural uncertainties, an improved DSM-SFA method was proposed in this paper. This method modifies the way of estimating the seismic response in DSM-SFA, so that the predicted error of the approximation can be obtained by the Kriging model. Based on the ability to predict errors, an AL sampling strategy is presented to adaptively select an appropriate DoE, thus improving the efficiency and accuracy of the DSM framework for fragility analysis. The proposed method is generic in nature and can be used for other structures under seismic load.

An example was analyzed by using different approaches to verify the effectiveness and feasibility of the proposed procedure. It has been observed that the calculation cost of this method is far less than that of MC simulation and the fragility curves of the two methods are very close. With the same number of calls to FEA, the errors of the fragility curves obtained by the DSM-SFA method are larger than those by the proposed method. The results show that the developed method is efficient and accurate.

The finite element model of MFVCS was established, and the proposed procedure was applied to its seismic fragility analysis. For the cases with different initial points, the analysis results were in good agreement, which indicates that this method can obtain the fragility curves stably. The comparison between the results of IDA and the developed method reveals that structural uncertainties have an effect on the fragility curves, especially the curve of the collapse prevention level. In addition, the fragility curves of MFS and MFVCS were compared in this paper, which shows that the damage probability of MFVCS is smaller than that of MFS under the same PGA and MFVCS has better seismic performance.

The proposed procedure provides a foundation for our future study on reliability-based design optimization of MFVCS considering the RTR variability of wind load.

Author Contributions: Conceptualization, Y.X.; methodology, Y.X.; software, Y.X.; validation, F.Y., B.F., M.M.S. and X.W.; formal analysis, Y.X.; investigation, Y.X.; resources, Y.X.; data curation, F.Y., B.F., M.M.S. and X.W.; writing—original draft preparation, Y.X.; writing—review and editing, F.Y.; visualization, Y.X.; supervision, X.Z.; project administration, X.Z.; funding acquisition, X.Z. All authors have read and agreed to the published version of the manuscript.

Funding: This work was supported by the National Natural Science Foundation of China under Grant No. 51078311.

Institutional Review Board Statement: Not applicable.

Informed Consent Statement: Not applicable.

Data Availability Statement: The data used to support the findings of this study are available from the corresponding author upon request.

Conflicts of Interest: The authors declare no conflict of interest.

Abbreviation

Abbreviation	Meaning
MFS	mega-frame structure
MFVCS	mega-frame with vibration control substructure
TMD	tuned mass damper
DSM	dual surrogate model
IDA	incremental dynamic analysis
IM	intensity measure
EDP	engineering demand parameter
AL	active learning
DoE	design of experiments
CoV	coefficient of variation
FEA	finite element analysis
LHS	Latin hypercube sampling
PGA	peak ground acceleration
NLTHA	nonlinear time history analysis

References

1. Zhang, X.A.; Qin, X.; Cherry, S.; Lian, Y.; Zhang, J.; Jiang, J. A new proposed passive mega-sub controlled structure and response control. *J. Earthq. Eng.* **2009**, *13*, 252–274. [CrossRef]
2. Feng, M.Q.; Mita, A. Vibration control of tall buildings using mega subconfiguration. *J. Eng. Mech.* **1995**, *121*, 1082–1088. [CrossRef]
3. Qin, X.; Zhang, X.A.; Sheldon, C. Study on semi-active control of mega-sub controlled structure by MR damper subject to random wind loads. *Earthq. Eng. Eng. Vib.* **2008**, *7*, 285–294. [CrossRef]
4. Jiang, Q.; Wang, H.Q.; Ye, X.G.; Chong, X.; Feng, Y.L.; Wang, J.; Yao, H.T. Shaking table model test of a mega-frame with a vibration control substructure. *Struct. Des. Tall Spec. Build.* **2020**, *29*, e1742. [CrossRef]
5. Abdulhadi, M.; Zhang, X.A.; Fan, B.; Moman, M. Design, Optimization and Nonlinear Response Control Analysis of the Mega Sub-Controlled Structural System (MSCSS) Under Earthquake Action. *J. Earthq. Tsunami* **2020**, *14*, 2050013. [CrossRef]
6. Liang, Q.; Li, L.; Yang, Q. Seismic analysis of the tuned—Inerter—Damper enhanced mega-sub structure system. *Struct. Control. Health Monit.* **2021**, *28*, e2658. [CrossRef]
7. Xiao, Y.J.; Yue, F.; Zhang, X.A.; Shahzad, M.M. Aseismic Optimization of Mega-sub Controlled Structures Based on Gaussian Process Surrogate Model. *KSCE J. Civ. Eng.* **2022**, *26*, 2246–2258. [CrossRef]
8. Kalehsar, H.E.; Khodaie, N. Wind-induced vibration control of super-tall buildings using a new combined structural system. *J. Wind. Eng. Ind. Aerodyn.* **2018**, *172*, 256–266. [CrossRef]
9. Fan, B.; Zhang, X.A.; Abdulhadi, M.; Moman, M.; Wang, Z. Seismic behavior of MSCSS based on story drift and failure path. *Lat. Am. J. Solids Struct.* **2021**, *18*, e411. [CrossRef]
10. Abdulhadi, M.; Zhang, X.A.; Fan, B.; Moman, M. Substructure design optimization and nonlinear responses control analysis of the mega-sub controlled structural system (MSCSS) under earthquake action. *Earthq. Eng. Eng. Vib.* **2021**, *20*, 687–704. [CrossRef]
11. Shahzad, M.M.; Zhang, X.A.; Wang, X.; Abdulhadi, M.; Wang, T.; Xiao, Y.J. Response control analysis of a new mega-subcontrolled structural system (MSCSS) under seismic excitation. *Struct. Des. Tall Spec. Build.* **2022**, *2022*, e1935. [CrossRef]
12. Park, J.; Towashiraporn, P. Rapid seismic damage assessment of railway bridges using the response-surface statistical model. *Struct. Saf.* **2014**, *47*, 1–12. [CrossRef]

13. Gentile, R.; Galasso, C. Gaussian process regression for seismic fragility assessment of building portfolios. *Struct. Saf.* **2020**, *87*, 101980. [CrossRef]
14. Wang, X.; Shahzad, M.M.; Shi, X. Fragility analysis and collapse margin capacity assessment of mega-sub controlled structure system under the excitation of mainshock-aftershock sequence. *J. Build. Eng.* **2022**, *49*, 104080. [CrossRef]
15. Nazri, F.M.; Kian Yern, C.; Kassem, M.M.; Farsangi, E.N. Assessment of structure-specific fragility curves for soft storey buildings implementing IDA and SPO approaches. *Int. J. Eng.* **2018**, *31*, 2016–2021. [CrossRef]
16. Bao, X.; Jin, L.; Liu, J.; Xu, L. Framework for the mainshock-aftershock fragility analysis of containment structures incorporating the effect of mainshock-damaged states. *Soil Dyn. Earthq. Eng.* **2022**, *153*, 107072. [CrossRef]
17. Asgarian, B.; Ordoubadi, B. Effects of structural uncertainties on seismic performance of steel moment resisting frames. *J. Constr. Steel Res.* **2016**, *120*, 132–142. [CrossRef]
18. Lin, D.K.; Tu, W. Dual response surface optimization. *J. Qual. Technol.* **1995**, *27*, 34–39. [CrossRef]
19. Fang, J.; Gao, Y.; Sun, G.; Xu, C.; Li, Q. Multi-objective robust design optimization of fatigue life for a truck cab. *Reliab. Eng. Syst. Saf.* **2015**, *135*, 1–8. [CrossRef]
20. Wu, Y.; Li, W.; Fang, J.; Lan, Q. Multi-objective robust design optimization of fatigue life for a welded box girder. *Eng. Optim.* **2018**, *50*, 1252–1269. [CrossRef]
21. Qiu, N.; Jin, Z.; Liu, J.; Fu, L.; Chen, Z.; Kim, N.H. Hybrid multi-objective robust design optimization of a truck cab considering fatigue life. *Thin-Walled Struct.* **2021**, *162*, 107545. [CrossRef]
22. Towashiraporn, P. Building Seismic Fragilities Using Response Surface Metamodels. Ph.D. Thesis, Georgia Institute of Technology, Atlanta, CA, USA, 2004.
23. Zhang, Y.; Wu, G. Seismic vulnerability analysis of RC bridges based on Kriging model. *J. Earthq. Eng.* **2019**, *23*, 242–260. [CrossRef]
24. Perotti, F.; Domaneschi, M.; De Grandis, S. The numerical computation of seismic fragility of base-isolated nuclear power plants buildings. *Nucl. Eng. Des.* **2013**, *262*, 189–200. [CrossRef]
25. Saha, S.K.; Matsagar, V.; Chakraborty, S. Uncertainty quantification and seismic fragility of base-isolated liquid storage tanks using response surface models. *Probabilistic Eng. Mech.* **2016**, *43*, 20–35. [CrossRef]
26. Xiao, Y.; Ye, K.; He, W. An improved response surface method for fragility analysis of base-isolated structures considering the correlation of seismic demands on structural components. *Bull. Earthq. Eng.* **2020**, *18*, 4039–4059. [CrossRef]
27. Datta, G.; Bhattacharjya, S.; Chakraborty, S. Efficient reliability-based robust design optimization of structures under extreme wind in dual response surface framework. *Struct. Multidiscip. Optim.* **2020**, *62*, 2711–2730. [CrossRef]
28. Datta, G.; Bhattacharjya, S.; Chakraborty, S. Robust design of offshore jacket platform structure under random wave in dual response surface framework. *Struct. Infrastruct. Eng.* **2021**, *17*, 887–901. [CrossRef]
29. Majumder, S.; Datta, G.; Bhattacharjya, S. A dual response surface-based efficient fragility analysis approach of offshore structures under random wave load. *Appl. Math. Model.* **2021**, *98*, 680–700. [CrossRef]
30. Datta, G.; Bhattacharjya, S.; Chakraborty, S. A metamodeling-based robust optimisation approach for structures subjected to random underground blast excitation. *Structures* **2021**, *33*, 3615–3632. [CrossRef]
31. Guo, Q.; Liu, Y.; Zhao, Y.; Li, B.; Yao, Q. Improved resonance reliability and global sensitivity analysis of multi-span pipes conveying fluid based on active learning Kriging model. *Int. J. Press. Vessel. Pip.* **2019**, *170*, 92–101. [CrossRef]
32. Guo, Q.; Liu, Y.; Chen, B.; Yao, Q. A variable and mode sensitivity analysis method for structural system using a novel active learning Kriging model. *Reliab. Eng. Syst. Saf.* **2021**, *206*, 107285. [CrossRef]
33. Xiao, Y.J.; Yue, F.; Zhang, X.A. Seismic Fragility Analysis of Structures Based on Adaptive Gaussian Process Regression Metamodel. *Shock Vib.* **2021**, *2021*, 7622130. [CrossRef]
34. Qin, S.; Zhou, Y.L.; Cao, H.; Wahab, M.A. Model updating in complex bridge structures using kriging model ensemble with genetic algorithm. *KSCE J. Civ. Eng.* **2018**, *22*, 3567–3578. [CrossRef]
35. Preuss, R.; Von Toussaint, U. Global optimization employing Gaussian process-based Bayesian surrogates. *Entropy* **2018**, *20*, 201. [CrossRef] [PubMed]
36. Phan, H.N.; Paolacci, F.; Corritore, D.; Tondini, N.; Bursi, O.S. A Kriging-Based Surrogate Model for Seismic Fragility Analysis of Unanchored Storage Tanks. In *Pressure Vessels and Piping Conference*; ASME: San Antonio, TX, USA, 2019; p. V008T08A024. [CrossRef]
37. Xiao, Y.J.; Yue, F.; Wang, X.; Zhang, X.A. Reliability-Based Design Optimization of Structures Considering Uncertainties of Earthquakes Based on Efficient Gaussian Process Regression Metamodeling. *Axioms* **2022**, *11*, 81. [CrossRef]
38. *GB 50011-2010*; Code for Seismic Design of Buildings. Ministry of Construction of the People's Republic of China: Beijing, China, 2010. (In Chinese)
39. Jin, S.; Du, H.; Bai, J. Seismic performance assessment of steel frame structures equipped with buckling-restrained slotted steel plate shear walls. *J. Constr. Steel Res.* **2021**, *182*, 106699. [CrossRef]
40. Barbato, M.; Gu, Q.; Conte, J.P. Probabilistic push-over analysis of structural and soil-structure systems. *J. Struct. Eng.* **2010**, *136*, 1330–1341. [CrossRef]
41. Porter, K.A.; Beck, J.L.; Shaikhutdinov, R.V. Sensitivity of building loss estimates to major uncertain variables. *Earthq. Spectra* **2002**, *18*, 719–743. [CrossRef]

42. Mehanny, S.S.F.; Ayoub, A.S. Variability in inelastic displacement demands: Uncertainty in system parameters versus randomness in ground records. *Eng. Struct.* **2008**, *30*, 1002–1013. [CrossRef]
43. Cai, T.; Zhang, X.A.; Lian, Y.D.; Li, B. Research on model design and shaking table experiment of mega-sub controlled structure system. *Ind. Constr.* **2016**, *46*, 139–143. (In Chinese) [CrossRef]
44. Limazie, T.; Zhang, X.A.; Wang, X. Vibration control parameters investigation of the mega-sub controlled structure system (MSCSS). *Earthq. Struct.* **2013**, *5*, 225–237. [CrossRef]
45. Dall'Asta, A.; Scozzese, F.; Ragni, L.; Tubaldi, E. Effect of the damper property variability on the seismic reliability of linear systems equipped with viscous dampers. *Bull. Earthq. Eng.* **2017**, *15*, 5025–5053. [CrossRef]
46. Altieri, D.; Tubaldi, E.; De Angelis, M.; Patelli, E.; Dall'Asta, A. Reliability-based optimal design of nonlinear viscous dampers for the seismic protection of structural systems. *Bull. Earthq. Eng.* **2018**, *16*, 963–982. [CrossRef]
47. Abdulhadi, M.; Zhang, X.A.; Fan, B.; Moman, M. Evaluation of Seismic Fragility Analysis of the Mega Sub-Controlled Structural System (MSCSS). *J. Earthq. Tsunami* **2020**, *14*, 2050025. [CrossRef]

 buildings

Article

Architectural Characteristics and Seismic Vulnerability Assessment of a Historical Masonry Minaret under Different Seismic Risks and Probabilities of Exceedance

Ercan Işık [1], Fatih Avcil [1], Ehsan Harirchian [2,*], Enes Arkan [3], Hüseyin Bilgin [4] and Hayri Baytan Özmen [5]

[1] Department of Civil Engineering, Bitlis Eren University, Bitlis 13100, Turkey
[2] Institute of Structural Mechanics (ISM), Bauhaus-Universität Weimar, 99423 Weimar, Germany
[3] Department of Architecture, Bitlis Eren University, Bitlis 13100, Turkey
[4] Department of Civil Engineering, Epoka Universtiy, 1001 Tirana, Albania
[5] Department of Civil Engineering, Uşak University, Uşak 64300, Turkey
* Correspondence: ehsan.harirchian@uni-weimar.de

Abstract: Masonry structures began to be built with the existence of human beings and are an inspiration for today's structures. Monumental historical buildings built according to people's religious beliefs have special importance among such structures. Despite being exposed to many natural disasters over time, such structures that have survived till today are an indispensable part of the historical heritage. Within the scope of this study, structural analyses were carried out for the historical Ulu Mosque's minaret in Bitlis (Turkey), located in the Van Lake basin, using both on-site measurements and finite element methods. Detailed historical and architectural features were given for the minaret and the mosque. In addition to four different earthquake ground motion levels of 2%, 10%, 50% and 68%, structural analyses were deployed separately for seven different geographical locations in the same seismic risk area. Moreover, time history analyses were conducted using the acceleration records of the Van earthquake that occurred in the region. The minaret performance levels were determined by using the displacement values obtained. The study examined the different probabilities of exceedance and the changes in the regions with the same seismic risk. As a result of each structural analysis, base shear forces, displacement, period and maximum stress values were obtained for the minaret. The displacement, base shear force, and stress values increased as the exceedance probability decreased. While the same seismic and structural analysis results were obtained for the selected settlements in the same earthquake zone in this study, remarkable differences were observed for these settlements using the geographical-location-specific design spectrum.

Keywords: architectural; historical heritage; seismic risk; probability of exceedance; minaret

Citation: Işık, E.; Avcil, F.; Harirchian, E.; Arkan, E.; Bilgin, H.; Özmen, H.B. Architectural Characteristics and Seismic Vulnerability Assessment of a Historical Masonry Minaret under Different Seismic Risks and Probabilities of Exceedance. *Buildings* **2022**, *12*, 1200. https://doi.org/10.3390/buildings12081200

Academic Editor: Fulvio Parisi

Received: 23 June 2022
Accepted: 6 August 2022
Published: 10 August 2022

Publisher's Note: MDPI stays neutral with regard to jurisdictional claims in published maps and institutional affiliations.

1. Introduction

Cultural heritage is defined as an expression of lifestyles that are inherited from the past generations, including traditions, practices, places, objects, artistic works, structures and other values developed by any community [1]. Cultural heritage can be classified as follows:

- Built environment (buildings, city scenes, archaeological remains).
- Natural environment (rural landscapes, beaches and shorelines, agricultural heritage).
- Other works (books and documents, objects, pictures) [2].

Historical monumental buildings built in different periods by following different faiths are also evaluated within this scope. At the same time, such structures appear as an indicator of societies' belief types, engineering backgrounds, understanding of art and economic status [3–7]. Such structures are invaluable cultural assets that strongly connect the past and the future [8,9]. Minarets, which have an important meaning in Islamic belief, are high, slender and elegant structures built in the form of towers in which the call to prayer is read. The behavior of such structures under the influence of horizontal loads has a special importance due to their structural features. While minarets used to be built with local stones and materials, they can also be built using different types of materials such as concrete, reinforced concrete and steel, depending on the developments in building technologies today. In general, the pulpit, transition segment, web (body), balcony, upper part of web, spire and end ornaments are the components of minaret structures. The parts of the Bitlis Ulu Mosque minaret, considered in the study, are shown in Figure 1.

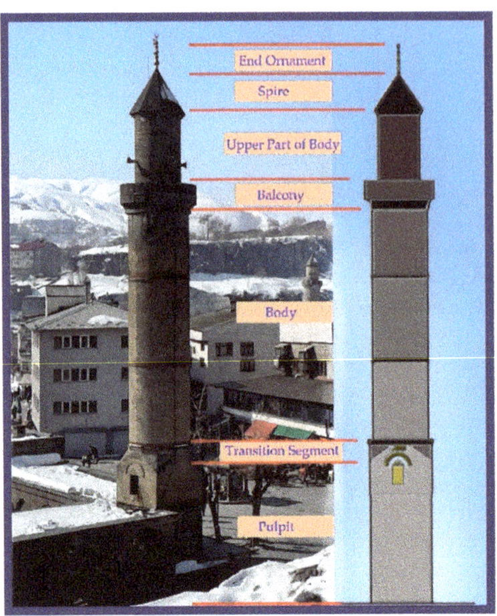

Figure 1. Bitlis Ulu Mosque's minaret parts.

Historical masonry minarets were generally built using local materials and craftsmanship without any engineering service. Even though such minarets have been exposed to many natural disasters over time, they have managed to survive till today. At the same time, such structures provide important information about the architecture, art, construction technologies and lifestyles of that day. Such structures, which are evaluated within the scope of historical and cultural heritage, are invaluable. Each interdisciplinary study to be deployed on such structures has a separate importance. Several studies have been carried out in the field of civil and earthquake engineering related to masonry minarets. Işık and Antep (2018) analyzed the earthquake behavior for the minaret of Kadı Mahmut Mosque in the Ahlat district according to Turkey's 2007 earthquake code and calculated the resulting stress and displacement values [10]. Çalık et al. (2012) investigated the static and dynamic behaviors of the masonry minaret of the Merkez Hacı Kasım Muhittin Mosque in Trabzon with analytical and experimental methods in order to determine its structural safety and to restore it [11]. Suliman et al. (2021) studied the behavior of the minaret of the Carol I Mosque in Constanta, Romania, under seismic loads [12]. Işık et al. (2022) carried out the structural analyses for the historical Five Minarets, one of the important symbols of Bitlis, according to the Turkish Building Earthquake Code-2018 only. In their studies, only one earthquake ground motion level and the location of the minaret were considered in structural analyzes [13]. Çoşgun and Türk (2012) investigated the dynamic behavior of a historical masonry minaret in Istanbul and proposed a strengthening method for the minaret [14]. Erdil et al. (2018) examined the behavior of the Van Ulu Mosque under seismic loads and compared the stress and crack locations for the Van Ulu Mosque, where the amount of damage increased as a result of the 2011 Van earthquakes [15]. Dogangun et al. (2008) carried out structural analyses for three masonry minarets using the 1999 Kocaeli and Düzce earthquake ground motions data set on the dynamic behavior of unreinforced masonry minarets. Modal analyses of the models showed that the structural periods and the overall structural response were affected by the minaret height and the spectral characteristics of the input motion [16]. Pekgökgöz et al. (2018) determined the elasticity module of the Şanlıurfa Ulu Mosque's minaret building stone with ultrasonic tests [17]. Livaoglu et al. (2018) revealed the effect of the change of geometric properties on the dynamic behavior of seven different masonry minarets in Bursa [18]. Basaran et al. (2016) conducted structural analyses of the minaret of the Hacı Mahmut mosque using the material properties obtained as a result of nondestructive and destructive tests [19]. Oliveira et al. (2012) compared the results of vibration tests and numerical modeling in historical minarets with different characteristics [20]. Hejazi et al. (2016) performed structural analyses for nine historical brick masonry minarets in Isfahan under the effects of weight, temperature, wind and earthquake [21]. Türkeli (2020) deployed dynamic analyses for the minaret of Iskender Pasha Mosque using the finite element method under the effects of wind and earthquake [22]. Muvafik (2014) presented the field survey results together with the seismic analyses of the masonry brick minaret of the historical Ulu Mosque, which was damaged in the 23 October (Erciş) and 9 November (Edremit) 2011 Van earthquakes [23]. Karaşin et al. (2016) stated the damages and solution proposals for the Bitlis Ulu Mosque together with a site survey and structural analyses [24]. In these studies, besides structural analyses under the influence of earthquake and wind forces, rehabilitation proposals are also presented. These studies can also be considered as case studies on the modeling and strengthening of masonry minarets with the finite element method and determining their seismic behavior using different seismic analysis methods.

While determining the behavior of engineering structures under the influence of earthquakes, seismic design regulations and the seismicity parameters of the geographical locations where the structures built should be considered [25,26]. Seismic parameters may vary with different probabilities of exceedance. Within the scope of this study, four different probabilities of exceedance were considered for the selected masonry minaret. While considering the probabilities, four different values included in the Turkish Building Earthquake Code-2018 (TBEC-2018) [27], with a probability of exceedance in 50 years with 2%, 10%, 50% and 68% were taken into account. These obtained values were compared with the standard earthquake ground motion level in the previous earthquake regulation (TSDC-2007) [28] in the country.

Developments and innovations in the scientific literature make changes in seismic design regulations inevitable [29,30]. Necessary updates and amendments were made to both earthquake regulations and earthquake hazard maps on different dates in Turkey. One of the important parameters that have changed with the current regulation has been the earthquake ground motion levels. While there was only one ground motion level in the previous regulation, there are four different ground motion levels with four different probabilities to be exceeded in the current regulation [31,32]. Design spectra obtained on a regional basis in the previous regulations have been replaced by site-specific design spectra for the first time with the current regulation [33]. In short, the concept of earthquake zone has been completely removed and the concept of earthquake hazard specific to a geographical location has started [34–37]. Another variable considered in this study is the effect of different geographical locations on the seismic behavior of masonry minarets. Structural analyses were carried out separately by considering seven different settlements located in the same region in the previous earthquake hazard map whose locations are in seven different geographical regions in Turkey.

The main purpose of the study was to reveal the effects of different earthquake ground motion levels and geographical-site-specific design spectra on a masonry minaret. For this purpose, the historical Ulu Mosque's minaret in the province of Bitlis (Turkey) was chosen as a case study. Structural analyses were carried out by creating a finite element model of the minaret using the macromodeling technique. First, analyses were carried out using the design spectra obtained for the different exceedance probabilities specified in the current earthquake code for the province of Bitlis, where the minaret is located. In order to make comparisons between the two earthquake codes, analyses were made using the standard design spectrum given for the probability of exceedance in the previous earthquake code. In the next stage of the study, considering the provinces of Çankırı, Aydın, Amasya, Kocaeli, Siirt, Bitlis and Osmaniye, which are in the same earthquake zone in the previous earthquake zone map, structural analyses were carried out for the same ground motion level in the last two earthquake codes and the effect of geographical-site-specific design spectra. At the same time, by using the acceleration records of the 2011 Van earthquake, which is the closest region to Bitlis, where the Ulu Mosque's minaret is located, analyses were performed in the time history domain and compared with the values predicted in the last two regulations. In addition, seismic parameters and design spectra were obtained for the settlements considered and comparisons were made. By giving detailed information about the Ulu Mosque and its minaret, its current structural status was determined based on site observations. The study aimed at revealing the two important amendments in the current regulation used in Turkey from the response of the selected masonry minaret.

Within the scope of this study, Ulu Mosque's minaret, which is one of the five historical minarets in Bitlis, was chosen as a sample masonry minaret. This study includes not only the minaret of the Great Mosque, but also a very detailed architectural and structural features of the mosque part. Structural analyzes were carried out taking into account not only the latest earthquake code used in Turkey, but also the previous earthquake code. Thus, it was possible to make a comparison between the last two earthquake codes of masonry structures in Turkey. In addition, comparisons were made with the recent earthquake acceleration values that occurred in the regions close to the minaret's location and the acceleration values predicted in the last two earthquake hazard maps in Turkey. With this, we tried to reveal whether the earthquake hazard was adequately specified in the structural analysis. Moreover, the novelty of this study was the structural analyzes according to different probabilities of exceedance used for the first time in the current earthquake code. In addition, it is one of the novelties in the study to perform structural analyzes by using the geographical-location-specific design spectra that were used for the first time with the current earthquake code in Turkey, taking into account five different geographical locations in the same earthquake hazard zone in the previous earthquake hazard map. With this, the effect of location-specific design spectra on masonry minarets was revealed. In addition, performance levels were determined based on the detailed analyses deployed.

2. Ulu (Grand) Mosque and Its Minaret

Although Bitlis is located on a strategic transition corridor in the Eastern Anatolia region of Turkey, it has been the cradle of many civilizations. There are many historical and monumental buildings and artifacts belonging to different civilizations in the province. One of the most important of them is the Bitlis Grand Mosque and its minaret. Different images of the Ulu (Grand) Mosque and its minaret are shown in Figure 2.

Figure 2. The Ulu (Grand) Mosque and its minaret.

Minarets are tall and slender structures in the form of towers, which are generally built adjacent to mosques and in a separate area, in which the call to prayer is made. For the scope of this study, the Ulu Mosque's minaret, which is located in the province of Bitlis built from unreinforced masonry, was chosen. Different images of the minaret are shown in Figure 3.

Figure 3. The minaret of the Ulu Mosque.

Bitlis Ulu Mosque was built in the Gazibey neighborhood, southeast of Bitlis Castle, on the edge of the Kömüs stream. The building, which forms a complex with a sanctuary in the south, a courtyard with a portico in the middle and a minaret in the northwest outer corner of the courtyard, is thought to have been a place of worship before it was built as a mosque (Figures 4 and 5). The typology of the windows and their position on the wall are important data supporting this argument.

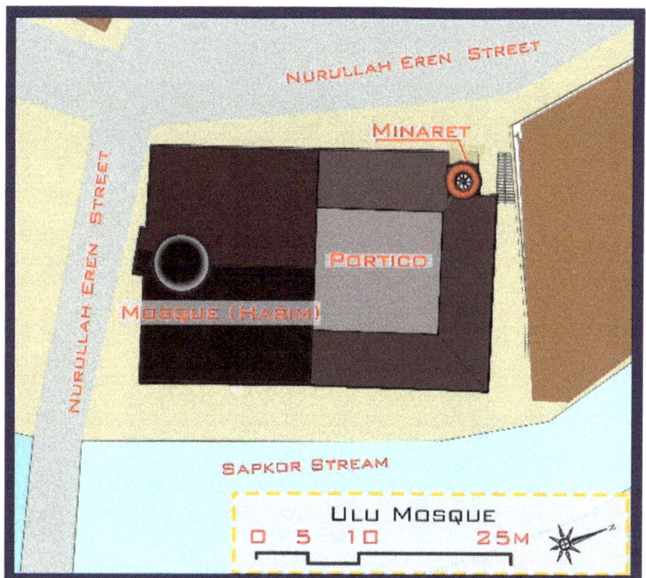

Figure 4. Bitlis Ulu (Grand) Mosque's layout plan.

Figure 5. Bitlis Ulu Mosque's plan and section: (A) sanctuary; (B) portico court; (C) minaret.

There is no definite information about the construction date of the mosque. Şen reports from Ibn-ul Esir in 2018 that it was built in the period of the first Islamic conquests, and it was one of the mosques that was destroyed in the Byzantine period in 928 M [38]. The mosque appears in the miniature of Bitlis, drawn by Matrakçı Nasuh around 1535. It is pictured with its minaret in the southeast of the inner castle, inside the outer castle walls. The south façade of the mosque is stylized with the protrusion of the sanctuary and the window layout placed on the upper level (Figure 6) [39–42]. The Seljuk period work, which is referred to as Cami-façade Köhne in the Şerefname, also has this mosque.

Figure 6. Matrakçı Nasuh Bitlis miniature and the Ulu (Grand) Mosque [39].

There are data regarding the construction period of the part of the building used as a place of worship. There are two written inscriptions, which are among the most important of them, in the structure. These are the inscriptions on the right side of the door from the middle of the north façade and above the door at the west entrance. In the inscription to the right of the middle door, there is information that the building was renovated by Ebu'n-Nazr Muhammed bin El-Muzaffer bin Rüstem in 1150. The inscription on the door at the western entrance was translated by Arık (1971) and it is written that it was repaired by Osman Ağa in 1651 [43]. The building typology and early inscription clearly reveal which period the existing building belongs to. This inscription is an old-dated one, which coincides with the period when Dilmaçoğulları Principality ruled, and the principality was under the influence of the Artuqids. The current state of the sanctuary section also constitutes the first example of the Artuqid period's cross-planned, domed and symmetrical mosque scheme in front of the mihrab [43–47]. The other inscription is related with its repair made during the Serefhans period.

On the exterior part of the rectangular prism-shaped structure, the mihrab protrusion in the direction of the qibla and the conical dome on this section overflow. The mosque, which has an east–west-oriented plan scheme, is divided into three transverse naves in this direction. The spatial part of the structure is supported by five cross-like columns in two horizontal rows. The roofing part of the mihrab is domed. The columns in this section were kept wider and thrusting arches were added to support the dome. There are four windows in the dome. Other sections are covered with east–west-oriented vaults. It is supported by hidden arches and pendentives in the wall joints. The building originally had a flat roof, but a metal hipped roof was added after restoration work. It has been raised 80 cm above the exterior wall to hide the roof. The windows on the three façades of the building are just below the ceiling level. Three windows in the east and west have circular arches, and two windows in the south are rectangular. On the facade where the entrance is located, there are three pointed arched doors in the middle, and two windows between the doors. Due to the sloppy ground level, a gallery was built up to the outer ground level, enclosing the western and southern facades of the building.

The sanctuary and the minaret appear as two separate structures, with approximately 11 m between them. During the restoration works initiated by Bitlis Regional Directorate of Foundations in 2012, traces of a courtyard with a fountain in the middle and the portico around it were unearthed. Yegin (2019) stated that the cruciform column parts were of the same size and typology as those in the mosque, so this part was built at the same time as the mosque [48,49]. The upper cover of the reconstructed portico was built with a pointed barrel vault like the one in the sanctuary. In the section where the minaret and the portico section meet, a load-bearing wall was also built.

The minaret, located in the northwest part of the courtyard with a portico, has three entrance gates at different levels. It is written on the inscription on the eastern entrance gate that it was built in the period of Serefhans in 1492/1493 [43,49]. There are two more inscriptions on the web of the body. The inscription on the western part of the lower part of the body cannot be read because it was destroyed during the 1916 Russian occupation. The inscription in the middle part of the body cannot be read due to similar reasons and deterioration, albeit slight. Its base consists of a square prism lectern and a cube section from the corners to a chamfered circular body. There are two rectangular gates at the ground level and at the 4 m level on the south face. A wall was built in front of the lower door during the restoration phase. The epitaph on the rectangular door on the upper level in the western part is decorated with an archivolt. There is one crenelated window on the southwest face of the lower part and one on the southeast face of the upper part of the body. There are also bullet marks around these windows. From the body, which does not have an ornamental element, to the balcony, a bracelet is used. Arınç (1991) stated that the cone section of the minaret was destroyed in 1975 because of a lightning strike [50]. These sections were renovated as a plain honeycomb-octagonal prism cone with a cylindrical body, such as that of the Meydan Minaret and the Şerefiye Külliyesi Minaret (Figure 7).

Within the scope of this study, the appearance and plans of the minaret of the Great Mosque, one of the most important historical structures of Bitlis Province, are shown in Figure 8.

Figure 7. Bitlis Ulu Mosque's minaret: (**a**) Arık (1971) [43]; (**b**) by authors (2022).

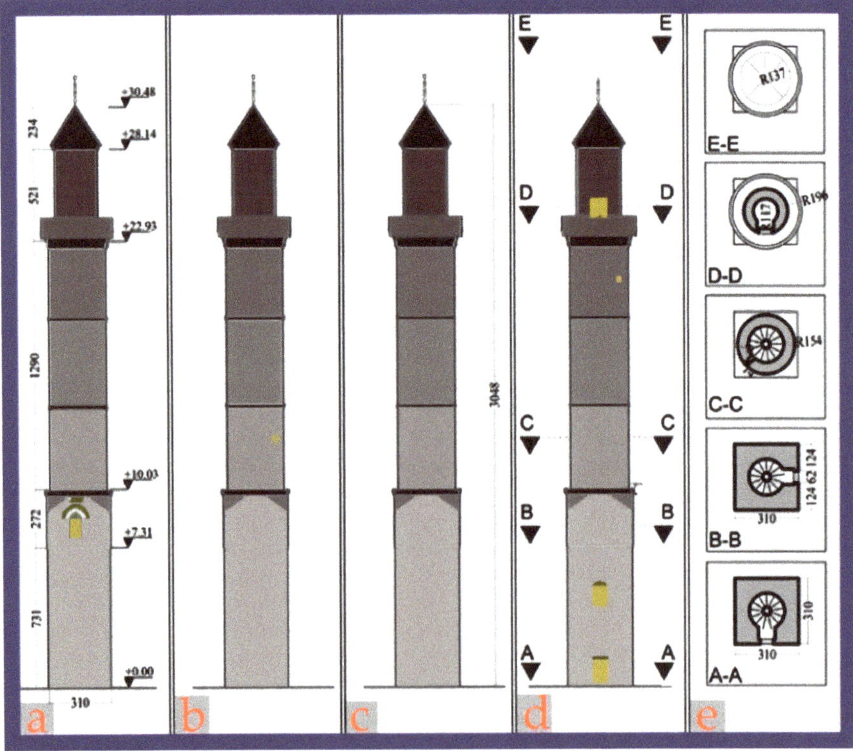

Figure 8. Bitlis Ulu Mosque's minaret view and plans: (**a**) east elevation; (**b**) west elevation; (**c**) north elevation; (**d**) south elevation; (**e**) plans.

The Ulu Mosque and its minaret were built with a single balcony by local masters and workers, using the Bitlis stone. The dimensions of the Ulu Mosque's minaret are 3.10 × 3.10 m and it was built on a square foundation. It has a cylindrical body shape and has a body diameter of 3.10 m. The body wall thickness of the minaret is 0.60 m. The total height for this minaret was determined as 30.48 m. These features of the Bitlis Ulu Mosque's minaret are shown in Table 1.

Table 1. Bitlis Grand Mosque and minaret features.

Parameter	Value
Date of construction	1492 or 1493
Material	Bitlis stone
Balcony	Single
Height	3048 cm
Footing dimensions	3.10 × 3.10 m
Body (web) diameter	3.10 m
Body wall thickness	0.60 m

3. Observation-Based Analysis on the Great Mosque and Minaret

In this part of the study, the current structural condition of the minaret and the mosque has been evaluated based on site observations. The city center of Bitlis, where the minaret is located, is prone to heavy snowfall in the region [51]. Day and night temperature differences are quite high in the city center, especially in winter. For these reasons, all the buildings in the province are under a freeze–thaw effect. In addition to the high groundwater level in the city center, natural disasters such as many earthquakes, rock falls, landslides, floods and avalanches are frequently experienced in and around Bitlis. Due to earthquakes and other natural disasters, various types of damage have occurred to the Ulu Mosque and its minaret over time. Moreover, environmental factors have also contributed significantly to these damages. The mosque and minaret were built with the material known locally as Bitlis stone. The minaret and the mosque were constantly monitored by the relevant institutions and organizations over time, and necessary interventions were made in a timely manner. The damages noted by the on-site observations of the mosque and minaret by the authors are shown in Table 2.

The locations of the damages observed in different parts of the Bitlis Ulu Mosque are shown in Figure 9.

Due to its location at the lowest elevation of the province, Ulu Mosque is also affected by the high groundwater level. This effect is an important factor causing a moisture problem with minor partial settlements in the structure. It is possible to see this effect in the lower parts of almost all vertical load-bearing elements in the mosque. In addition, calcifications are clearly observed on the exterior walls of the mosque due to rain and snow waters. The mosque has recently been repaired and renovated. Negligible mass losses and wall joint losses occurred on the outer walls of the mosque. There are plant formations in some parts of the dome of the mosque. Precipitation water coming from the canals, which were built to remove rain and snow water on the roof of the mosque, known as coratan in its local name, creates a moisture effect in these parts by making a collision effect on the walls of the exterior part. The fact that the mosque is very close to the Bitlis stream bed causes groundwater to be influential inside the mosque. It can be said that the biggest problem for the mosque today is the underground and precipitation waters. A drainage system surrounding the structure can be suggested to prevent possible damage induced by ground surface water that may cause partial settlements. No cracks or damage were observed on the elements of the mosque load-bearing system. The visuals of the damages observed in different parts of the Bitlis Ulu Mosque are shown in Figure 10.

Based on the visual inspections on the Ulu (Grand) Mosque's minaret, the following damages were observed as shown in Figure 11.

Table 2. Damages observed in Bitlis Ulu (Grand) Mosque and its minaret.

Parameters	Yes	No
Time-dependent distortions	X	
Effect of natural conditions	X	
The status of whether the building is actively used or not	X	
Lack of maintenance		X
Random repairs and alterations		X
Has its originality been preserved?	X	
Deterioration of mosque and minaret facades		X
Movement of the stones that make up the structure		X
Fragmentation and rupture		X
Calcification on surfaces	X	
Wear of joints	X	
Scratches in the foundation of the structure		X
Algae and vegetative formations	X	
Repaired or not	X	
Is there a layer of germinated soil?		X
Cracks and features		X
Surface rot		X
Darkening on the surface	X	
Loss of mass	X	
Are there any inconsistencies in the joints?		X
Cracks due to rooting of plants		X
Horizontal and vertical deformations		X
Are the protective measures sufficient?		X
Presence of water entering the structure	X	
Freeze–thaw effect	X	
The effect of construction in the surrounding area		X
Discoloration	X	
Natural disaster effects		X

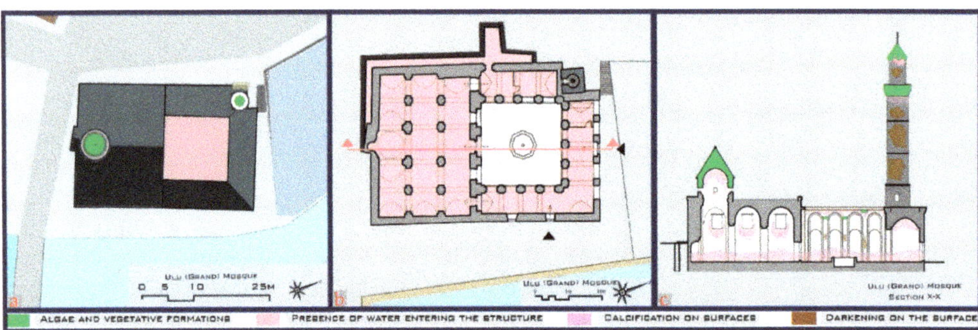

Figure 9. The locations of observed damages; (**a**) layout plan; (**b**) plan; (**c**) section.

Recently, the minaret of the Ulu Mosque has also undergone some repairs in order to preserve its originality. It was determined that the most obvious damage observed on the minaret was traces of bullets. There are numerous traces of bullets on the minaret. Calcifications have been observed especially in the balcony part of the minaret due to precipitation. There is mass loss and joint loss in very few parts. There are vegetal formations in places of the cone part. There is a darkening on some stones that form the minaret. The Bitlis Ulu Mosque and its minaret is one of the Five Minarets that are one of the beauties of Bitlis Province. In this respect, the necessary interventions and repairs to the mosque and minaret were made by the relevant institutions and organizations over time. However, it is inevitable that there will be deteriorations and damages due to the characteristics of the Bitlis stone used, subjected to the excessive and long duration of snowfall. It is known that the Bitlis stone, which was used in the construction of the minaret and the mosque, causes erosion and destruction due to its soft spongy feature. This situation is affected by temperature differences, frost and humid environments, causing fragmentation and rupture. The water movement seen at the ground level of the building induced by the ground water and rainwater must be prevented by installing a drainage system. Moisture-induced deterioration is observed in the parts of the stone texture that met water. Installing the necessary water insulation to the areas where water is a problem at points of the building floor would limit the possible future damages.

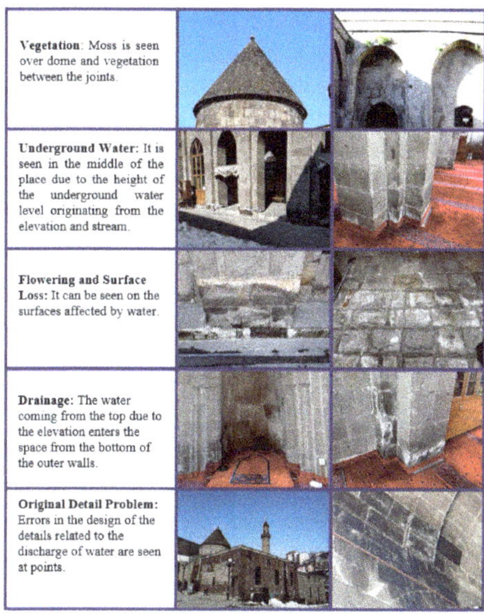

Figure 10. Damages observed in the Ulu Mosque.

Figure 11. Damages observed on the minaret of the Ulu Mosque.

4. Determination of Seismic Parameters for the Selected Settlements

Earthquake zone and hazard maps were prepared at different dates in Turkey. The last two maps differ from the others in that they were prepared using a probabilistic approach. The map of 1972 was prepared using the deterministic method and was amended in 1996 depending on scientific developments. The 1996 Turkey Earthquake Zones Map was prepared by considering the 10% probability of exceedance within 50 years. In this map, five different earthquake zones were taken into account, with the first degree being the most dangerous. All values on the map were on a regional basis and required the use of the same values in the same earthquake zone. In this map, seismicity parameters specific to each geographical location were not considered. Using the recent developments in earthquake and civil engineering, an up-to-date fault database and earthquake catalogues, the map in 1996 was replaced by the Turkey Earthquake Hazard Map, which was completed in 2018. With the current map, the earthquake hazard on a microscale was considered. The last two earthquake maps used in Turkey considered in this study are shown in Figures 12 and 13, respectively. The Turkey Earthquake Hazard Map Interactive Web Earthquake Application (TEHMIWA) [52] was used in order to obtain the seismic parameters of any location with the help of today's technology along with the updated map.

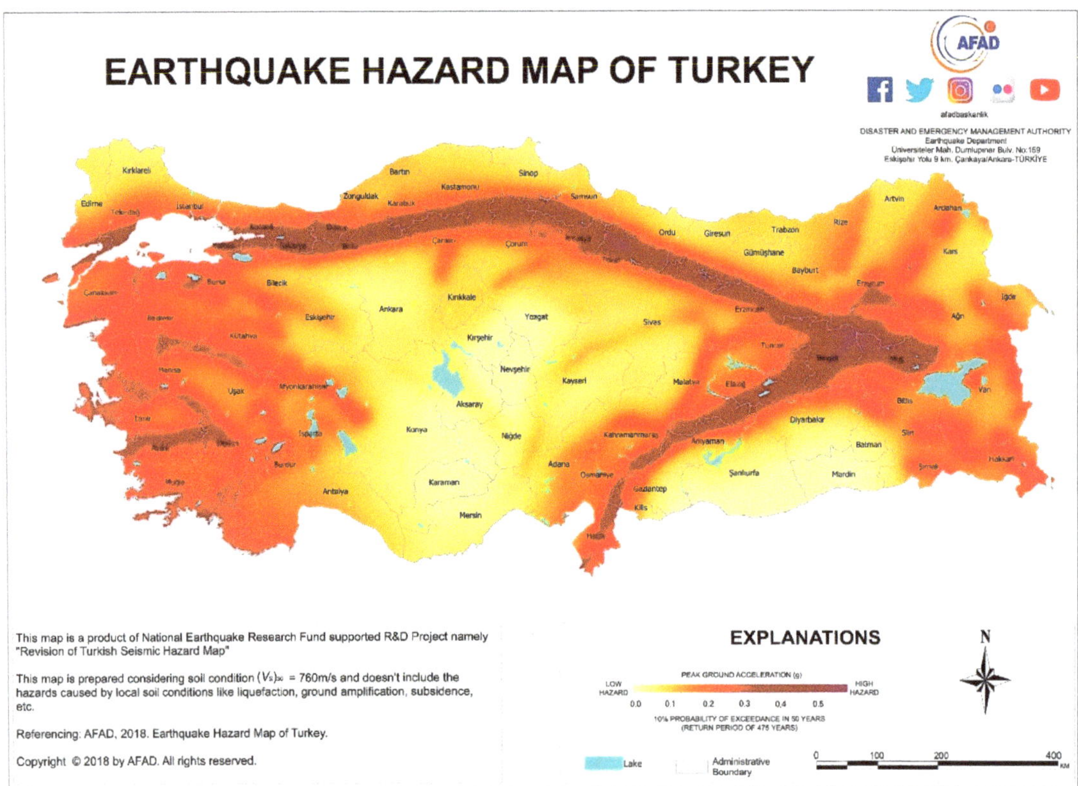

Figure 12. The current earthquake hazard map of Turkey [52].

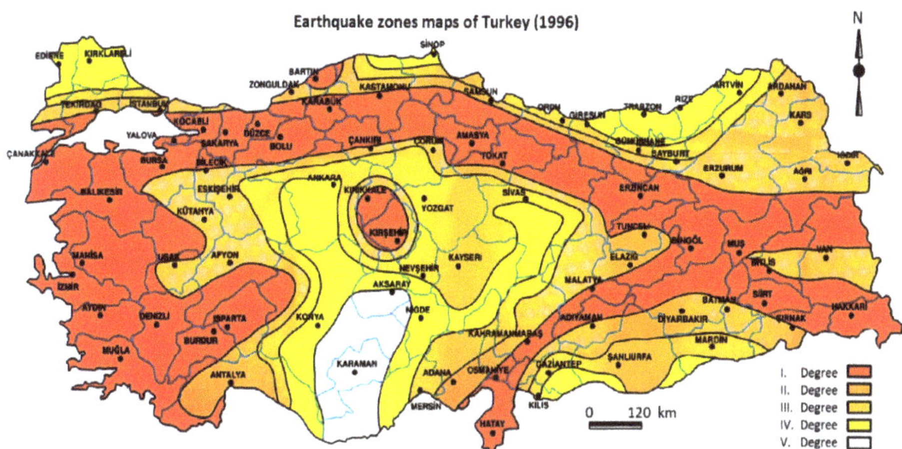

Figure 13. The previous earthquake zone and hazard maps used in Turkey [53].

Within the scope of this study, in order to reveal the effect of the transition from regional to site-specific earthquake hazards on the seismic behavior of masonry minarets, seven different settlements, each one from seven different geographical regions of Turkey, with the same earthquake zone (1. °) on the previous map were taken into account. The considered settlements are shown in Figure 14.

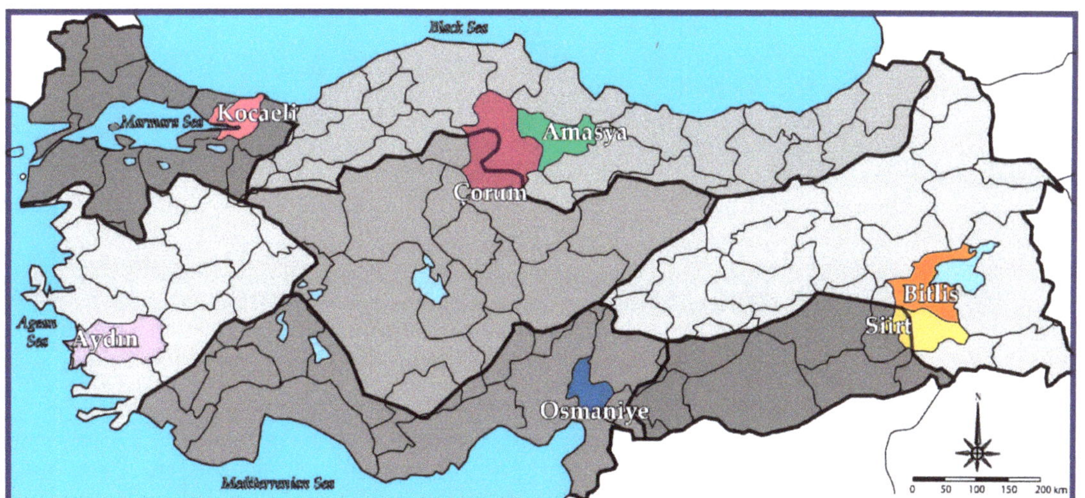

Figure 14. Settlements considered in the study.

Kocaeli (Marmara region), Bitlis (Eastern Anatolia region), Aydın (Aegean region), Siirt (Southeast Anatolia region), Amasya (Black Sea region), Çankırı (Central Anatolia region) and Osmaniye (Mediterranean region) provinces are located in the same earthquake zone. A random geographic location was chosen from each of the centers. For these settlements, earthquake parameters were obtained with the help of TEHMIWA, considering four different ground motion levels with different exceedance probabilities. Different earthquake ground motion levels specified in TBEC-2018 are given in Table 3 to be considered in this study. In the previous regulation, TSDC-2007, only the standard design ground motion level, which has a 10% probability of being exceeded in 50 years, used to be taken into account.

Table 3. Earthquake ground motion levels [27].

Earthquake Level	Repetition Period (Year)	Probability of Exceedance (in 50 Years)	Description
DD-1	2475	0.02	Largest earthquake ground motion
DD-2	475	0.10	Standard design earthquake ground motion
DD-3	72	0.50	Frequent earthquake ground motion
DD-4	43	0.68	Service earthquake movement

In order to make a comparative assessment, the local soil class ZB in TBEC-2018 determined in the site survey reports conducted by the relevant public institutions for the location of the Bitlis Great Mosque's minaret was taken into account. The peak ground acceleration (PGA) and the peak ground velocity (PGV) values obtained for different probability of exceedance for the selected settlements are shown in Table 4.

Table 4. PGA and PGV values for selected settlements for different ground motion levels.

Province	Peak Ground Acceleration (g)—PGA				Peak Ground Velocity (cm/s)—PGV			
	Probability of Exceedance in 50 Years				Probability of Exceedance in 50 Years			
	2%	10%	50%	68%	2%	10%	50%	68%
Çankırı	0.533	0.283	0.111	0.079	34.409	18.877	7.456	5.301
Aydın	1.090	0.593	0.214	0.149	69.915	36.949	11.618	7.948
Amasya	0.809	0.447	0.182	0.131	54.265	29.458	11.404	8.063
Kocaeli	1.132	0.673	0.275	0.142	96.795	56.699	16.345	7.891
Siirt	0.456	0.244	0.093	0.064	24.078	12.828	5.448	4.004
Bitlis	0.490	0.260	0.106	0.077	28.215	15.081	6.508	4.847
Osmaniye	0.598	0.310	0.115	0.079	36.886	18.251	6.542	4.539

Among the selected settlements, the highest PGA and PGV values were obtained for Kocaeli, while the lowest values were obtained for the province of Siirt. The ratio of the highest and lowest PGA values was obtained as 2.48 for DD-1. The comparison of the PGA values and design spectral acceleration coefficients predicted in the last two regulations and the maps according to the DD-2 ground motion level is also presented in Table 5.

Table 5. PGA and S_{DS} values and comparison over the last two maps.

Location	TSDC-2007 Seismic Zone	TSDC-2007 PGA (g)	TBEC-2018 PGA (g)	PGA_{2018}/PGA_{2007}	S_{DS2007}	S_{DS2018}	S_{DS2018}/S_{DS2007}
Çankırı	1	0.400	0.283	0.71	1.000	0.605	0.61
Aydın	1	0.400	0.593	1.48	1.000	1.308	1.31
Amasya	1	0.400	0.447	1.12	1.000	0.968	0.97
Kocaeli	1	0.400	0.673	1.68	1.000	1.482	1.48
Siirt	1	0.400	0.244	0.61	1.000	0.510	0.51
Bitlis	1	0.400	0.260	0.65	1.000	0.553	0.55
Osmaniye	1	0.400	0.310	0.78	1.000	0.651	0.65

PGA and S_{DS} values for all settlements considered in this study for the same earthquake zone in the former map were completely represented with different values in the current regulation. However, since the previous regulation and the map were prepared on a regional basis, these two values were assumed to be the same. The fact that each location had its own seismicity parameters indicated that it was more realistic to base the current regulation on a site-specific seismic hazard analysis. While PGA values increased for Aydın, Amasya and Kocaeli within the selected settlements, lower values were obtained for other provinces than the values predicted in the previous regulation. While the biggest increase was observed in Kocaeli, the biggest decrease was in Siirt. This situation remained valid for S_{DS} as well. Another variable used within the scope of this study was the different probability of exceedance levels. Other seismic parameters considered within the scope of this study are given in Table 6.

Table 6. Comparison of earthquake parameters for different probability of exceedance levels.

Ground Motion	S_S	S_1	F_S	F_1	S_{DS}	S_{D1}	T_A	T_B	T_{AD}	T_{BD}
DD-1	1.193	0.312	0.900	0.800	1.074	0.250	0.046	0.232	0.015	0.077
DD-2	0.614	0.172	0.900	0.800	0.553	0.138	0.050	0.249	0.017	0.083
DD-3	0.243	0.076	0.900	0.800	0.219	0.061	0.056	0.278	0.019	0.093
DD-4	0.176	0.055	0.900	0.800	0.158	0.044	0.056	0.278	0.019	0.093

There was no change in the local ground effect coefficients obtained for the different probability of exceedance levels for the province of Bitlis. However, as the probability of exceedance levels increased, the S_S, S_1, S_{DS} and S_{D1} values decreased. This is due to the fact that very rare earthquakes have higher magnitudes. The design spectra, on the other hand, were obtained by fitting a smoothed envelope curve to the response spectra calculated from different ground motion acceleration records in order to take into account the maximum earthquake effects that may occur, by taking into account the seismic characteristics and local site conditions of any region [54,55]. The change in the design spectrum curves also affected the displacement demands in the buildings. It is obvious that damage estimations and building performances would deviate from the realistic values in buildings when displacement demands are not met [56,57]. A comparison of the obtained horizontal and vertical elastic design spectra is given in Figure 15.

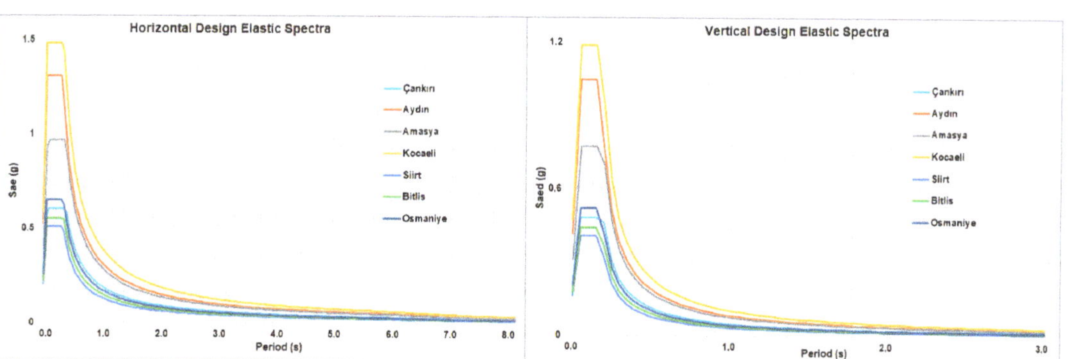

Figure 15. Comparison of horizontal and vertical elastic design spectra.

Since the vertical elastic design spectrum was used for the first time with the current regulation, it was not possible to compare it with the previous regulation. While a single design spectrum was used for the settlements selected in the previous map, a design spectrum specific to each geographical location was used together with the current regulation. A comparison of the obtained horizontal elastic design spectra revealed this difference. The comparison of the horizontal and vertical elastic design spectra obtained by considering the different probabilities of exceedance for the minaret is given in Figure 16.

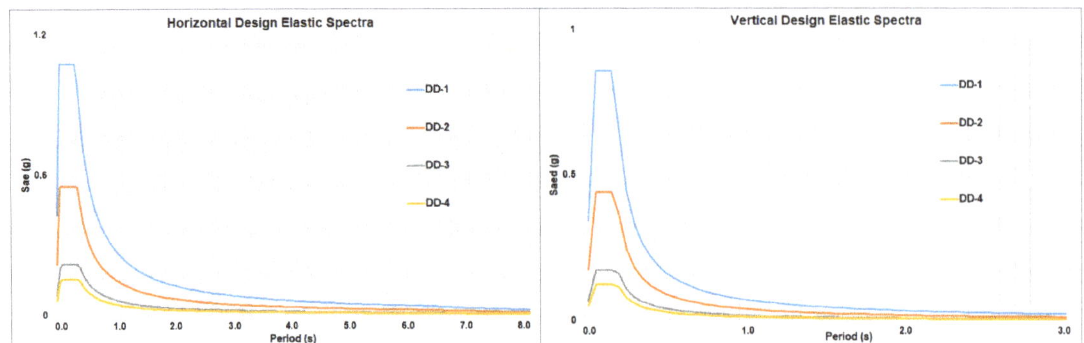

Figure 16. Comparison of horizontal and vertical elastic design spectra obtained for different probabilities of exceedance.

The horizontal and vertical elastic design spectra with the largest amplitude for ground motion levels with different recurrence periods were obtained for the largest earthquake with a recurrence period of 2475 years. Table 7 shows the comparison of the earthquake data that occurred recently around the province of Bitlis, where the minaret is located, and the values obtained for the location of the minaret.

Table 7. Comparison of measured and obtained PGAs for minaret location.

Earthquake No.	Date	Location	Measured Values	PGA_{2018} (g)				PGA_{2007} (g)
			PGA (g)	DD-1	DD-2	DD-3	DD-4	DD-2
1	24 January 2020	Sivrice	0.298					
2	23 October 2011	Van	0.182					
3	9 November 2011	Van	0.251					
4	8 March 2010	Elazığ	0.068	0.490	0.260	0.106	0.077	0.400
5	1 May 2003	Bingöl	0.511					
6	27 January 2003	Tunceli	0.011					
7	13 March 1992	Erzincan	0.485					
8	30 October 1983	Erzurum	0.175					

It is seen that the PGA values measured for the Van earthquakes, which are the recent earthquakes that have occurred in and around Bitlis, where the Ulu Mosque is located, are lower than the values predicted by the current regulation. Except for the Bingöl (2003) earthquake, the PGA values measured for all earthquakes were lower than the predictions for DD-1. For DD-2, the values measured for the Bingöl (2003) earthquake as well as the Sivrice (2020) and Erzincan (1992) earthquakes were larger than the predicted one. Bitlis Province is approximately 200 km from Bingöl and 350 km from Erzincan and Sivrice. In addition, the measured value of the 2011 Van earthquake at the earthquake station in Bitlis is known to be 0.104 g. Therefore, the values obtained for the largest earthquake affecting Bitlis recently were considerably lower than the PGA value foreseen in the last two earthquake codes.

5. Structural Analyses

For the analysis and design of today's modern engineering structures, many computer software packages have been developed that facilitate data transfer and transfer the results to application projects in an integrated manner. The load-bearing systems of masonry buildings differ from today's modern engineering structures. For this reason, the finite element method is preferred in the structural analysis of such structures. The first step in this method is to create a numerical model of the structure to be examined. Numerical modeling can be defined as the reliable and compatible conversion of structural system elements made of different materials and having a variable cross-section geometry into mathematical terms according to the fundamental principles of mechanics. The finite element analysis of masonry structures is computationally a demanding process [58,59].

The realization of the analysis in the created structural model roughly consists of the stages of creating the geometry, mesh production, physical setup, numerical solution and obtaining the results. ABAQUS is software for both the modeling and analysis of mechanical components and the visualization of the finite element analysis result. Preprocessing and postprocessing stages can also be performed with all the components of modeling such as monitoring the solution, intervening in the process and examining the results [60]. In some cases, simple finite element models can be created by developing some approaches and simplifications regarding the geometric properties of structures. Despite such minor changes, FE models can be used to evaluate the seismic performance of such building typologies [61]. The sign convention and directional assumptions of the elements used in the structural models created using finite elements are shown in Figure 17, adhering to the assumptions stipulated by the software program (ABAQUS 2022) [60] in which the numerical modeling was made.

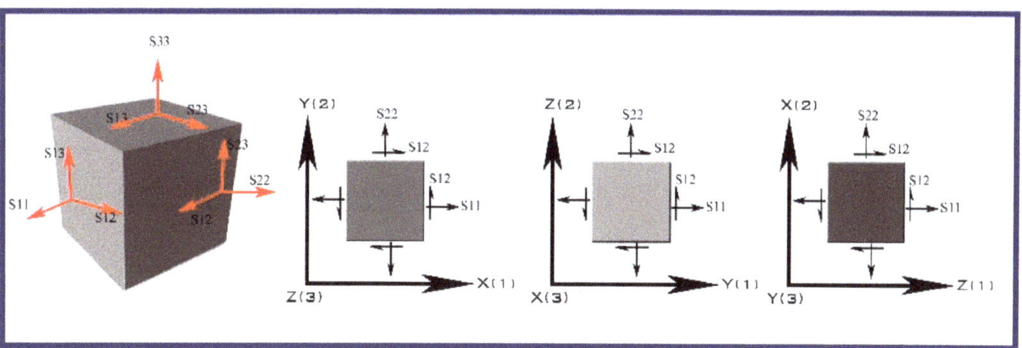

Figure 17. Sign convention in the analyses.

As indicated in Figure 17: S11: vertical stress in (x) direction, S22: axial stress in (y) direction, S33: axial stress in (z) direction, (S12 = S21): shear stresses in x–y plane. While determining the material properties for the minaret, the following literature works were followed Işık et al. (2020) [62] and Işık et al. (2022) [13]. The elastic modulus (E) and unit weight (γ) values of Bitlis stone were taken as a single value in all structures. The properties of the material used are given in Table 8.

Table 8. Material properties [13,62].

	Unit Volume Weight (kN/m^3)	Elastic modulus, E (MPa)	Poisson Ratio
Bitlis stone	14.60	4006	0.22

The modeling of masonry walls is extremely important in the evaluation and design of historical and modern masonry structures. Masonry walls can be modeled using three different modeling techniques such as a detailed micro modeling, simplified micro modeling and macro modeling. These models can be seen in Figure 18.

In this study, a finite element model of the minaret was created using the macro modeling technique. The macro modeling technique is one of the commonly used masonry structural modeling techniques in the literature. The historical masonry bell towers in South-East Lombardy in Italy [63], historical fortified masonry palaces in Switzerland and Northern Italy [64], Emir Bayındır Bridge in Turkey [65], San Pietro and San Benedetto churches in Italy [66], Torre De la Vela in Spain [67], Gaskar brick minaret in Iran [68] and five historical masonry minarets in Antalya (Turkey) [69] are some of the studies examining the seismic behavior of different types of structures using this type of structural modeling technique. While doing this type of modeling, analyses are carried out without making any distinction between the binding material (mortar, etc.) used in the building and the structural elements. In this modeling, the masonry unit and the properties of the mortar are homogenized and considered as masonry composite material. The mechanical properties of this model are the values obtained as a result of the homogenization process. Macro modeling is more convenient in practice because it requires less computational cost. However, with macro modeling, stress distributions in masonry units and mortar can be obtained accurately [70–72]. The Bitlis stone used in the minaret and the mortar that connects them were considered as a single material, and the material properties of the Bitlis stone were taken into account in the analyses for these two materials.

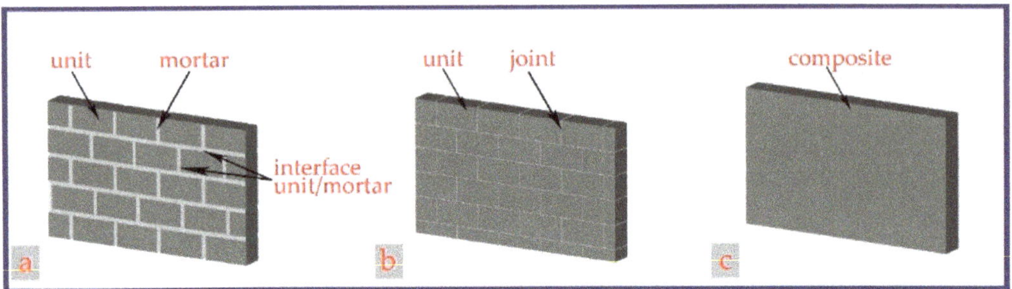

Figure 18. Modeling methods of masonry: (**a**) detailed micro modeling; (**b**) simplified micro modeling; (**c**) macro modeling.

5.1. Structural Analysis Results According to Different Probability of Exceedance Levels

While modeling with the ABAQUS program, stairs were taken into account and drawn together with the outer wall to avoid possible problems in the meshing system and intersection. The macro modelling presupposed that the masonry structure was a homogenous continuum that could be discretized with a finite element mesh that did not replicate the wall texture but rather met the criteria established by the method itself. This modeling approach stroke a balance between accuracy and simplicity by aiming to produce results at a global level and keeping computational work at a manageable level. The minaret was analyzed with FE models consisting of ten-node quadratic tetrahedral elements (C3D10) with four integration points chosen as the mesh type. In order to further improve the outcomes, several mesh sizes were examined, beginning with the coarse ones, until stable results were obtained. The masonry minaret consisted of a total of 19,663 elements and 39,557 nodes. The dimensions based on site measurements of the minaret and the 3D models developed by software are shown in Figure 19.

A modal analysis is a dynamic analysis method that enables the determination of free vibration periods, frequency values, mass participation rates and mode shapes of the structure. In order to determine the dynamic properties of the minaret, first, modal analyses were performed. The natural vibration periods of the minaret were added as a result of the eigenvalue analysis. While performing the modal analysis, the first 10 modes formed in the structure were taken into account. According to the modal analysis results of the model, its effective modes, natural vibration periods and frequency values are shown in Table 9. The calculated mass participation of the modes shows the contribution of the mode effects to the overall dynamic response. The calculated mass participations of the first and second modes have a significant proportion of the overall dynamic behavior of the structure. In the selected minaret, the mass participation rates in the first two modes were around 44%. It was seen that the mass participation rates in the X and Y directions were above 80% and considering additional modes did not have much effect.

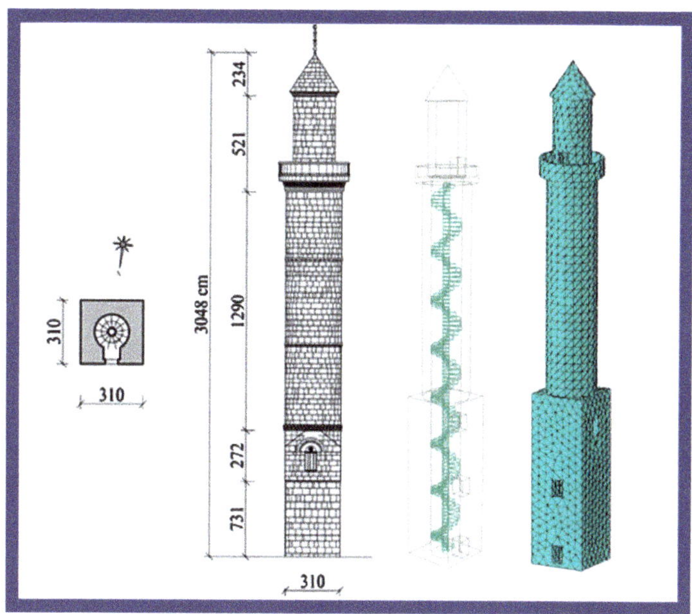

Figure 19. Bitlis Ulu Mosque's minaret and 3D minaret models developed by the software program.

Table 9. Modal analysis results of the minaret model.

Mode	Frequency (Hz)	Period (s)	Mass Participation (X) (%)	Mass Participation (Y) (%)	Total Mass Participation (X) (%)	Total Mass Participation (Y) (%)
1	1.499	0.667	0.01	43.68	0.01	43.68
2	1.515	0.660	43.32	0.01	43.33	43.68
3	6.418	0.156	0	23.26	43.34	66.94
4	6.482	0.154	24.07	0	67.41	66.94
5	11.758	0.085	0.04	0	67.45	66.94
6	14.005	0.071	0.00	8.45	67.45	75.39
7	14.245	0.070	8.90	0	76.35	75.39
8	17.026	0.059	0	0.03	76.35	75.42
9	23.818	0.042	0	5.47	76.36	80.90
10	24.182	0.041	4.91	0	81.27	80.90

The first natural period value obtained for the Ulu Mosque's minaret was checked by averaging the four differential empirical period relations suggested in the literature. The empirical relations considered are shown in Table 10. The natural period of the minaret is in agreement with that of the literature. In addition, the obtained period value for the Ulu Mosque's minaret remains between the lowest and highest values obtained from the empirical formulas taken into account.

The mode shapes obtained while performing the analysis in the software program of the Bitlis Ulu Mosque's minaret are also shown in Figure 20. As can be seen from Figure 20, torsion occurred in the fifth mode.

Table 10. Suggested empirical formula for fundamental period.

Empirical Formula	Period of the Sample Minaret	Description	References
$T_1 = C_t H^{0.75}$	$T_1 = 0.646$ s	$C_t = 0.05$ and H = total height	[69,73,74]
$f_1 = Y. (H/B)^{-z}$	$T_1 = 0.88$ s	H: total height; B: minimum base width of the minaret. Y = 8.03 and z = 0.86 for minarets	[75]
$T_1 = 0.0187$ H	$T_1 = 0.570$ s	H: total height	[76]
$T_1 = 0.0113$ $H^{1.138}$	$T_1 = 0.552$ s	H: total height	[77]
Mean	$T_1 = 0.662$ s		

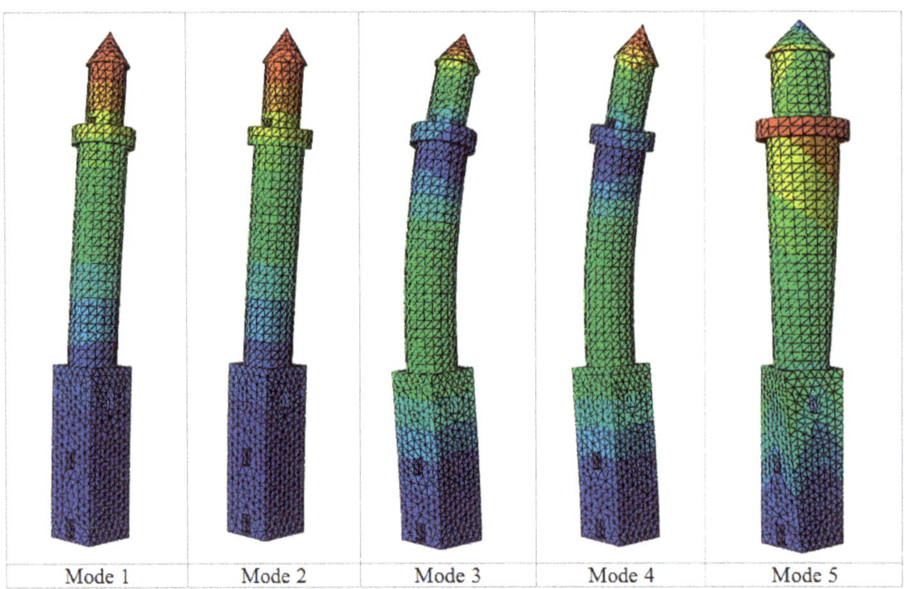

Mode 1   Mode 2   Mode 3   Mode 4   Mode 5

Figure 20. Mode shapes fort the first five modes.

An Abaqus response spectrum analysis was used to estimate the peak response parameters (displacement, stress and base shear force) of the minaret. The stresses distribution in the minaret is shown in Figure 21. Since the same design spectrum curve was used for seven different settlements considered in the study and located in the same earthquake zone, the same stresses would be obtained. The stress diagrams obtained for the largest earthquake ground motion level (DD-1) with a 2% probability of exceedance in 50 years are shown in Figure 22. The stress diagrams obtained for the standard design earthquake ground motion level (DD-2) with a 10% probability of exceedance in 50 years and a recurrence period of 475 years are shown in Figure 23. The stress diagrams obtained for the frequent earthquake ground motion level (DD-3), which has a 50% probability of exceedance in 50 years, are shown in Figure 24. The stress diagrams obtained for the service earthquake ground motion level (DD-4), which has a 68% probability of exceedance in 50 years, are shown in Figure 25.

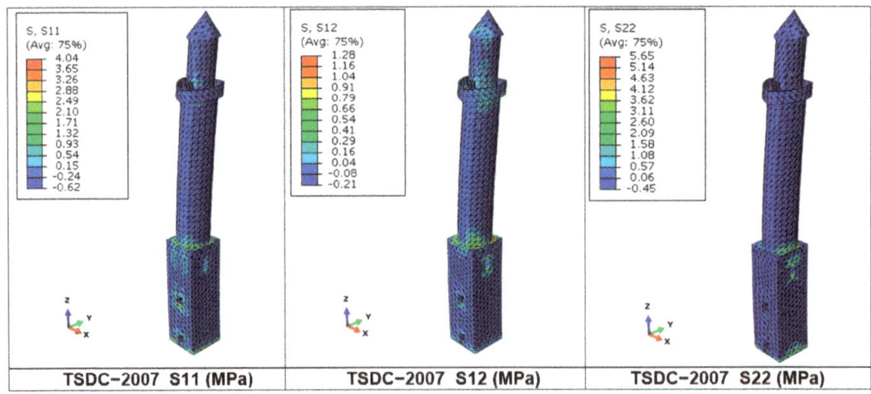

TSDC−2007 S11 (MPa)   **TSDC−2007 S12 (MPa)**   **TSDC−2007 S22 (MPa)**

Figure 21. Stress distribution according to TSDC−2007.

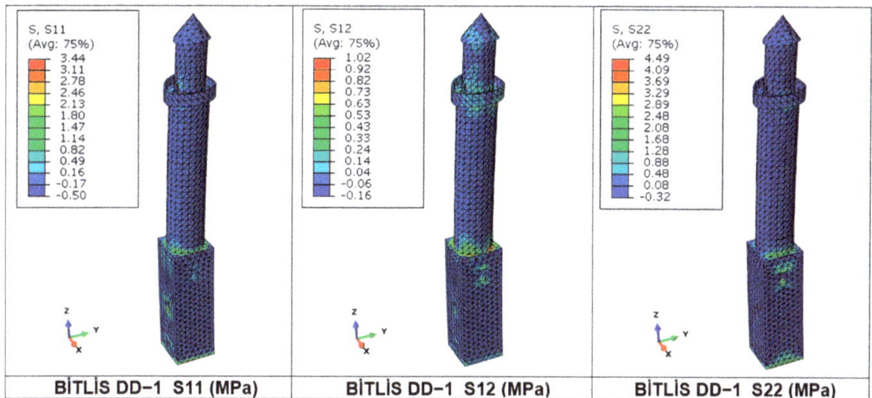

Figure 22. Relative stresses for DD−1 ground motion level.

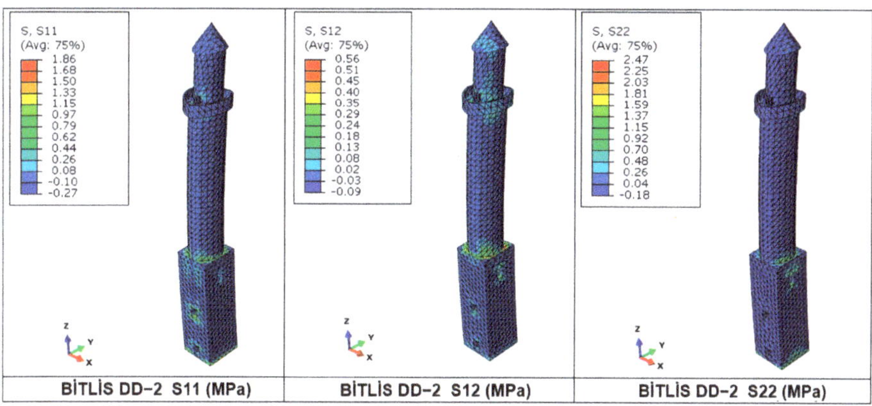

Figure 23. Stresses for DD−2 ground motion level.

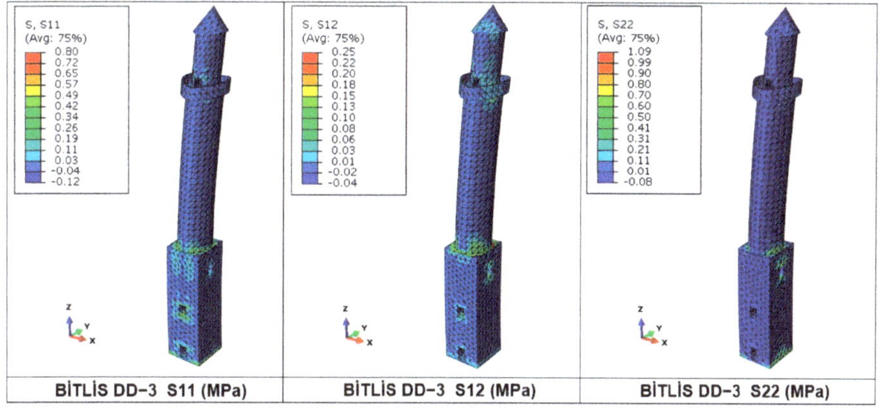

Figure 24. Stresses for the DD−3 ground motion level.

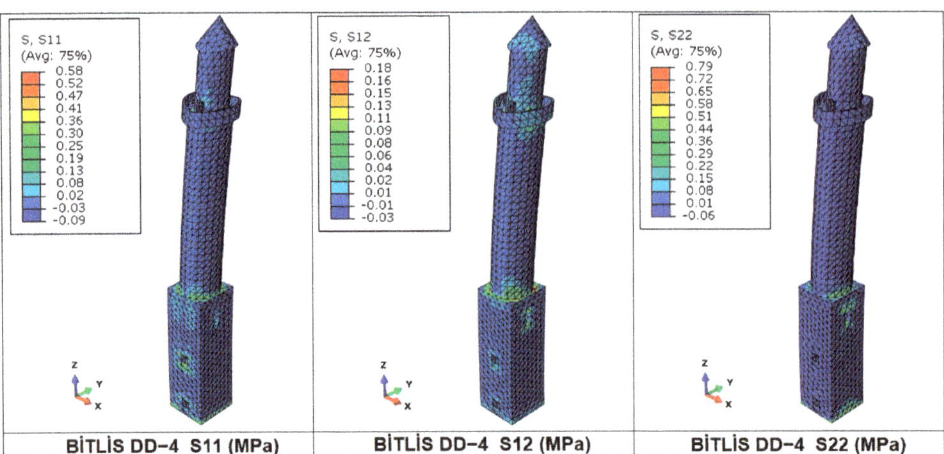

Figure 25. Stresses for DD−4 ground motion level.

The comparison of the maximum displacement, base shear force and stress values obtained for the ground motion levels with different probabilities for the minaret of Bitlis Ulu Mosque is given in Table 11.

Table 11. The highest values obtained according to different exceedance probabilities.

Ground Motion Level	Displacement (mm)	Base Shear Force (N)	S11 (MPa)	S12 (MPa)	S22 (MPa)
TSDC-2007	380.9	4.25×10^6	3.44	1.02	4.49
Bitlis DD-1	481.3	4.36×10^6	4.04	1.28	5.65
Bitlis DD-2	210.1	2.23×10^6	1.86	0.56	2.47
Bitlis DD-3	92.9	0.91×10^6	0.80	0.25	1.09
Bitlis DD-4	67.2	0.66×10^6	0.58	0.18	0.79

The largest displacement, base shear force and stress values were obtained for DD-1 and the lowest values were obtained for DD-4. Different values were obtained for the standard design ground motion level (DD-2) in both earthquake codes. For this ground motion level, the largest displacement value for the previous regulation was 0.38 m, while it was 0.21 m with the current regulation. The stresses obtained for the 2007 regulation decreased by approximately 45% compared to those of the current regulation. The displacement value obtained for the DD-1 ground motion level increased by 129% compared to that of the DD-2 ground motion level. In the previous regulation, only one ground motion level (DD-2) was taken into account, and since all selected settlements were located in the same earthquake zone, the same displacement, base shear and same stresses were obtained. With the current regulation, different values were obtained for all selected settlements. A comparison of the obtained result values is shown in Figure 26.

Performance levels were also determined by using the displacement values obtained in the study. For this purpose, the limit value assumptions specified in the Earthquake Risk Management Guide for Historical Buildings (TYDRYK-2017) [78] were used and these values are presented in Figure 27. It is stated that it is sufficient to use linear calculation for the immediate occupancy (IO) performance level, and one of the linear or nonlinear calculation methods for the life safety (LS) and collapse prevention (CP) performance levels. If one of these calculation methods is selected, the minimum limit values that must be provided are shown.

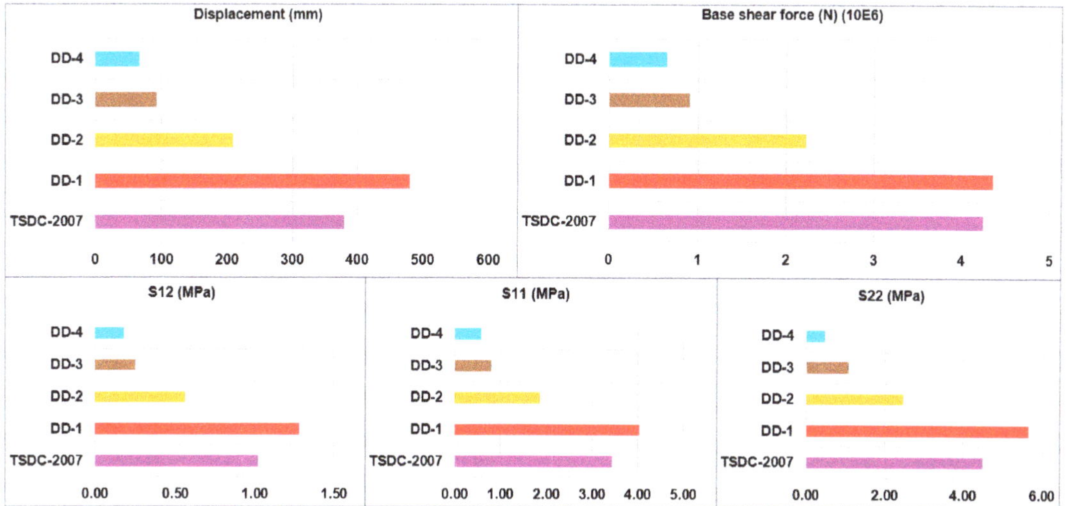

Figure 26. Comparison of result values for different probabilities.

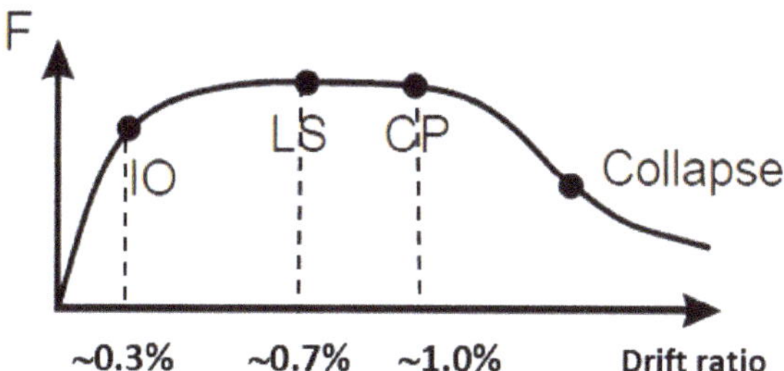

Figure 27. Pushover curve and limit states.

The performance levels obtained for the DD-2 earthquake as the standard ground motion level, which is predicted in TSDC-2007 and TBEC-2018, are shown in Table 12.

Table 12. Design earthquake drift check.

Code	Maximum Displacement (mm)	Maximum Drift (%)	Immediate Occupancy < 0.3%		Life Safety < 0.7%		Collapse Prevention < 1%	
TSDC-2007	380.90	1.25	91.44	✘	213.36	✘	304.80	✘
TBEC-2018	210.10	0.69	91.44	✘	213.36	✔	304.80	✔

The values obtained for the standard design earthquake ground motion level (DD-2) were taken into account as the maximum displacement values. The maximum drift (%) was obtained by dividing the obtained maximum displacement value by the minaret height. The target displacement values for the performance levels were obtained by multiplying the marginal percentages of the minaret height under the unmitigated earthquake effect seen in TYDRYK-2017 (Figure 27). Considering the displacement value obtained for TSDC-2007, it was determined that it did not satisfy the limit states for three different performance levels, but the value obtained for TBEC-2018 satisfies the life safety and collapse prevention performance levels.

5.2. *Influence of the Site-Specific Design Spectrum*

In this part of the study, the effect of the site-specific design spectrum, which is one of the important changes between the last two earthquake codes, is examined. In the previous regulation, site-specific seismicity effects were neglected by using the same design spectrum for the same earthquake zone. The fact that each geographical location has unique seismic parameters requires a differentiation of the design spectra. In order to show the difference, the stresses occurring for Kocaeli Province, which has the highest PGA value among the settlement units considered within the scope of the study, are shown in Figure 28, while the stresses occurring in the minaret of the Ulu Mosque, which was chosen as an example for Siirt with the lowest PGA value, are shown in Figure 29. During this comparison, no changes were made in the structural characteristics, material properties and local soil class of the minaret. The only variable was the design spectrum specific to each settlement. The importance of using the site-specific design spectra is shown.

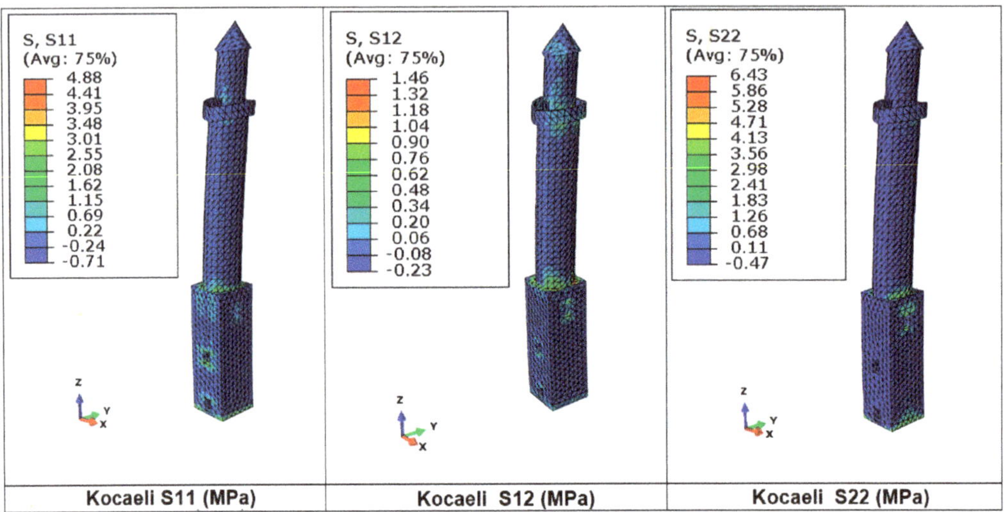

Figure 28. Stresses for Kocaeli (Central) Province.

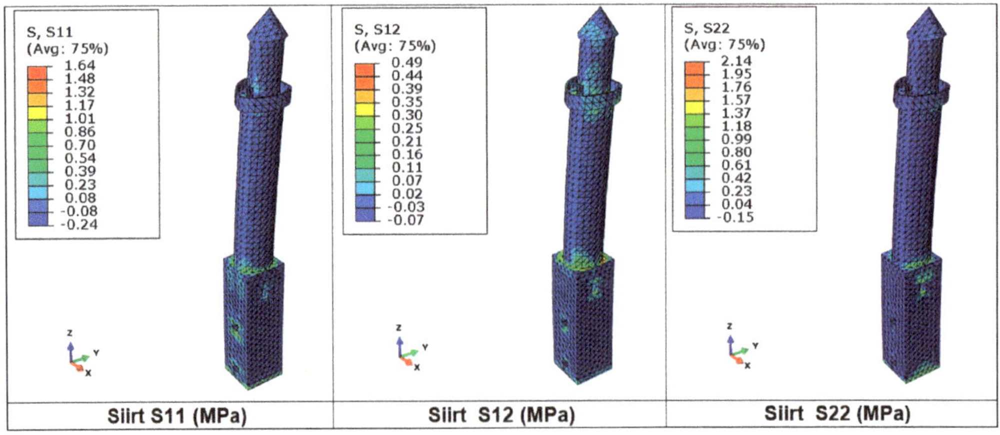

Figure 29. Stresses for the province of Siirt (Center).

The comparison of the result values obtained for all settlements considered in the study is shown in Table 13. The comparison of the displacement values obtained for the settlements is shown in Figure 30.

Table 13. Comparison of the results obtained for different settlements.

Location	Displacement (mm)	Base Shear Force (N)	S11 (N/mm^2)	S12 (N/mm^2)	S22 (N/mm^2)
Çankırı	264.1	2.53×10^6	2.26	0.71	3.10
Aydın	439.7	5.11×10^6	4.04	1.18	5.19
Amasya	398.1	3.98×10^6	3.45	1.06	4.68
Kocaeli	546.2	5.93×10^6	4.88	1.46	6.43
Siirt	181.6	2.02×10^6	1.64	0.49	2.14
Bitlis	210.1	2.23×10^6	1.86	0.56	2.47
Osmaniye	238.9	2.60×10^6	2.14	0.64	2.81
TSDC-2007	380.9	4.25×10^6	3.44	1.02	4.49

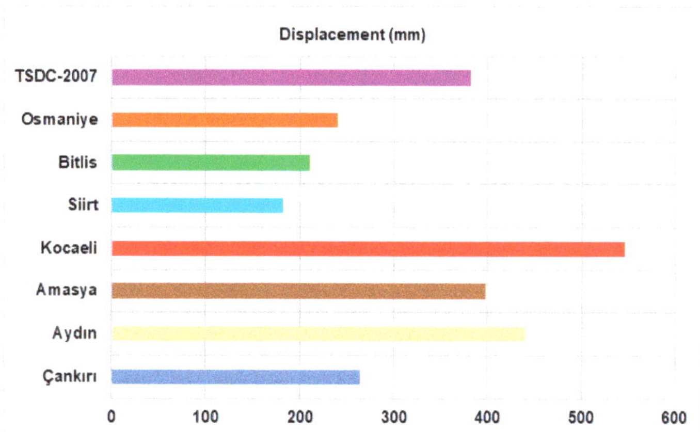

Figure 30. Comparison of displacement values for each settlement.

With the current regulation, the design spectra obtained on a regional basis were replaced by the geographical-site-specific design spectra. While the same values used to be obtained for all the settlements in that earthquake zone, different values can be obtained for the settlements within the region. This clearly shows that structural analyses should be performed using site-specific seismic parameters. Using the design spectra stipulated in the previous regulation, the same values were obtained within the same region, whereas each geographic location has its unique seismic parameters and will be subjected to different levels of shakings. The highest displacement, base shear force and stress values were obtained for Kocaeli, which has the highest PGA value, while the lowest values were obtained for the province of Siirt, which had the lowest PGA value. While higher values were obtained for Aydın, Amasya and Kocaeli in the new regulation, lower values were obtained for the other settlements. The comparison of the performance levels for the settlements considered in the study is shown in Table 14.

Table 14. Design earthquake drift controls.

Province	Maximum Drift (mm)	Maximum Drift (%)	Immediate Occupancy (IO) < 0.3%		Life Safety (LS) < 0.7%		Collapse Prevention (CP) < 1%	
Çankırı	264.1	0.87	91.44	✗	213.36	✗	304.8	✓
Aydın	439.7	1.44	91.44	✗	213.36	✗	304.8	✗
Amasya	398.1	1.31	91.44	✗	213.36	✗	304.8	✗
Kocaeli	546.2	1.79	91.44	✗	213.36	✗	304.8	✗
Siirt	181.6	0.6	91.44	✗	213.36	✓	304.8	✓
Bitlis	210.1	0.69	91.44	✗	213.36	✓	304.8	✓
Osmaniye	238.9	0.78	91.44	✗	213.36	✗	304.8	✓
TSDC-2007	380.9	1.25	91.44	✗	213.36	✗	304.8	✗

The values in Table 13 were obtained by taking into account the calculation principles considered in Table 11. For the settlements located in the same earthquake zone, the same values were obtained in the previous regulation, and it was determined that the limit conditions foreseen for the three performance levels were exceeded. For Siirt, Bitlis, Osmaniye and Çankırı, it was concluded that the limit states were exceeded for the other settlements, while the CP level was achieved.

5.3. Dynamic Time History Analysis

Time history analysis is the numerical analysis of the equation of motion, which is created by considering the mass, damping and stiffness properties of the structure, under a selected ground motion. In this part of the study, time history analyses were carried out by using the acceleration records of the 2011 Van earthquake, which occurred recently in the closest region to the Bitlis Ulu Mosque's minaret. The Van Earthquake south–north acceleration–time curve of this considered earthquake is shown in Figure 31. Solid elements were used in the construction of the finite element model under the presumption that the structure's materials were homogeneous, and a linear elastic material model was used for the time history analysis.

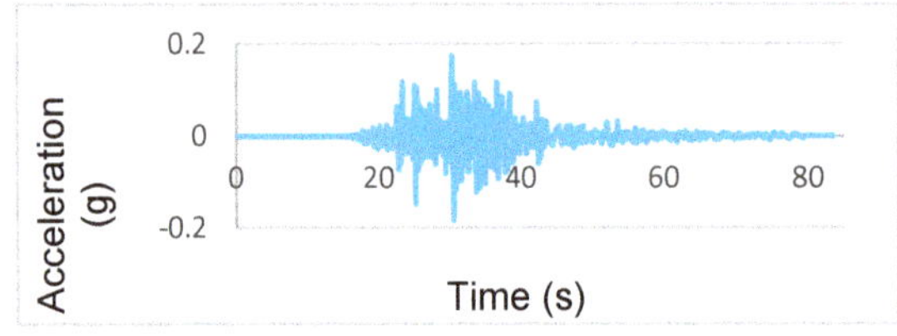

Figure 31. 2011 Van earthquake south–north acceleration–time graph.

The displacement–time graph obtained as a result of the time history analysis is shown in Figure 32 by using the 2011 Van earthquake south–north acceleration–time curve.

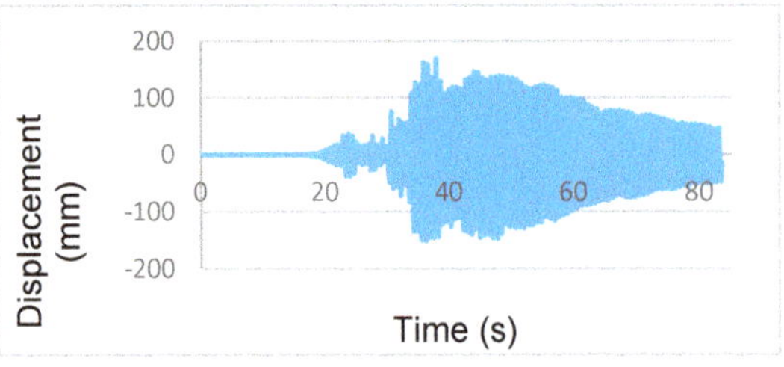

Figure 32. Time history analysis (Van earthquake) (max = 169.64 mm).

As a result of the 2011 Van earthquake acceleration record, it was obtained that the minaret of the Ulu Mosque's minaret had a maximum displacement of 0.17 m. This value was lower than the displacement values of 0.38 m and 0.21 m determined for the last two earthquake codes. Considering the displacement value obtained for the 2011 Van earthquake, the performance levels are shown in Table 15.

Table 15. Analysis translation results and limit values for the 2011 Van earthquake.

Material Type	Maximum Drift (mm)	Maximum Drift (%)	Immediate Occupancy (IO) < 0.3%		Life Safety (LS) < 0.7%		Collapse Prevention (CP) < 1%	
TBEC-2018	169.64	0.56	91.44	✗	213.36	✓	304.80	✓

The minaret performance level was obtained as life safety according to the largest displacement value obtained by considering the 2011 Van earthquake acceleration record.

6. Conclusions

Within the scope of this study, ground motion levels with different exceedance probabilities and different design spectra in masonry structures were chosen as variables. As a case study, the minaret of the historical Ulu Mosque in the province of Bitlis (Turkey) was chosen. This minaret is one of the Five Minarets, one of the symbolic structures of Bitlis Province, and is an invaluable part of the historical heritage. For this minaret, a structural model was created with the finite element method and analyses were carried out separately for different settlements. Finally, the current situation of the mosque and minaret was determined based on on-site measurements and observations, and solution suggestions were discussed.

Using the seismicity parameters and predicted design spectra in the last two earthquake hazard maps and regulations in Turkey, the seismic behavior of this minaret under the influence of earthquakes was compared. In recent years, with the use of site-specific design spectra, the variation of the amplitude of the design spectra and the PGA value significantly changed, resulting in different response estimations in many countries [36,57,79,80]. While the same seismic and structural parameters were used for the selected seven different settlements with the same seismic risk in the previous earthquake hazard map, completely different seismic and structural parameters were used together with the current earthquake hazard map and regulation for Turkey. As a result, the results of the structural analysis differed completely. In the seismic parameters, along with the current map, increases in some settlements and decreases in others were observed. This allowed for more realistic earthquake hazard and structural analyses of the seismic and structural parameters to be obtained according to the geographical location.

Both seismic and structural parameters for different ground motion levels were obtained, which is one of the innovations in the current earthquake code and one of the variables of this study. As the probability of exceedance increases, the earthquake effect that the structure will be exposed to decreases and as a result, the displacement, base shear force and stress values decrease.

Maximum stress levels occurred in the transition zones between the parts of the minaret in different settlements and different probability of exceedance. The resulting stress values were lower than the allowable stress values suggested in the literature for both Bitlis stone and natural stones. This proved once again that the engineering knowledge and experience at the time the minarets were built were very high. When necessary, monitoring the Ulu Mosque and its minaret by the relevant institutions/organizations and the timely engineering interventions minimized the destruction and damage that may occur to the structure. From this point of view, the originality of this minaret has been preserved by the necessary works and procedures until today. The continuity of monitoring and intervention processes related to such structures, which are an important part of the historical and cultural heritage, is very important for transferring such structures to the next generations.

The maximum displacement value obtained by considering the 2011 Van earthquake, located in the closest environment to Bitlis Province, was lower than the values predicted in the last two regulations. This can be considered as an indication that the last two regulations provide a certain level of safety.

It is known that Bitlis stone has low strength and is more damaged by abrasion due to its soft spongy feature. This situation is affected by temperature differences, frost and humid environments, causing fragmentation and rupture. The freeze–thaw effect, which is cyclical especially in cold periods, is one of the most important factors in the deterioration of the building blocks in the region. This poses a risk to the minaret and mosque. Therefore, the loss and destruction of the Bitlis stone, which is the main element of the minaret, over time due to the natural process may affect the load-bearing feature of the structure.

The surrounding of the monument should be considered together with the city infrastructure in order to drain the ground water from the building. Chemical and mechanical cleaning should be done in the planted parts. After this stage, the damages caused by moisture in the building should be monitored periodically by the relevant public institutions and organizations.

The almost disappearance of traditional construction methods and building construction has similarly affected the number of qualified personnel who understand this business. The use of contemporary materials and techniques is also required from time to time. As a result of these two situations, faulty productions occur in the details of the repairs. In this context, masters should be trained through local governments, and contemporary materials and techniques should be reconsidered with original details.

Timely interventions by the relevant public institutions and organizations did not allow us to obtain data on cracks or significant damage in the minaret. Therefore, the effect of the resulting stresses on the minaret could not be clearly demonstrated. The continuity of these processes is important for the structure.

Interventions on damage that may occur over time in a timely manner and in a way that preserves the structure's originality are important in terms of transferring our historical heritage to the next generations. In this context, it is necessary to examine and monitor the structural health of historical buildings. The best way to protect our cultural heritage is to make necessary interventions in the light of experimental and numerical studies to be carried out on historical buildings before they are damaged.

The minaret of Bitlis Ulu Mosque, one of the first mosques of the Anatolian Principalities period, is a unique structure with multiple entrances at different elevations. The fact that Bitlis contains traces of the occupation years has an important place in terms of the memory of the city, but it is an important landmark of the city in terms of its location. This study can be used as a source for the structural health of the minaret.

In future studies, similar structural analyses will be carried out by using the micromodeling technique and by determining the material properties in more detail using experimental methods. The modal frequencies and mode shapes for the minaret of the Ulu Mosque were determined by a numerical model. In addition, modal parameters can be obtained by performing different techniques and field ambient vibration tests on the minaret. This study can be a source for such studies.

Author Contributions: Conceptualization, E.I., E.A., F.A. and E.H.; methodology, E.H., E.I., E.A, H.B., H.B.Ö. and E.H.; software, F.A., H.B.Ö. and E.I.; validation, H.B., E.I., H.B.Ö., F.A. and E.H.; formal analysis, E.H. and H.B.; investigation, E.A., F.A., E.I. and E.A.; resources, E.A., E.I., F.A. and E.H.; data curation, E.I. and F.A., writing—original draft preparation, E.H., H.B., H.B.Ö. and E.A.; writing—review and editing, E.H., E.I. and F.A.; visualization, E.A.; supervision, E.H and E.I.; project administration, E.I.; funding acquisition, E.H. All authors have read and agreed to the published version of the manuscript.

Funding: This research received no external funding.

Institutional Review Board Statement: Not applicable.

Informed Consent Statement: Not applicable.

Data Availability Statement: Most data are included in the manuscript.

Conflicts of Interest: The authors declare no conflict of interest.

References

1. ICOMOS. *International Cultural Tourism Charter, Principles and Guidelines for Managing Tourism at Places of Cultural and Heritage Significance*; ICOMOS International Cultural Tourism Committee: Melbourne, Australia, 2002.
2. Feather, J. Managing the documentary heritage: Issues from the present and future. In *Preservation Management for Libraries, Archives and Museums*; Gorman, G.E., Sydney, J.S., Eds.; Facet: London, UK, 2006; pp. 1–18.
3. Karasin, I.B.; Isik, E. Protection of Ten-Eyed Bridge in Diyarbakır. *Bud. I Archit.* **2016**, *15*, 87–94. [CrossRef]
4. Bilgin, H.; Ramadani, F. Numerical study to assess the structural behavior of the Bajrakli Mosque (Western Kosovo). *Adv. Civ. Eng.* **2021**, *2021*, 4620916. [CrossRef]
5. Hadzima-Nyarko, M.; Ademovic, N.; Pavic, G.; Sipos, T.K. Strengthening techniques for masonry structures of cultural heritage according to recent Croatian provisions. *Earthq. Struct.* **2018**, *15*, 473–485.
6. Cosgun, T.; Sayin, B.; Gunes, B.; Osman Avşar, A.; Şengün, R.; Gümüşdağ, G. Rehabilitation of historical ruined castles based on field study and laboratory analyses: The case of Bigalı Castle in Turkey. *Rev. Constr.* **2020**, *19*, 52–67. [CrossRef]
7. Isik, E.; Antep, B.; Buyuksarac, A.; Isik, M.F. Observation of behavior of the Ahlat Gravestones (TURKEY) at seismic risk and their recognition by QR code. *Struct. Eng. Mech.* **2019**, *72*, 643–652.
8. Betti, M.; Vignoli, A. Modelling and analysis of a Romanesque church under earthquake loading: Assessment of seismic resistance. *Eng. Struct.* **2008**, *30*, 352–367. [CrossRef]
9. Gunes, B.; Cosgun, T.; Sayin, B.; Ceylan, O.; Mangir, A.; Gumusdag, G. Seismic assessment of a reconstructed historic masonry structure: A case study on the ruins of Bigali castle mosque built in the early 1800s. *J. Build. Eng.* **2021**, *39*, 102240. [CrossRef]
10. Işık, E.; Antep, B. Structural analysis of historical masonry minaret in Ahlat. *BEU. J. Sci.* **2018**, *7*, 46–56.
11. Çalık, İ.; Demirtaş, B.; Bayraktar, A.; Türker, T. Yığma taş minarelerin analitik ve deneysel yöntemlerle güvenliğinin belirlenmesi: Trabzon Muhittin Camii Minaresi örneği. *Vakıflar Derg.* **2012**, *38*, 121–139.
12. Suliman, S.; Gramescu, A.M.; Gelmambet, S. Modelling the structure of Carol I Mosque Minaret taken into account the seismic evaluation. *IOP Conf. Ser. Mater. Sci. Eng.* **2021**, *1138*, 012040. [CrossRef]
13. Işık, E.; Harirchian, E.; Arkan, E.; Avcil, F.; Günay, M. Structural analysis of Five Historical Minarets in Bitlis (Turkey). *Buildings* **2022**, *12*, 159. [CrossRef]
14. Cosgun, C.; Turk, A.M. Seismic behavior and retrofit of historic masonry minaret. *Gradevinar* **2012**, *64*, 39–45.
15. Erdil, B.; Tapan, M.; Akkaya, İ.; Korkut, F. Effects of structural parameters on seismic behaviour of historical masonry minaret. *Period. Polytech. Civ. Eng.* **2018**, *62*, 148–161. [CrossRef]
16. Dogangun, A.; Acar, R.; Sezen, H.; Livaoglu, R. Investigation of dynamic response of masonry minaret structures. *Bull. Earthq. Eng.* **2008**, *6*, 505–517. [CrossRef]
17. Pekgökgöz, R.K.; İzol, G.; Avcil, F.; Gürel, M.A. Şanlıurfa Ulu Cami Minaresi yapı taşının elastisite modülünün ultrasonik test cihazı kullanılarak belirlenmesi. *Harran Üniversitesi Mühendislik Derg.* **2018**, *3*, 35–45.
18. Livaoğlu, R.; Baştürk, M.H.; Doğangün, A.; Serhatoğlu, C. Effect of geometric properties on dynamic behavior of historic masonry minaret. *KSCE J. Civ. Eng.* **2016**, *20*, 2392–2402. [CrossRef]
19. Basaran, H.; Demir, A.; Ercan, E.; Nohutcu, H.; Hokelekli, E.; Kozanoglu, C. Investigation of seismic safety of a masonry minaret using its dynamic characteristics. *Earthq. Struct.* **2016**, *10*, 523–538. [CrossRef]
20. Oliveira, C.S.; Çaktı, E.; Stengel, D.; Branco, M. Minaret behavior under earthquake loading: The case of historical Istanbul. *Earthq. Eng. Struct. Dyn.* **2012**, *41*, 19–39. [CrossRef]
21. Hejazi, M.; Moayedian, S.M.; Daei, M. Structural analysis of Persian historical brick masonry minarets. *J. Perform. Constr. Facil.* **2016**, *30*, 04015009. [CrossRef]
22. Türkeli, E. Dynamic seismic and wind response of masonry minarets. *Period. Polytech. Civ. Eng.* **2020**, *64*, 353–369. [CrossRef]
23. Muvafik, M. Field investigation and seismic analysis of a historical brick masonry minaret damaged during the Van Earthquakes in 2011. *Earthq. Struct.* **2014**, *6*, 457–472. [CrossRef]
24. Karaşin, İ.B.; Işık, E.; Eren, B. Structural analysis of Bitlis Grand Mosque. In Proceedings of the International Conference on Natural Science and Engineering (ICNASE–2016), Kilis, Turkey, 19–20 March 2016.
25. Shkodrani, N.; Bilgin, H.; Hysenlliu, M. Influence of interventions on the seismic performance of URM buildings designed according to pre-modern codes. *Res. Eng. Struct. Mater.* **2021**, *7*, 315–330. [CrossRef]
26. Avcil, F.; Işık, E.; Bilgin, H.; Özmen, H.B. Tbdy-2018'de verilen tasarım spektrumlarının anıtsal yığma yapı sismik davranışına etkisi. *Adıyaman Üniversitesi Mühendislik Bilimleri Derg.* **2022**, *16*, 165–177.
27. TBEC. *Turkish Building Earthquake Code*; T.C. Resmi Gazete: Ankara, Turkey, 2018.
28. TSDC. *Turkish Seismic Design Code*; T.C. Resmi Gazete: Ankara, Turkey, 2007.
29. Işık, E. A comparative study on the structural performance of an RC building based on updated seismic design codes: Case of Turkey. *Challenge* **2021**, *7*, 123–134. [CrossRef]
30. Ozmen, H.B. A view on how to mitigate earthquake damages in Turkey from a civil engineering perspective. *Res. Eng. Struct. Mater.* **2021**, *7*, 1–11. [CrossRef]

31. Akkar, S.; Kale, Ö.; Yakut, A.; Ceken, U. Ground-motion characterization for the probabilistic seismic hazard assessment in Turkey. *Bull. Earthq. Eng.* **2018**, *16*, 3439–3463. [CrossRef]
32. Akkar, S.; Azak, T.; Çan, T.; Çeken, U.; Tümsa, M.D.; Duman, T.Y.; Kale, Ö. Evolution of seismic hazard maps in Turkey. *Bull. Earthq. Eng.* **2018**, *16*, 3197–3228. [CrossRef]
33. Işık, E. Comparative investigation of seismic and structural parameters of earthquakes (M $\geq$ 6) after 1900 in Turkey. *Arab. J. Geosci.* **2022**, *15*, 971. [CrossRef]
34. AFAD. Available online: https://www.afad.gov.tr/ (accessed on 20 May 2022).
35. Çeken, U.; Dalyan, İ.; Kılıç, N.; Köksal, T.S.; Tekin, B.M. Türkiye Deprem Tehlike Haritaları İnteraktif Web Uygulamasi. In Proceedings of the International Earthquake Engineering and Seismology Conference, Bucharest, Romania, 12–14 June 2017.
36. Karaşin, İ.B.; Işık, E.; Demirci, A.; Aydın, M.C. The effect of site-specific design spectra for geographical location on reinforced-concrete structure performance. *Duje* **2020**, *11*, 1319–1330.
37. Peker, F.U.; Işık, E. A study on the effect of local soil conditions in TBDY-2018 on earthquake behavior of steel structure. *BEU J. Sci.* **2021**, *10*, 1125–1139.
38. Şen, K. Two important inscriptions belong to the Bitlis grand mosque and Bitlis castle. *ASEAD* **2018**, *5*, 147–156.
39. Nasûhü's-Silâhî, N.M. *Beyân-ı Menâzil-i Sefer-i Irâkeyn-i Sultan Süleyman Han*; Yurdaydın, H.G., Ed.; Türk Tarih Kurumu: Ankara, Türkiye, 2014; Volume 1537, p. 38.
40. Talasoğlu, A. Yüzyıl Osmanlı Minyatür Sanatının Şehir Kuruluşları ve Mimarlıkla Ilişkisi. Ph.D. Thesis, İstanbul Üniversitesi Sosyal Bilimler Enstitüsü, İstanbul, Turkey, 2000.
41. Sayan, Y.; Öztürk, Ş. *Bitlis Evleri*; Kültür Bakanlığı Yayınları: Ankara, Turkey, 2001.
42. Gelişkan, N.N. Matrakçı Nasuh gözüyle Bitlis. Master's Thesis, İstanbul Teknik Üniversitesi, Fen Bilimleri Enstitüsü, İstanbul, Turkey, 2015.
43. Arık, O. *Bitlis Yapılarında Selçuklu Rönesansı*; Selçuklu Tarih ve Medeniyeti Enstitüsü Yayını: Ankara, Türkiye, 1971.
44. Erkan, S. *Türkiye'de Vakıf Abideler ve Eski Eserler-II*; Vakıflar Genel Müdürlüğü Yayınları: Ankara, Türkiye, 1977.
45. Sinclair, T.A. *Eastern Turkey: An Architectural and Archaeological Survey*; Pindar Press: London, UK, 1989; Volume I.
46. Uluçam, A. *Ortaçağ ve Sonrasında Van Gölü Çevresi Mimarlığı-II-Bitlis*; Kültür Bakanlığı Yayınları: Ankara, Turkey, 2002.
47. Aslanapa, O. *Anadolu'da Türk sanatı Anadolu'da Büyük Selçuklulara Bağlanan Camiler*; Atatürk Kültür Merkezi Yayinlari: Ankara, Turkey, 2007.
48. Yeğin, M. Bitlis Ulu Cami restorasyonunda onarım ve güçlendirme çalışmaları. In Proceedings of the Uluslararası Türkiye Vizyonu Kongresi (UTVİK), Adana, Türkiye, 18–20 March 2019; pp. 487–498.
49. Ülkü, C.; Yeğin, M. Courtyard in 11–12. century mosques/The newly discovered courtyard in Bitlis Ulu Mosque. *J. Social Sci.* **2017**, *17*, 17–37.
50. Arınç, K. Bitlis Çayı havzasının Coğrafi Etüdü. Ph.D. Thesis, Atatürk Üniversitesi, Erzurum, Turkey, 1991.
51. Aydın, M.C.; Işık, E. Evaluation of ground snow loads at the micro-climate regions. *Russ. Meteorol. Hydrol.* **2015**, *40*, 741–748. [CrossRef]
52. AFAD. Available online: https://tdth.afad.gov.tr (accessed on 2 February 2020).
53. Gunes, O. Turkey's grand challenge: Disaster-proof building inventory within 20 years. *Case Stud. Constr. Mater.* **2015**, *2*, 18–34. [CrossRef]
54. Kale, Ö.; Akkar, S. An auxiliary tool to build ground-motion logic-tree framework for probabilistic seismic hazard assessment, 3. In Proceedings of the Türkiye Deprem Mühendisliği ve Sismoloji Konferansı, İzmir, Turkey, 14–16 October 2015.
55. Koçer, M.; Nakipoğlu, A.; Öztüztürk, B.; Al-hagri, M.G.; Arslan, M.H. Comparison of TBSC 2018 and TSC 2007 through the values of seismic load related spectral acceleration. *J. Selcuk-Tech.* **2018**, *17*, 43–58.
56. Kutanis, M.; Ulutaş, H.; Işık, E. PSHA of Van province for performance assessment using spectrally matched strong ground motion records. *J. Earth Syst. Sci.* **2018**, *127*, 99. [CrossRef]
57. Işık, E.; Harirchian, E.; Bilgin, H.; Jadhav, K. The effect of material strength and discontinuity in RC structures according to different site-specific design spectra. *Res. Eng. Struct. Mater.* **2021**, *7*, 413–430. [CrossRef]
58. Giordano, A.; Mele, E.; De Luca, A. Modelling of historical masonry structures: Comparison of different approaches through a case study. *Eng. Struct.* **2002**, *24*, 1057–1069. [CrossRef]
59. Akan, A.E.; Başok, G.; Er, A.; Örmecioğlu, H.T.; Koçak, S.Z.; Cosgun, T.; Uzdil, O.; Sayin, B. Seismic evaluation of a renovated wooden hypostyle structure: A case study on a mosque designed with the combination of Asian and Byzantine styles in the Seljuk era (14th century AD). *J. Build. Eng.* **2021**, *43*, 103112. [CrossRef]
60. Abaqus, G. *Abaqus 6.11*; Dassault Systemes Simulia Corporation: Providence, RI, USA, 2011.
61. Valente, M.; Milani, G. Effects of geometrical features on the seismic response of historical masonry towers. *J. Earthq. Eng.* **2018**, *22*, 2–34. [CrossRef]
62. Işık, E.; Büyüksaraç, A.; Avşar, E.; Kuluöztürk, M.F.; Günay, M. Characteristics and properties of Bitlis ignimbrites and their environmental implications. *Mater. Construcción* **2020**, *70*, 214. [CrossRef]
63. Valente, M. Seismic vulnerability assessment and earthquake response of slender historical masonry bell towers in South-East Lombardia. *Eng. Fail. Anal.* **2021**, *129*, 105656. [CrossRef]
64. Valente, M. Seismic behavior and damage assessment of two historical fortified masonry palaces with corner towers. *Eng. Fail. Anal.* **2022**, *134*, 106003. [CrossRef]
65. Işık, E.; Antep, B.; Karaşin, İ.B. Structural Analysis of Ahlat Emir Bayındır Bridge. *Bitlis Eren Uni. J. Sci. Techn.* **2018**, *8*, 11–18. [CrossRef]

66. Valente, M.; Milani, G. Advanced numerical insights into failure analysis and strengthening of monumental masonry churches under seismic actions. *Eng. Fail. Anal.* **2019**, *103*, 410–430. [CrossRef]
67. Vuoto, A.; Ortega, J.; Lourenço, P.B.; Suárez, F.J.; Núñez, A.C. Safety assessment of the Torre de la Vela in la Alhambra, Granada, Spain: The role of on site works. *Eng. Struct.* **2022**, *264*, 114443. [CrossRef]
68. Pouraminian, M. Multi-hazard reliability assessment of historical brick minarets. *J. Build. Path. Rehab.* **2022**, *7*, 10. [CrossRef]
69. Usta, P. Assessment of seismic behavior of historic masonry minarets in Antalya, Turkey. *Case Stud. Constr. Mater.* **2021**, *15*, e00665. [CrossRef]
70. Pande, G.N.; Middleton, J.; Kralj, B. Computer methods in structural masonry. In Proceedings of the Fourth International Symposium on Computer Methods in Structural Masonry, London, UK, 3–5 September 1998.
71. Lourenco, P.B.; Rots, J.G.; Blaauwendraad, J. Two approaches for the analysis of masonry structures: Micro and macro-modeling. *Heron* **1995**, *40*, 1995.
72. Lourenço, P.B. Computations on historic masonry structures. *Prog. Struct. Eng. Mater.* **2002**, *4*, 301–319. [CrossRef]
73. Norme Tecniche per le Costruzioni (NTC), Ministero delle Infrastrutture e dei Trasporti, Norme Tecniche per le Costruzioni, NTC 2008, (English: Technical Standards for Construction), D.M. del Ministero delle Infrastrutture e dei Trasporti del 14/01/2008. Official Gazette n. 29, del 04.02.2008, S.O. n. 30, Roma. Available online: https://www.studiopetrillo.com/files/ntc2008-completa.pdf (accessed on 20 May 2022). (In Italian).
74. Adam, M.A.; El-Salakawy, T.S.; Salama, M.A.; Mohamed, A.A. Assessment of structural condition of a historic masonry minaret in Egypt. *Case Stud. Constr. Mater.* **2020**, *13*, e00409. [CrossRef]
75. Shakya, M.; Varum, H.; Vicente, R.; Costa, A. Empirical formulation for estimating the fundamental frequency of slender masonry structures. *Int. J. Archit. Herit.* **2016**, *10*, 55–66. [CrossRef]
76. Faccio, S.; Podestà, A.; Saetta, A. Venezia Campanile della Chiesa di Sant'Antonio, Esempio 5. In *Linee Guida per la Valutazione e riduzione del Rischio Sismico del Patrimonio Culturale Allineate Alle Nuove Norme Tecniche per le Costruzioni (DM 14/01/2008)*; Ministero per i Beni e le Attività Culturali (a cura di): Roma, Italy, 2009. (In Italian)
77. Rainieri, C.; Fabbrocino, G. Estimating the elastic period of masonry towers. In *Topics in Modal Analysis I*; Springer: New York, NY, USA, 2012; Volume 5, pp. 243–248.
78. Vakıflar Genel Müdürlüğü. *Tarihi Yapılar için Deprem Risklerinin Yönetimi Kılavuzu (TYDRYK-2017)*; Vakıflar Genel Müdürlüğü: Ankara, Türkiye, 2017.
79. Wang, X.L.; Lu, D.G. MCS-based PSHA procedure and generation of site-specific design spectra for the seismicity characteristics of China. *Bull. Seismol. Soc. Am.* **2018**, *108*, 2408–2421. [CrossRef]
80. Ansal, A.; Tönük, G.; Kurtuluş, A. A simplified approach for site-specific design spectrum. *Geot. Geol. Earthq.* **2018**, *44*, 73–86.

 *buildings*

Article

Mechanical Steel Stitches: An Innovative Approach for Strengthening Shear Deficiency in Undamaged Reinforced Concrete Beams

Ceyhun Aksoylu [1], Yasin Onuralp Özkılıç [2,*] and Musa Hakan Arslan [1]

[1] Faculty of Engineering and Natural Sciences, Department of Civil Engineering, Konya Technical University, Konya 42130, Turkey
[2] Faculty of Engineering, Department of Civil Engineering, Necmettin Erbakan University, Konya 42000, Turkey
* Correspondence: yozkilic@erbakan.edu.tr

Abstract: In this study, reinforced concrete beams with insufficient shear capacity were strengthened on both sides of the beam along the shear openings by a novel approach: Mechanical Steel Stitches (MSS). This innovative method facilitates the application of strengthening the beams with a low-cost solution. In this concept, six specimens were experimentally investigated under vertical load. While one of the specimens was tested as a reference, the others were strengthened with MSS application at different ratios (ρ_{MS}), ranging from 0.2% to 1% at both the beams' shear span. MSS were applied with the angle of 90° considering stirrup logic. The diameter, anchorage depth and mechanical properties of the MSSs were kept constant, and their effects on the strengthening of the beams in terms of ductility, strength, stiffness, and energy dissipation capacities were investigated by changing the spacing of the MSSs. The results revealed that increasing MSS ratio caused a dramatic positive change in the behavior in terms of both strength and energy dissipation capacity. MSSs to be made at appropriate intervals (($\%$1) MSS ratio or ($d/5$) MSS spacing) significantly improved the shear capacity. However, a 43% loss in stiffness occurred with the increase in ρ_{MS} since the MSSs are applied to the beams by drilling and anchoring from the outside.

Keywords: strengthening; steel; reinforced concrete; mechanical steel stitches; shear

Citation: Aksoylu, C.; Özkılıç, Y.O.; Arslan, M.H. Mechanical Steel Stitches: An Innovative Approach for Strengthening Shear Deficiency in Undamaged Reinforced Concrete Beams. *Buildings* **2022**, *12*, 1501. https://doi.org/10.3390/ buildings12101501

Academic Editors: Tom Lahmer, Ehsan Harirchian and Viviana Novelli

Received: 26 August 2022
Accepted: 16 September 2022
Published: 21 September 2022

Publisher's Note: MDPI stays neutral with regard to jurisdictional claims in published maps and institutional affiliations.

1. Introduction

Reinforced concrete members are expected to have a sufficient level of the three most important parameters, such as ductility, strength, and rigidity, to carry the loads safely. Some of the existing reinforced concrete structures do not have enough of these parameters for various reasons. For this reason, even under the influence of service loads, damages are observed in the structures. The beam is usually the first element to experience damage in a reinforced concrete building [1,2]. Damages in beams are observed as shear or bending damage (in some cases, both). In order to prevent cracks caused by principal tensile stresses in beams, transverse reinforcement (stirrup) is placed perpendicular to the cracks. No matter how small the shear stress may be, in all construction regulations used today, it is obligatory to use a minimum level of transverse reinforcement to prevent possible shear damage. If there will be damage to the beams, it must first be bending damage. For this purpose, the transverse reinforcements placed in sufficient amounts prevent shear fracture and reach the bending capacity of the beam to ensure bending power depletion.

Structural elements may need to be strengthened during their service life due to design and application errors, time-related capacity losses due to corrosion and durability problems, changing the purpose of use and being insufficient according to new regulations [3,4]. In general, the low compressive strength of the concrete and the insufficient stirrup reinforcement to meet the shear stress of the reinforced concrete elements are the most common deficiencies in existing reinforced concrete structures [5]. For this reason, conventional and innovative strengthening methods have been investigated by many researchers in the literature to increase the bearing capacity, bending and shear strength of the structural members [6,7]. Reinforced concrete jacketing, wrapping with steel plate and fiber reinforced polymers (FRP) etc., form conventional reinforcements. While methods such as these have been examined, relatively new reinforcement methods such as reinforcement with FRP have been examined by many researchers on beams due to the decrease in their cost in recent years [8–11]. It is seen that some of the studies in the literature are for the repair of damaged beams, and some of them are on the strengthening of undamaged beams. Table 1 presents a review of current studies for pre-damaged and undamaged reinforced concrete beams.

Table 1. Current studies on reinforced pre-damaged and undamaged beams.

Strengthening Type		Pre-Damaged Beams	Undamaged Beams
	AFRP *	Raza et al. (2019) [12]	More and Kulkarni (2014) [13]; Wu et al. (2016) [14]; Zhang and Wu (2019) [15]; Raval et al. (2020) [16];
	BFRP *	Ma et al. (2017) [17]; Ma et al. (2018) [18]; Qin et al. (2019) [19];	Duic et al. (2018) [20]; Joyklad et al. (2019) [21]; Pham et al. (2020) [22]; Shen et al. (2021) [23]
FRP	CFRP *	Prado et al. (2016) [24]; Karzad et al. (2017) [25]; Karam et al. (2017) [26]; Karzad et al. (2019) [27]; Yu et al. (2020) [28]; Yu et al. (2020) [29]; Yu et al. (2021) [30]; Bahij et al. (2020) [31]	Gemi et al. (2019) [11]; Zaki et al. (2019) [32]; Aksoylu et al. (2021) [33]; Al-Khafaji et al. (2021) [34]; Kotynia et al. (2021) [35]; Abed et al. (2021) [36]; Al-Fakih et al. (2021) [37]; Jahami et al. (2021) [38]; Alhassan et al. (2021) [39]; Samb et al. (2021) [40]; Mukhtar and Shehadah (2021) [41]; Mansour (2021) [42]; Gemi et al. (2022) [43];
	GFRP *	Siddika et al. (2019) [44]; Capozucca et al. (2021) [45];	Panigrahi et al. (2014) [46]; Boumaaza et al. (2017) [47]; Aksoylu (2021) [48]; Rahman (2021) [49]; Kumari ve Nayak (2021) [50]; Ali et al. (2021) [51]; Abbas et al. (2021) [52]; Miruthun et al. (2021) [53]; Al-Shalif et al. (2022) [54];
Steel Plate		Peng et al. (2017) [55]; Kazem (2018) [56]; Alam et al. (2020) [57]	Aykaç ve Özbek (2011) [58]; Acar (2014) [59]; Aykaç and Acar (2014) [60]; Abdul-Razzaq et al. (2017) [61]; Demir et al. (2018) [62]
Mechanical Connections		Osman et al. (2017) [63]; Xu et al. (2018) [64]; Xu et al. (2019) [65]; Alshlash et al. (2019) [66]	Hamoush and Ahmad (1997) [67]; Altin et al. (2004) [68]; Rizal et al. (2019) [69]; Chalioris et al. (2019) [70]; Aldhafairi et al. (2020) [71]; Di Trapani et al. (2020) [72]; Yuan et al. (2020) [73];

Table 1. Cont.

Strengthening Type	Pre-Damaged Beams	Undamaged Beams
Jacketing	Murthy et al. (2019) [74]; Hassan et al. (2021) [75]; Ganesh and Murthy (2021) [76]	Chandrakar ve Singh (2017) [77]; Rodrigues et al. (2018) [78]

* A: Aramid, B: Bazalt C: Carbon and G: Glass.

As seen in Table 1, there are many alternatives for strengthening beam members. However, it is crucial and challenging to determine which damage type, damage level, and strengthening/repair method will be more effective in practice. An inexperienced engineer often has difficulty choosing an effective retrofit/repair method based on existing or potential damage. Effective strengthening depends on the extent to which the chosen reinforcement has increased the structural parameters of the beam, such as ductility, strength and rigidity, as well as the method being easy, applicable and economical.

A new strengthening method has emerged in the literature in recent years. This method, which emerged as a Mechanical Steel Stitch (MSS) or Cracks Locking System (CLS), is used to strengthen reinforced concrete beams that are insufficient in shear. When the literature is examined comprehensively, it is seen that the first study aimed at repairing bending cracks in small-scale unreinforced concrete beams was carried out by Hamoush and Ahmad in 1997. The study is on a limited number of experimental studies and their analytical confirmation. In this way, the study was created to represent a guide prepared for the use of seams as a crack repair method in concrete structures. The study presents an analytical method to determine the effectiveness of crack suturing as a repair tool.

For MS repair/reinforcement of pre-damaged beams, Alshlash et al. [66] carried out a series of experimental studies. As a result of the experimental test, 50%, 65% and 85% of pre-damaged shear beams were strengthened with MSSs, resulting in 17%, 43% and 50% increases in shear capacity, respectively, compared to the reference beam. As a result, it was stated that it would be among the main methods that can be used in practice in the future. On top of that, Aksoylu [48] performed a series of experimental studies on undamaged front beams using the similar beam geometry of Alshlash et al. [66]. The MSSs in the study were placed systematically with 45° angles to prevent this damage, considering the shear damage in the reference beam. The observation of ductile behavior at the end of the experiment also proved that MSSs with 45° angles are effective reinforcement alternatives that can be used directly for both strength increase and sufficient ductility.

In this study, the strengthening technique with MSSs, which was proposed as an innovative strengthening alternative, was preferred because it is easy to apply, economical and effective. However, unlike the two studies [48,66] in the literature, MSs were applied to beams at 90° considering stirrup logic in this study. For this, six reinforced concrete MSSs, one of which is a reference, were applied to reinforced concrete beams with insufficient shear, considering the different volumetric ratios. With the study, the effectiveness of MSSs in the case of applying the stirrup logic to the beams was investigated.

2. Materials and Method

2.1. Preparation of Shear Beam

Shear deficient reinforced concrete beams were tested in Konya Technical University Construction and Earthquake Laboratory. The beams produced have a geometric scale of $1/2$ and a cross-sectional area of 125×250 mm, and their length was 2500 mm. Details of the beam are shown in Figure 1. Although concrete compressive strength targets 20–25 MPa, the 28-day cylinder concrete compressive strength of the beams was calculated as 29 MPa (between C25 and C30 grade). The splitting tensile strength was determined as 1.3 MPa. B420c type ribbed $3\phi12$ ($\rho = 0.0117$) longitudinal reinforcement was placed in the bending zone of the beam and $2\phi8$ ($\rho' = 0.00347$) longitudinal reinforcement was placed in the compression zone. The reinforcement ratio considered in the design of the beams is higher than the minimum reinforcement ratio (ρ_{min} = 0.00306), according to the Turkish Reinforced Concrete Building Code [79]. In addition, the longitudinal reinforcements were selected in accordance with the under-balance ductile design. The transverse reinforcements were applied as φ5/350 mm. On the other hand, stirrup hooks were bent at 90°, which is common and contrary to the regulation. Additionally, 25 mm placers are used in the beams.

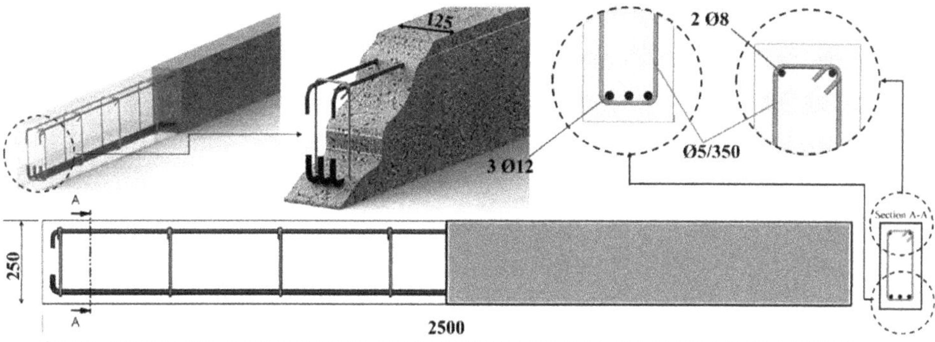

Figure 1. The reinforcement details of the specimens, dimensions in mm.

2.2. Strengthening Technic by Mechanical Steel Stitches (MSS)

In reinforcement applications, U-type mechanical steel stitches (MSS) were prepared by using a φ6 mm diameter cold-formed transmission steel. Each stitch length is 150 mm, and the squares are designed as 60 mm (10φ). The points determined on the beam surface to be applied were drilled to a depth of 60 mm with an 8 mm diameter drill (φ8), and the holes were filled with Dubell-F.1311 brand epoxy after cleaning with compressed air. Prepared U-type MSSs were placed in the holes drilled on both sides of the beam. According to the manufacturer's recommendations, after the MSs were fixed to the beams with epoxy, they were left to cure for one day at room temperature (25 °C). The MSSs applied to the shear opening of the beams were applied at 90° in stirrup logic (Figure 2). MS with five different placements was applied to the shear span of the beams (700 mm part). One is the reference (S0), and the other five represent the specimens strengthened by the placement of MSSs with different spacings (Figure 3).

The placed MSSs were applied considering the increasing volumetric ratio ($\rho_{MS} = \frac{n \times A_{0,MS}}{b_w \times S}$). Here n; number of stirrup arms, A_0; stirrup cross-sectional area, b_w; beam width and s; MSS represents the application range. Details depending on the MSS diameter (φ), spacing (s) and volumetric ratio (ρ_{MS}) for beams reinforced with MSSs applied with different spacing are given in Table 2. All the MSSs made were placed perpendicular to the expected crack, taking into account the crack mechanism of the S0 beam at the end of the experiment. MSSs transmit force with the friction force they create between the surfaces to which they are attached. For this, the number and size of MSS are significant for effective strengthening. Therefore, MSSs were prepared considering the remaining height after deducting the rust, longitudinal reinforcement and transverse reinforcement allowances of the MSS beam. MSS application has very important advantages compared to other strengthening methods. First of all, since it is very easy to apply, fewer workers are needed compared to other reinforcement alternatives, and this process can even be carried out with a single worker. It is also the most cost-effective compared to other strengthening methods.

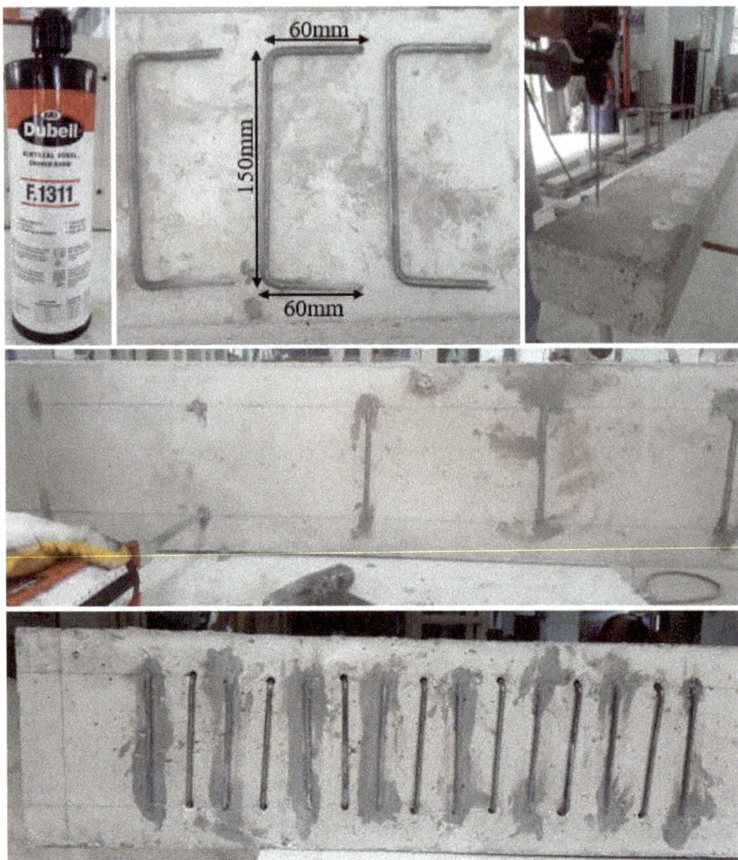

Figure 2. Strengthening beams with MSS.

Table 2. Details of specimens.

Specimen	MS Diameter (φ) (mm)	MS Number	MS Spacing (s) (mm)	MS Volumetric Ratio (ρ_{MS})
S0	-	-	-	---
S1	6	4	220	0.0020
S2	6	5	165	0.0027
S3	6	6	130	0.0034
S4	6	7	110	0.0041
S5	6	15	45	0.0100

2.3. Test Setup

The reference and strengthened specimens were tested under vertical loads in a four-point bending setup in the rigid steel loading frame shown in Figure 4. In Figure 5, the pre-experimental views of each specimen are shown. The beams are simply supported so that 100 mm of them fits on the supports. Since shear damage in the reference beam was desired, the a_v/d ratio (shear span to depth ratio (70/22.5) (mm/mm)) was chosen as 3.11 [80,81]. In this way, the formation of shear damage was observed in reference S0. The load cell used for loading the beam has a capacity of 300 kN. A load-displacement curve was obtained for each specimen by considering two displacement meters (LVDT) located in the middle of the beam, and the load and the distance between them are 160 mm. All records were recorded with the data collection system. As a result of splitting the load from the vertical piston into two over the spreader beam, a single load transfer is achieved. With vertical monotonic loading, 10 kN increments were continued until the end of the experiments. The experiments continued with displacement control depending on the behavior of the beams at the time of yielding. In the experiments, each monotonic loading was waited for a short time to mark the cracks on the test.

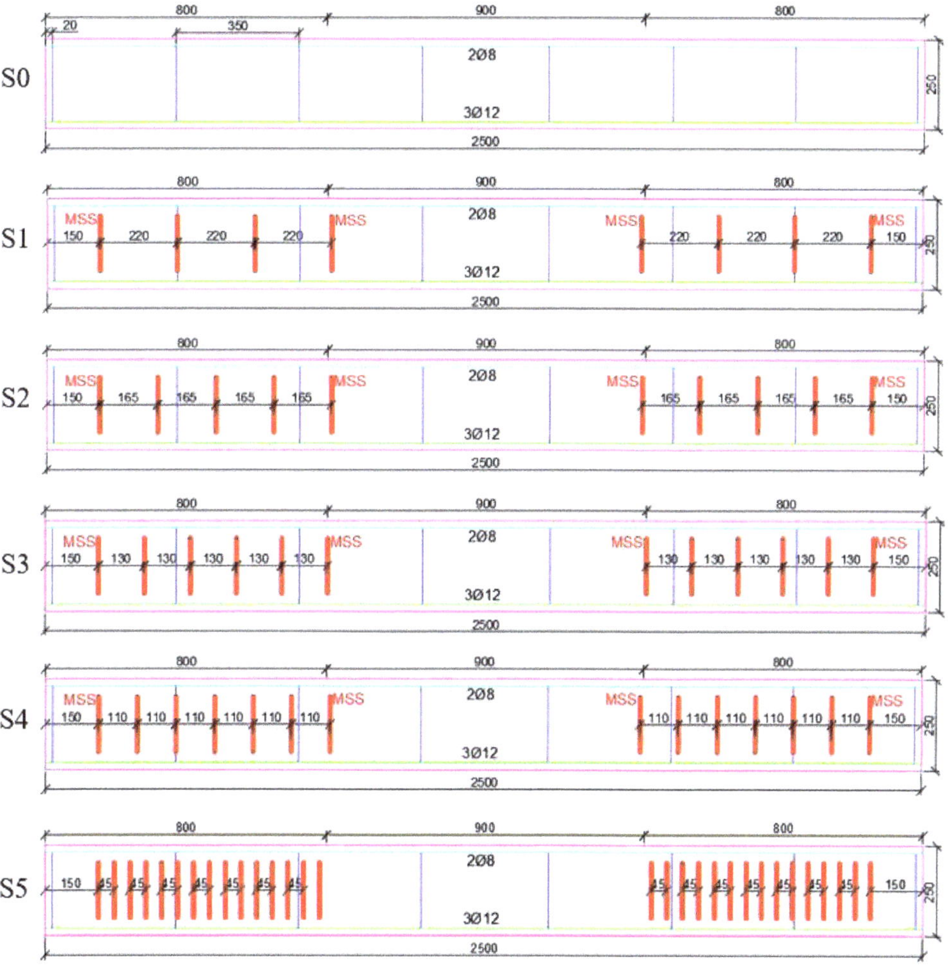

Figure 3. Appearance of test specimens after strengthening in mm.

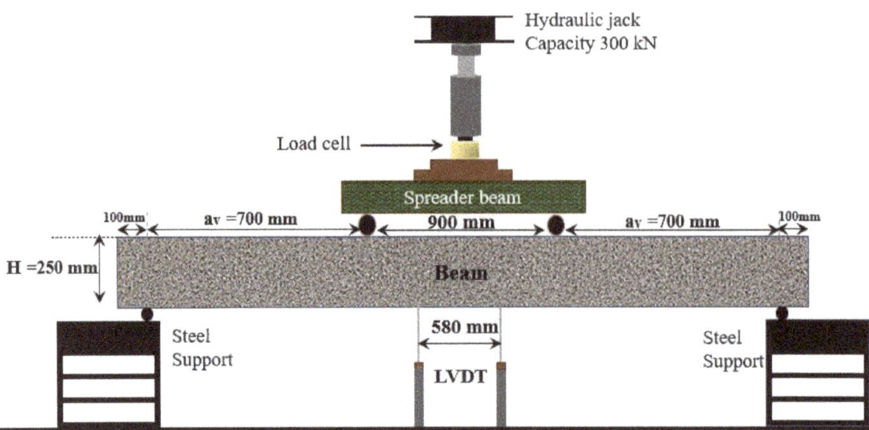

Figure 4. Four-point bending test setup in accordance with the standards used in the experiments.

Figure 5. Pre-experiment view of reference and reinforced beams.

3. Experimental Results and Discussion

In the experimental study, vertical load-mid-point displacement graphs of each specimen were drawn. By comparing the reference specimen (S0) with the strengthened specimens (S1, S2, S3, S4 and S5), deep discussions on MSS were reached. Comparisons were made step by step with the systematic application of MSSs to the beam. Graphical comparisons were made by considering the increasing volumetric ratio of MSS (ρ_{MS}) in strengthening applications. This way, behavior changes and suggestions could be put forward clearly. In the experiments, the number of MSS (4, 5, 6, 7 and 15) applied to the beams was increased step by step. Each experiment was compared with previous experiments, and a cumulative evaluation was made at the end. In other words, the number and range of MSSs were determined by making evaluations before each reinforcement design. All MS applications were performed along the shear span on both sides of the beam. Experimental studies, respectively, are explained in detail and interpreted by comparison. Comparisons are limited by the load carrying capacity, stiffness, ductility and energy dissipation capacity. The test results of all specimens are summarized in Table 3. In addition, the comparison of all specimens is shown in Figure 6. During the whole experiment, no peeling and rupture damage was observed in the MSSs, and shear, bending and adherence damage were observed in the concrete. This also showed that the anchor length ($10\phi = 60$ mm) of the applied MSSs was sufficient. At the end of the experiment, comprehensive damage analyzes were performed for each specimen.

Table 3. Experimental test results and observed damage.

Specimen No	First Crack				Beam Damage Type		MSS Damage Type	Special Cases
	Load (kN)	Design Type	Angle	Place	Load (kN)	Failure Type		
S0	30	Bending	90°	Bending zone	73.00	Shear	---	Experiment ended up shear failure on the left side
S1	10	Bending	90°	Bending zone	74.59	Shear	No damage observed on MS	Experiment ended up shear failure on the left side
S2	20	Bending	90°	Bending zone	75.79	Shear	No damage observed on MS	Experiment ended up shear failure on the right side
S3	20	Bending	90°	Bending zone	76.90	Shear	No damage observed on MS	Experiment ended up bending failure on the right side
S4	30	Bending	90°	Bending zone	78.10	Shear	No damage observed on MS	Experiment ended up bending failure on the right side
S5	20	Bending	90°	Bending zone	95.74	Shear	No damage observed on MS	Experiment ended up bending failure on the left side

3.1. Reference Specimen: S0

The S0 specimen was tested as a reference beam with insufficient shear reinforcement. The vertical load was applied in increments of 10 kN. The first cracks in the experiment were observed in the middle span of the beam (in the bending region) and at the load level of 30 kN. These first cracks represent linear elastic cracks at the minor level. With the load reaching 60 kN, shear cracks observed in beam shear openings occurred below the neutral axis (midpoint of the beam section). With the vertical load of 70 kN and the current displacement of 8.3 mm, the shear cracks extended to the support point under the spreader beam. Finally, when the S0 specimen reached the maximum load level of 73 kN at 8.83 mm displacement, sudden and brittle shear damage occurred in the left shear opening. This showed that the beam design with insufficient shear and the selected a_v/d ratio were sufficient. A damage analysis view of the beam is given in Figure 7. When the damage analysis is examined, it is seen that typical shear damage occurs. Choosing the stirrup spacing as 350 mm caused the cracks formed in the shear opening of the beam to extend along the beam height with increasing load. The absence of stirrups to limit the propagation of cracks made the occurrence of shear damage inevitable.

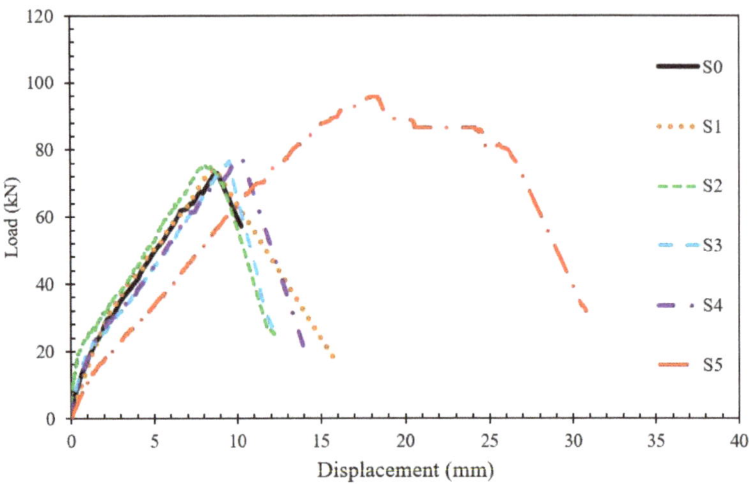

Figure 6. Comparison for load-displacement, S0–S5.

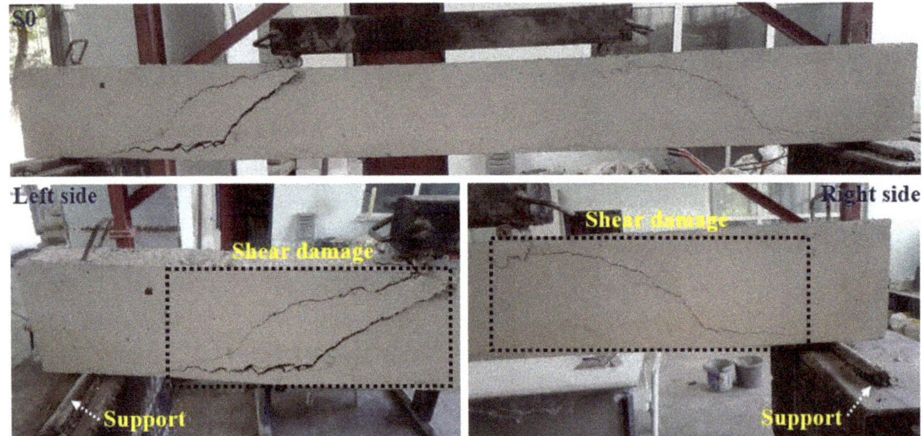

Figure 7. S0 specimen.

3.2. Strengthened Specimens: S1, S2, S3, S4 and S5

Specimen S1 was strengthened by epoxy application of 4 MSSs with a diameter of 6 mm, a length of 150 mm and an anchor length of 60 mm on one side of the shear opening of the reference beam (S0). In other words, the S1 beam was strengthened by applying a total of sixteen MSS to the beam shear openings. In the experiment, the first minor bending cracks started to appear in the middle span of the beam at a load level of 10 kN. When the load reached 30 kN, a displacement of 2.68 mm occurred in the middle region of the beam. At this load, the minor cracks extended towards the neutral axis. Additionally, as shown in Figure 8, the first capillary shear cracks occurred in the lower left part of MSS 2, just below MSS 5, and finally in the middle of MSS 5 and 6. With the load reaching 40 kN, a new crack occurred in the bending region. In addition, elongation was observed in the shear crack between 5 and 6 MSS at this load value. After this load value, a new shear crack was observed between MSS 1-2 and MSS 6-7 in the shear zone at 50 kN load level. With the load reaching 60 kN, no further elongation or new cracks were observed in the cracks in the bending region. With the load reaching 74.59 kN and the displacement reaching 8.67 mm, shear damage occurred in the left shear opening of the beam, passing through the middle of the MSS 2-3, and the experiment was terminated. Up to a certain value of the load, the crack propagation, which reaches 6 MSS, is prevented by the mechanical stitches. When the damage observed between MSS 2 and 3 in the left shear opening is examined, it is understood that the cracks try to reach the beam pressure zone by the shortest route. As a matter of fact, the angle of the shear damage in the reference sample with the horizontal is less. Here, the presence of MSSs 1 and 2 caused a partial change in the location of the shear damage. In addition, when compared to the S0 specimen, there was an increase of approximately 2.17% in the load-carrying capacity. This shows that the MSSs that are located makes a very small contribution to the load-carrying capacity. Finally, as in the S0 specimen, sufficient ductility could not be achieved in the S1 specimen.

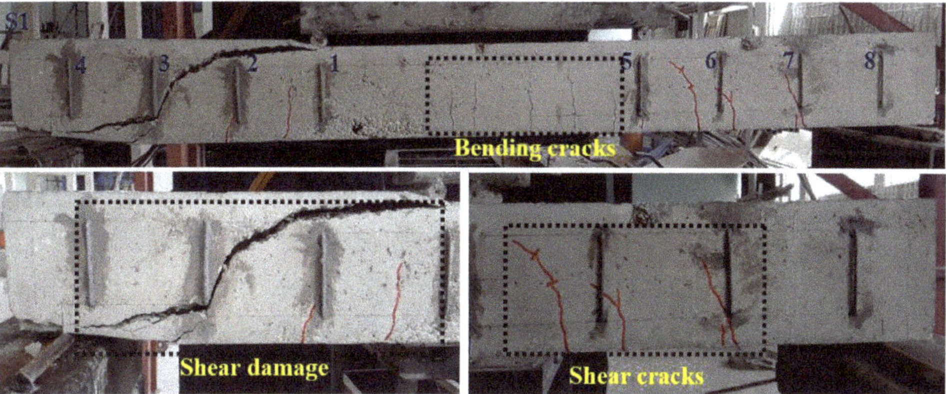

Figure 8. Damages at S1.

In the S2 specimen, the number of MSS was increased to 5. In this way, the distance between the MSSs has been reduced to 165 mm. The total number of MSSs placed on the beam shear zone is twenty. In the S2 specimen, the first microcracks were observed at the 20 kN load level, that is, at the vertical displacement level of 1.65 mm. These cracks represent elastic cracks. With the vertical load of 30 kN and the displacement value of 2.92 mm, the first shear cracks occurred between the MSSs 6-7 and 7-8. At this load value, it was observed that the cracks in the middle region of the beam were elongated towards the neutral axis, and new cracks were formed. Since the vertical load on the beam reached 50 kN and the displacement value reached 5.26 mm, shear cracks occurred between MSSs 1-2 and 2-3. In addition, bending cracks observed in the middle of the beam at this load value were limited to the neutral axis level. With the load reaching 60 kN and the displacement 6.23 mm, new shear cracks were observed between the MSSs 3-4 and 8-9. Finally, with the load reaching 75.79 kN and the vertical displacement 8.07 mm, shear damage occurred in the right shear opening and the experiment was terminated. It cannot be said that sufficient MSS number and spacing are provided in terms of ductility in the S2 specimen, where shear behavior was observed. However, the load-carrying capacity of the S2 specimen increased by 3.8% and 1.6%, respectively, compared to the S0 and S1 specimens. MSSs applied to the S2 specimen did not significantly contribute to the bending stiffness of the beams. This indicates that due to the fact that the shear capacity does not increase significantly, it can respond to less rotational demand and that the longitudinal reinforcements in the tension zone do not yield. The damage of the S2 specimen at the end of the test is shown in Figure 9. When Figure 9 is examined, the failure has migrated between MSSs 8 and 9. The shear damage in S2 is quite similar to the damage in S1. Reducing the MS gap in the S2 specimen partially changed the path of the crack. However, if it is taken into account, the occurrence of shear damage passing through the upper part of the MSS 8 and the lower point of the MSS 9 showed that the cracks were directed to the damaged points by drilling holes beforehand. This shows that there should be MSSs that prevent the propagation of cracks. Since this situation is thought to be possible only by reducing the MSS interval, the following test specimen was prepared.

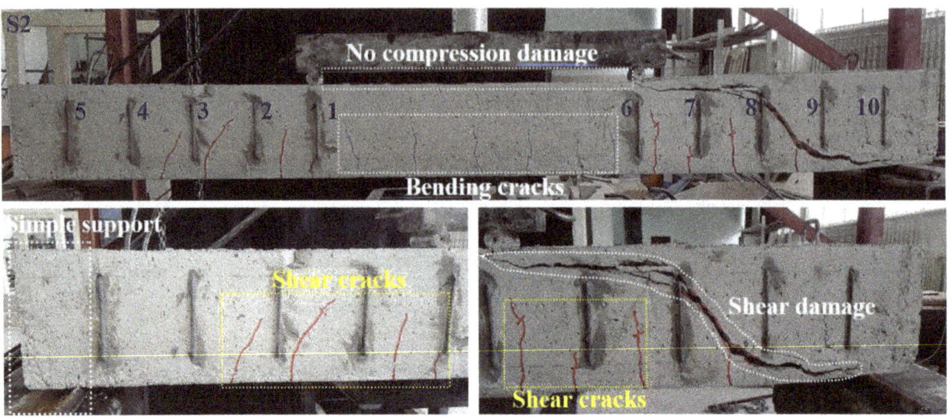

Figure 9. Damages at S2.

Although the MSSs placed in the S2 increased the load carrying capacity slightly, the S3 was created because they could not prevent shear damage. The total number of MSS placed in the S3 sample is twenty-four. The distance between each MSS was set to 130 mm. The aim here is to prevent the shear cracks to be formed by more MSS. For this, the first three bending cracks were observed in the middle of the beam with monotonic increasing loads and a load of 20 kN and a displacement of 1.95 mm (Figure 10). Cracks continued to increase in the bending zone as the load reached 30 kN and the displacement reached 3.98 mm. In addition, the first shear crack occurred at this load value between the 1-2 and 2-3 numbered MSSs in the left shear span. With a load of 50 kN and a displacement of 5.74 mm, minor shear cracks were observed at the lower points of MSSs 4, 7, 8 and 9. No propagation was observed in bending cracks at the neutral axis level at this load level. This indicates that the shear capacity is more difficult. With a load of 70 kN and a displacement of 7.34 mm, a minor shear crack started from the lower part of the MSS 4 and along the height of the MSS. This crack did not propagate in subsequent loadings. At this load level, the increase in minor shear cracks, especially in the right and left shear span, indicates that the shearing capacity is approached. Finally, with the load of 76.90 kN and the displacement of 9.53 mm, shear damage occurred with the propagation of the crack between the MSS 9-10 in the left shear span. Particularly, the weak areas at the upper point of MSS 9 and the lower point of MSS 10 accelerated the progression of the crack. In terms of load carrying capacity, the S3 increased by 5.34%, 3.08% and 1.45%, respectively, compared to S0, S1 and S2. The fact that sufficient ductility value could not be obtained with the increase in load carrying capacity showed that the number of MSS should be increased.

Since the desired ductile behavior could not be obtained in the S3, the MSS spacing was reduced to 110 mm. In this way, more MSS was applied to prevent possible shear damage that may occur in the shear zone. In other words, shear damage was tried to be prevented with seven MSSs placed on one side of the shear opening of the designed S4 specimen. In this way, the behavior change was investigated with a total of twenty-eight MSSs applied to the beam. In this loading, the first crack similarly occurred in the bending region at a load of 30 kN and a displacement of 3.63 mm. When the load is 40 kN and the displacement is 5.23 mm, the first shear crack was observed between MSS 3 and 4. When the load is 60 kN and the displacement is 7.16 mm, a shear crack occurred between MSS 1-2, 2-3 and the lower right of the MSS 10. The crack, which started under the MSS 13 with a vertical load of 70 kN and a displacement of 8 mm, progressed between MSS 11 and 12 and reached the upper cap (Figure 11). Finally, shear damage occurred when the vertical displacement reached 10.17 mm and the vertical load reached 78.10 kN. The load carrying capacity of the S4 increased by 6.98%, 4.69%, 3.03% and 1.56%, respectively, compared to the S0, S1, S2 and S3. However, since the ductility ratio was calculated as 1.36, it can also be said for S4 that sufficient ductility could not be obtained according to the literature. Although the load carrying capacity of the S1, S2, S3 and S4 obtained by strengthening the S0 up to this stage was relatively increased, the inability to obtain ductile behavior indicates that the MSSs do not work in the stirrup logic existing in the beam. The number of stitches was applied as 4 (range 220 mm), 5 (range 165 mm), 6 (range 130 mm) and 7 (range 110 mm) on one side in the shear area until this stage, but it was thought that more frequent MSS should be applied since shear damage could not be prevented. Since the crack formed in the shear zone at each step reached the beam's upper head between the two MSS in the shortest way and caused the formation of shear damage.

Figure 10. Damages at S3.

Finally, the S5 was prepared in order to strengthen the S0 beam. The difference between S-5 from other reinforcement types was that the most frequent (range 45 mm) MSS application was made along the shear opening. In this way, the behavior of MSSs placed in the stirrup logic in the most common situation was clearly seen. For this, fifteen MSS were applied to one side of the shear opening. In this way, the beam was strengthened using a total of 60 MSS. Initial bending cracks were observed for the S5 specimen at a load of 20 kN and a displacement of 2.56 mm. With the load of 30 kN and the displacement of 4.43 mm, the elongation of the cracks in the bending region was observed. In addition, at this load value, the first shear crack occurred in the right shear span of the beam, just below the 26 numbered MSS. With the increase in the load, new cracks were formed in the beam bending region with a load of 40 kN and a displacement of 6.16 mm and the elongation in the existing cracks continued. This was evaluated as a sign that the applied MSSs increased the shear capacity and forced the bending region of the beam. In addition, a shear crack was observed under the MSS 7 in the left shear opening at this load value. With the load reaching 50 kN and the displacement 7.94 mm, new cracks were formed in the bending region and elongation was observed in the existing cracks. When the load reached 60 kN and 9.47 mm, the cracks in the bending zone reached the neutral axis level. In addition, a shear crack was observed just below MSSs 1 and 22. With the increase in the vertical load, the 19-20-21 MSSs with 70 kN load and 11.27 mm displacement tried to prevent the propagation of shear cracks. At this load, new bending cracks also formed and moved towards the neutral axis. This showed that bending reinforcements started to yield at this load value. When the load reached 80 kN and the displacement reached 13.69 mm, the crack reaching MSS 19 advanced towards MSS 18. The propagation of cracks in the right shear span indicates that the shear capacity was forced. In addition, the fact that it continues to elongate in the cracks in the bending region at this load value shows that the longitudinal reinforcements are also forced. As the load reached 90 kN and the displacement reached 16.45 mm, the propagation of the cracks in the bending zone stopped. The crack reaching the number 18 MSS progressed and advanced to the bottom of the spreader beam right support. In addition, shear cracks were observed under MSSs 7 and 8 in the left shear opening at this load value. Finally, with the load reaching 95.74 kN and the displacement 17.89 mm, the load carrying capacity suddenly decreased and the experiment was terminated. It was observed in Figure 12 that cracks progressed on a horizontal line in the upper and lower parts of the MSSs in the left shear span. Due to the application of the applied MSSs between the lower and upper reinforcement, cracks developed from the weakest link. The MSSs present in the left shear span prevented the cracks from expanding and causing shear damage. Therefore, the support area, which was not reinforced, was broken by remaining weaker. When the specimen was examined, the test was terminated by the fracture of the weaker shell concrete than the point where the longitudinal reinforcements in the support area just ended. As a result, it can be said that the applied MSSs prevent the beam from direct shear damage. Considering the load carrying capacity, the S5 showed an increase of 31.15%, 28.33%, 26.30%, 24.49% and 22.58%, respectively, compared to the S0, S1, S2, S3 and S4 specimens.

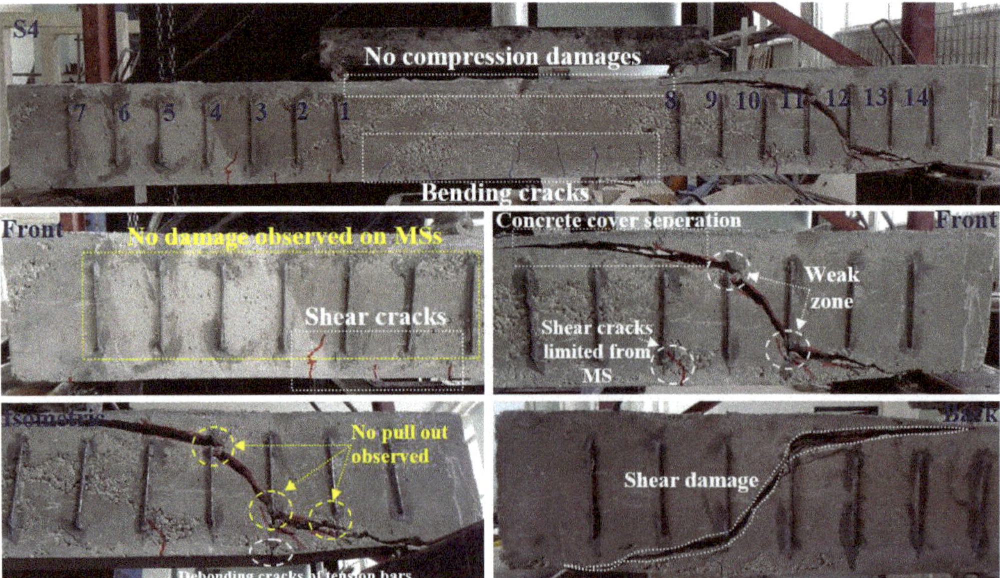

Figure 11. Damages at S4.

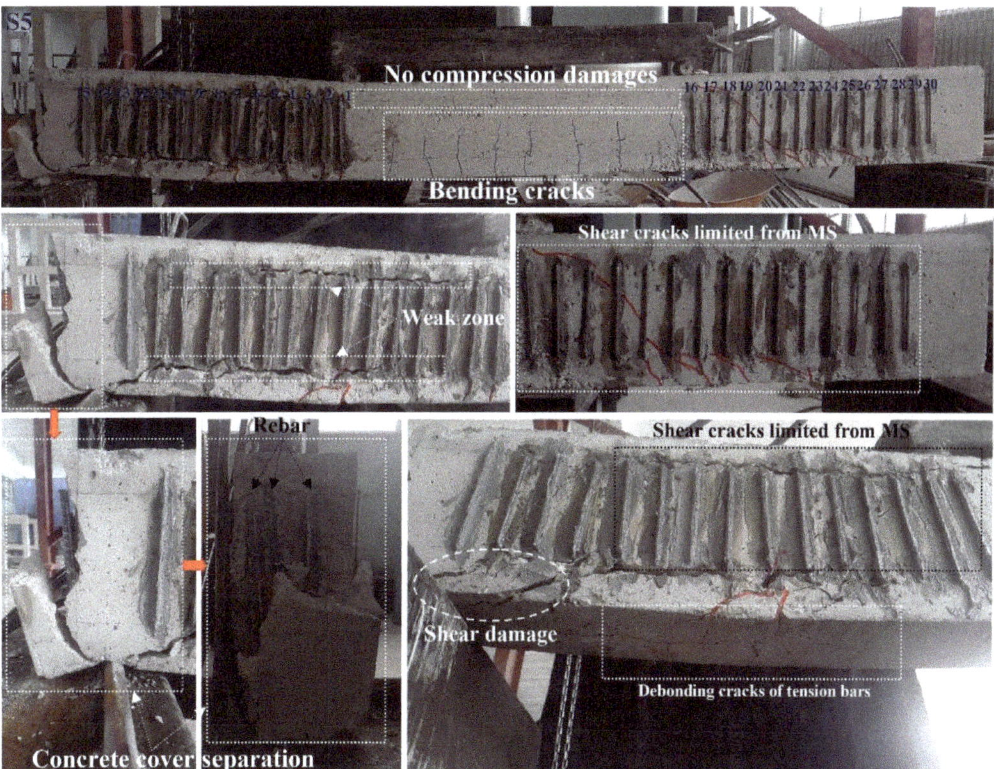

Figure 12. Damages at S5.

The energy dissipation capacities of each specimen are given in Figures 13 and 14 and Tables 4 and 5. When compared in terms of elastic energy dissipation, the S5 specimen showed an increase of 142.30%, 133.33%, 142.30%, 110% and 103.22%, respectively, compared to the S0, S1, S2, S3 and S4 specimens. This shows that the energy dissipation temporarily stored in the linear elastic behavior of the S5 specimen is the highest compared to other strengthening methods. In other words, it shows that the elastic energy dissipation capacity of the S5 specimen is better under sudden vertical load effects. The energy dissipated after damage to the building elements, especially under forced effects such as earthquakes, is known as plastic energy. In this respect, the S5 specimen has a higher plastic energy dissipation capacity. In other words, the S5 specimen has 8.62 times, 3.17 times, 3.85 times, 4.52 times and 3.85 times more plastic energy dissipation capacity than S0, S1, S2, S3 and S4 specimens, respectively. Although the S5 specimen has the highest plastic energy dissipation capacity among the reinforcement alternatives, it can be said that this is not at a sufficient level when evaluated with ductility, especially the high plastic energy dissipation capacity in S5 is due to the frequency of applied MSS. If this situation is considered to be applied to the beams in the stirrup logic, it should be applied at maximum 45 mm intervals. It should be noted that otherwise, direct shear damage to the beam will occur.

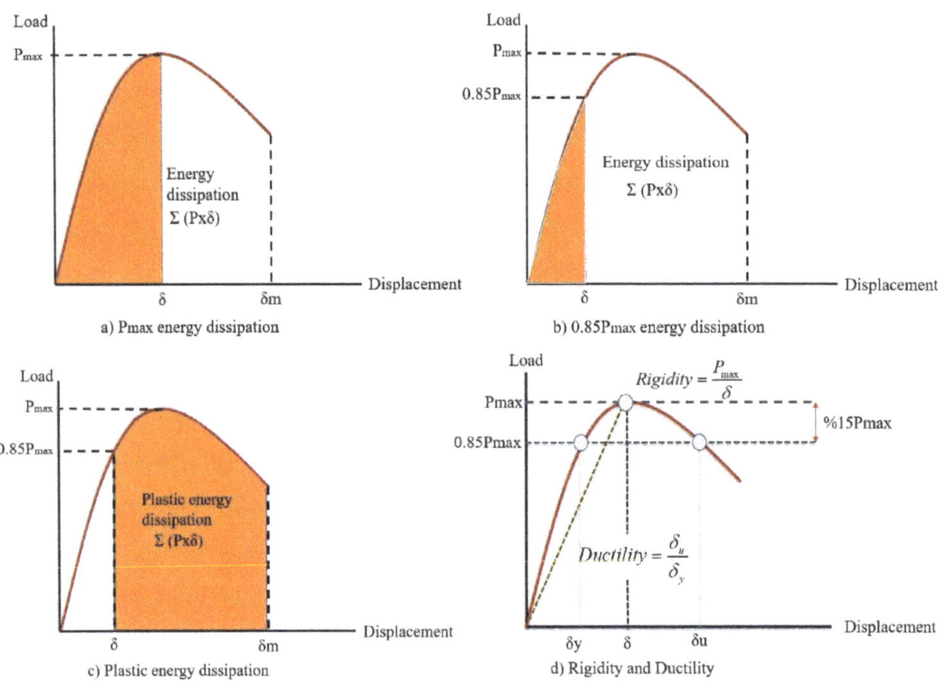

Figure 13. Energy dissipation capacity, ductility, and rigidity calculation of specimens.

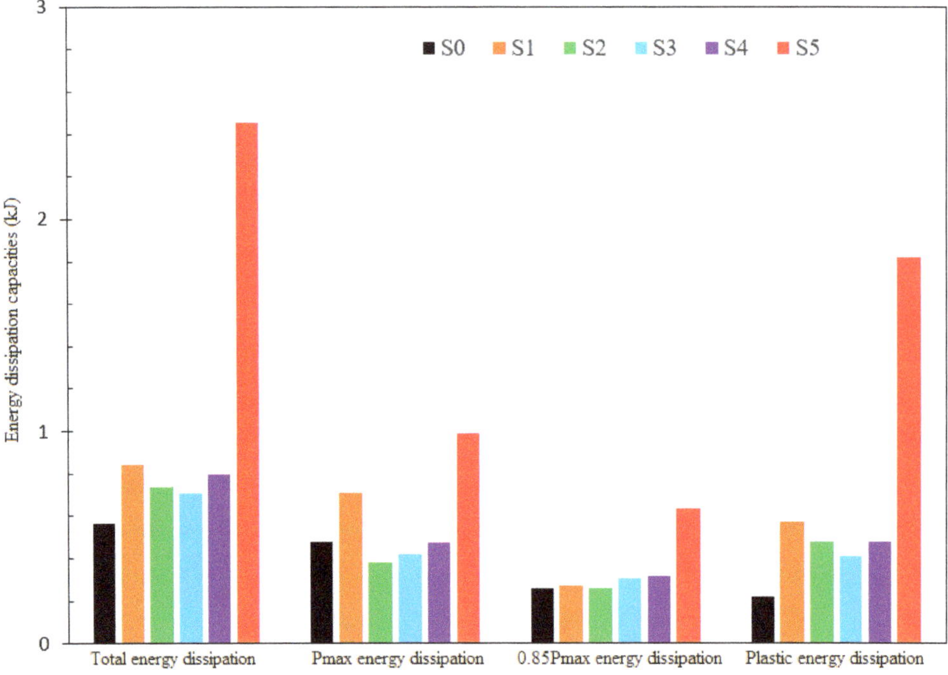

Figure 14. Energy dissipation values for specimens.

Table 4. Experimental results for load and displacement values.

Test Specimens	P_{max} (kN)	Rate of Increase at Max Load (%)	Displacement at Maximum Load (mm)	Stiffness at Maximum Load (P_{max}) (kN/mm)	P_u (0.85P_{max}) (kN)	Displacement at Yield, δ_y (mm)	At Yield (0.85P_{max}) Stiffness (kN/mm)	δ_u (mm)	Ductility Ratio
S0	73.00	1.00	8.83	8.26	62.00	6.62	9.36	9.83	1.49
S1	74.59	2.17	8.67	8.60	63.40	7.02	9.02	10.06	1.43
S2	75.79	3.82	8.07	9.39	64.42	6.45	9.98	9.49	1.47
S3	76.90	5.34	9.53	8.06	65.36	7.87	8.30	10.02	1.27
S4	78.10	6.98	10.17	7.68	66.38	8.07	8.22	11.02	1.36
S5	95.74	31.15	17.89	5.35	81.38	14.00	5.81	25.66	1.83

Table 5. Experimental test results for energy dissipation capacities.

Test Specimens	Maximum Displacement (mm)	Energy Dissipation at P_{max} (kJ)	Energy Dissipation at 0.85P_{max} (kJ)	Plastik Energy Dissipation (kJ)	Total Energy Dissipation (kJ)	Failure Type	Ductility Level
S0	10.26	0.47	0.26	0.21	0.56	Shear	Deficient
S1	15.84	0.70	0.27	0.57	0.84	Shear	Deficient
S2	12.21	0.37	0.26	0.47	0.73	Shear	Deficient
S3	12.09	0.41	0.30	0.40	0.71	Shear	Deficient
S4	14.01	0.47	0.31	0.47	0.794	Shear	Deficient
S5	30.77	0.98	0.63	1.81	2.45	Shear	Deficient

4. Conclusions

There are many different strategies (including the use of external steel reinforcement, section enlargement, internal steel or FRP reinforcement, supplemental members, FRP plates and strips, both steel and FRP NSMR, and external pre-stressing) in the conventional strengthening or retrofitting of existing reinforced concrete buildings. The strategies chosen vary in relation to the expected behavior of the existing reinforced concrete member. In this study, U-shaped Mechanical Steel Stitches (MSS) have been tested for the first time in the literature, especially for reinforced concrete beams where brittle fracture is expected under shear. The performance of MSSs applied over the cracks of damaged reinforced concrete elements, which were previously conducted in the literature, was tested on undamaged reinforced concrete beams in this study. In the experimental study carried out on six reinforced concrete beams, while the mechanical properties of the existing beam and MS were kept constant, the only variable was the application range (spacing) of MSSs. The findings obtained from the experimental study are as follows;

(1) As expected, shear failure occurred in the reference S0 beam. On the other hand, shear failure could not be prevented in S1, S2, S3 and S4 beams, where the MSS spacing gradually changes between d and d/2. It has been observed that the cracks formed in the range of 45°–60°. In the S5 specimen, where the MSS range was d/5 ((1%) MSS ratio), crack formation did not occur with this angle. Therefore, it can be concluded that tightening the spacing of MS would be helpful in preventing the shear fracture of the beams.

(2) Since the MSs are attached to the existing reinforced concrete beam with anchors, some losses in section due to the drilling have occurred in the stiffness of the existing beams. For example, a 38% loss in initial stiffness occurred in S5 compared to S0. This situation slightly increased the amount of deflection occurring in the span of the beam. Especially in MSS application, micro-cracks formed during the drilling of existing beams merged due to the close proximity of the holes, and a damage mechanism similar to an adherence crack was observed.

(3) While the capacity increase in S1, S2, S3 and S4 beams was limited compared to S0, a gain of nearly 31% occurred in the S-5 beam. However, a load carrying capacity increase depending on the d/s amount (s is spacing between MS) was not observed in the experiments. This situation is also related to the formation of cracks in the d to d/2 range without coinciding with the MSSs.

(4) The energy consumed (absorbed) by the beams S1, S2, S3 and S4 increased gradually compared to the reference beam S0. In addition, with the considerable increase in

strength, the energy consumption of the S5 beam increased approximately 4 times compared to the S0 beam. The increase in displacement due to the decrease in stiffness of the S-5 beam had an effect on this increase.

(5) Experimental results show that the RC beams strengthened with different MSS configurations as S1, S2, S3, and S4 have a modest increase in failure load. It would also seem that in terms of ductility the arrangement of the pins up to a spacing of 110 mm is negligible. On the other hand, the S5 MSS configuration allows a considerable increase in the ultimate load. Therefore, it is concluded that a certain level of spacing is quite critical in this novel external strengthening method.

(6) It has been seen that the method proposed in this study can be used for strengthening purposes, especially in the RC members under the effect of shear, when traditional strengthening methods are not suitable in terms of cost, application, and time. Therefore, the outcomes of this study will be frontiers for new studies to be carried out for the optimum design of MSSs, which is not in the existing codes and is a fairly new retrofit/strengthening alternative for the literature.

(7) In this study, MSSs applied angle, MS diameter, anchorage depth and mechanical properties were kept constant. Therefore, the effect of these parameters on the behavior of beams reinforced with MS should be investigated in future studies. Similarly, the mechanical properties of the beams, stirrups and longitudinal reinforcement amounts, beam' geometric shapes, loading patterns, etc., are also waiting as an important research topic in MSS-reinforced beams.

(8) In addition to the above-mentioned positive features, it is quite possible that MSSs will be exposed to corrosion over time due to their properties. For this, it is very important that the outside of the material is covered with a corrosion inhibitor in the strengthening to be made. In addition, in future studies, MSS applications can also be made with FRP materials. In this way, the corrosion situation is eliminated.

Author Contributions: Conceptualization, C.A., Y.O.Ö. and M.H.A.; methodology, C.A., Y.O.Ö. and M.H.A.; formal analysis, C.A., Y.O.Ö. and M.H.A.; investigation, C.A., Y.O.Ö. and M.H.A.; data curation, C.A., writing—original draft preparation, C.A., writing—review and editing, C.A., Y.O.Ö. and M.H.A.; funding acquisition, C.A, Y.O.Ö. and M.H.A. All authors have read and agreed to the published version of the manuscript.

Funding: The study was supported by the Scientific and Technological Council of Turkey (TÜBİTAK) through grant number 122M091. The opinions expressed in this paper are those of the authors and do not reflect the views of the sponsor.

Institutional Review Board Statement: Not applicable.

Informed Consent Statement: Not applicable.

Data Availability Statement: Not applicable.

Conflicts of Interest: The authors declare no conflict of interest.

References

1. Bousias, S.N.; Biskinis, D.; Fardis, M.N.; Spathis, A.-L. Strength, stiffness, and cyclic deformation capacity of concrete jacketed members. *ACI Struct. J.* **2007**, *104*, 521.
2. Altun, F. An experimental study of the jacketed reinforced-concrete beams under bending. *Constr. Build. Mater.* **2004**, *18*, 611–618. [CrossRef]
3. Bayülke, N. *Structural Damage on 27 June 1998 Adana-Ceyhan Earthquake*; General Directorate of Disaster Affairs, ERD: New Delhi, India, 1998.
4. Korkmaz, H.H.; Dere, Y.; Özkılıç, Y.O.; Bozkurt, M.B.; Ecemiş, A.S.; Özdoner, N. Excessive snow induced steel roof failures in Turkey. *Eng. Fail. Anal.* **2022**, *141*, 106661. [CrossRef]
5. Özkılıç, Y.O.; Aksoylu, C.; Arslan, M.H. Numerical evaluation of effects of shear span, stirrup spacing and angle of stirrup on reinforced concrete beam behaviour. *Struct. Eng. Mech. Int'l. J.* **2021**, *79*, 309–326.
6. Gemi, L.; Madenci, E.; Özkılıç, Y.O.; Yazman, Ş.; Safonov, A. Effect of Fiber Wrapping on Bending Behavior of Reinforced Concrete Filled Pultruded GFRP Composite Hybrid Beams. *Polymers* **2022**, *14*, 3740. [CrossRef]

7. Gemi, L.; Madenci, E.; Özkılıç, Y.O. Experimental, analytical and numerical investigation of pultruded GFRP composite beams infilled with hybrid FRP reinforced concrete. *Eng. Struct.* **2021**, *244*, 112790. [CrossRef]
8. Özkılıç, Y.O.; Aksoylu, C.; Yazman, Ş.; Gemi, L.; Arslan, M.H. Behavior of CFRP-strengthened RC beams with circular web openings in shear zones: Numerical study. *Structures* **2022**, *41*, 1369–1389. [CrossRef]
9. Arslan, M.H.; Yazman, Ş.; Hamad, A.A.; Aksoylu, C.; Özkılıç, Y.O.; Gemi, L. Shear strengthening of reinforced concrete T-beams with anchored and non-anchored CFRP fabrics. *Structures* **2022**, *39*, 527–542. [CrossRef]
10. Özkılıç, Y.O.; Yazman, Ş.; Aksoylu, C.; Arslan, M.H.; Gemi, L. Numerical investigation of the parameters influencing the behavior of dapped end prefabricated concrete purlins with and without CFRP strengthening. *Constr. Build. Mater.* **2021**, *275*, 122173. [CrossRef]
11. Gemi, L.; Aksoylu, C.; Yazman, Ş.; Özkılıç, Y.O.; Arslan, M.H. Experimental investigation of shear capacity and damage analysis of thinned end prefabricated concrete purlins strengthened by CFRP composite. *Compos. Struct.* **2019**, *229*, 111399. [CrossRef]
12. Raza, S.; Khan, M.K.; Menegon, S.J.; Tsang, H.-H.; Wilson, J.L. Strengthening and repair of reinforced concrete columns by jacketing: State-of-the-art review. *Sustainability* **2019**, *11*, 3208. [CrossRef]
13. More, R.U.; Kulkarni, D. Flexural behavioural study on RC beam with externally bonded aramid fiber reinforced polymer. *Int. J. Res. Eng. Technol.* **2014**, *3*, 316–321.
14. Wu, B.; Zhang, S.; Liu, F.; Gan, T. Effects of salt solution on mechanical behaviors of aramid fiber–reinforced polymer (AFRP) sheets and AFRP-to-concrete joints. *Adv. Struct. Eng.* **2016**, *19*, 1855–1872. [CrossRef]
15. Zhang, S.; Wu, B. Effects of salt solution on the mechanical behavior of concrete beams externally strengthened with AFRP. *Constr. Build. Mater.* **2019**, *229*, 117044. [CrossRef]
16. Raval, C.; Shah, S.; Machhi, C. Experimental Study on Shear Behaviour of RC Beam Strengthened by AFRP Sheet. *Int. J. All Res. Writ.* **2020**, *3*, 62–68.
17. Ma, C.-K.; Apandi, N.M.; Sofrie, C.S.Y.; Ng, J.H.; Lo, W.H.; Awang, A.Z.; Omar, W. Repair and rehabilitation of concrete structures using confinement: A review. *Constr. Build. Mater.* **2017**, *133*, 502–515. [CrossRef]
18. Ma, G.; Li, H.; Yan, L.; Huang, L. Testing and analysis of basalt FRP-confined damaged concrete cylinders under axial compression loading. *Constr. Build. Mater.* **2018**, *169*, 762–774. [CrossRef]
19. Qin, Z.; Tian, Y.; Li, G.; Liu, L. Study on bending behaviors of severely pre-cracked RC beams strengthened by BFRP sheets and steel plates. *Constr. Build. Mater.* **2019**, *219*, 131–143. [CrossRef]
20. Duic, J.; Kenno, S.; Das, S. Flexural rehabilitation and strengthening of concrete beams with BFRP composite. *J. Compos. Constr.* **2018**, *22*, 04018016. [CrossRef]
21. Joyklad, P.; Suparp, S.; Hussain, Q. Flexural response of JFRP and BFRP strengthened RC beams. *Int. J. Eng. Technol.* **2019**, *11*, 203–207. [CrossRef]
22. Pham, T.M.; Chen, W.; Elchalakani, M.; Karrech, A.; Hao, H. Experimental investigation on lightweight rubberized concrete beams strengthened with BFRP sheets subjected to impact loads. *Eng. Struct.* **2020**, *205*, 110095. [CrossRef]
23. Shen, D.; Li, M.; Kang, J.; Liu, C.; Li, C. Experimental studies on the seismic behavior of reinforced concrete beam-column joints strengthened with basalt fiber-reinforced polymer sheets. *Constr. Build. Mater.* **2021**, *287*, 122901. [CrossRef]
24. Prado, D.M.; Araujo, I.D.G.; Haach, V.G.; Carrazedo, R. Assessment of shear damaged and NSM CFRP retrofitted reinforced concrete beams based on modal analysis. *Eng. Struct.* **2016**, *129*, 54–66. [CrossRef]
25. Karzad, A.S.; Al Toubat, S.; Maalej, M.; Estephane, P. Repair of reinforced concrete beams using carbon fiber reinforced polymer. In Proceedings of the MATEC Web of Conferences: EDP Sciences 2017, Sharjah, United Arab Emirates, 18–20 April 2017; p. 01008.
26. Karam, E.C.; Hawileh, R.A.; El Maaddawy, T.; Abdalla, J.A. Experimental investigations of repair of pre-damaged steel-concrete composite beams using CFRP laminates and mechanical anchors. *Thin-Walled Struct.* **2017**, *112*, 107–117. [CrossRef]
27. Karzad, A.S.; Leblouba, M.; Al Toubat, S.; Maalej, M. Repair and strengthening of shear-deficient reinforced concrete beams using Carbon Fiber Reinforced Polymer. *Compos. Struct.* **2019**, *223*, 110963. [CrossRef]
28. Yu, F.; Zhou, H.; Jiang, N.; Fang, Y.; Song, J.; Feng, C.; Guan, Y. Flexural experiment and capacity investigation of CFRP repaired RC beams under heavy pre-damaged level. *Constr. Build. Mater.* **2020**, *230*, 117030. [CrossRef]
29. Yu, F.; Fang, Y.; Zhou, H.; Bai, R.; Xie, C. A Simplified Model for Crack Width Prediction of Flexural-Strengthened High Pre-Damaged Beams with CFRP Sheet. *KSCE J. Civ. Eng.* **2020**, *24*, 3746–3764. [CrossRef]
30. Yu, F.; Fang, Y.; Guo, S.; Bai, R.; Yin, L.; Mansouri, I. A simple model for maximum diagonal crack width estimation of shear-strengthened pre-damaged beams with CFRP strips. *J. Build. Eng.* **2021**, *41*, 102716. [CrossRef]
31. Bahij, S.; Omary, S.; Feugeas, F.; Faqiri, A. Structural Strengthening/Repair of Reinforced Concrete (RC) Beams by Different Fiber-Reinforced Cementitious Materials-A State-of-the-Art Review. *J. Civ. Environ. Eng.* **2020**, *10*. [CrossRef]
32. Zaki, M.A.; Rasheed, H.A.; Alkhrdaji, T. Performance of CFRP-strengthened concrete beams fastened with distributed CFRP dowel and fiber anchors. *Compos. Part B Eng.* **2019**, *176*, 107117. [CrossRef]
33. Aksoylu, C.; Özkılıç, Y.O.; Yazman, Ş.; Gemi, L.; Arslan, M.H. Experimental and Numerical Investigation of Load Bearing Capacity of Thinned End Precast Purlin Beams and Solution Proposals. *Tek. Dergi* **2021**, *3*, 10823–10858.
34. Al-Khafaji, A.; Salim, H.; El-Sisi, A. *Behavior of RC Beams Strengthened with CFRP Sheets under Sustained Loads*; Elsevier: Amsterdam, The Netherlands, 2021; pp. 4690–4700.
35. Kotynia, R.; Oller, E.; Marí, A.; Kaszubska, M. Efficiency of shear strengthening of RC beams with externally bonded FRP materials–State-of-the-art in the experimental tests. *Compos. Struct.* **2021**, *267*, 113891. [CrossRef]

36. Abed, M.J.; Fayyadh, M.M.; Khaleel, O.R. Effect of web opening diameter on performance and failure mode of CFRP repaired RC beams. *Mater. Today Proc.* **2021**, *42*, 388–398. [CrossRef]
37. Al-Fakih, A.; Hashim, M.H.M.; Alyousef, R.; Mutafi, A.; Sabah, S.H.A.; Tafsirojjaman, T. *Cracking Behavior of Sea Sand RC Beam Bonded Externally with CFRP Plate*; Elsevier: Amsterdam, The Netherlands, 2021; pp. 1578–1589.
38. Jahami, A.; Temsah, Y.; Khatib, J.; Baalbaki, O.; Kenai, S. The behavior of CFRP strengthened RC beams subjected to blast loading. *Mag. Civ. Eng.* **2021**, *3*, 10309.
39. Alhassan, M.; Al-Rousan, R.; Ababneh, A. Anchoring of the main CFRP sheets with transverse CFRP strips for optimum upgrade of RC Beams: Parametric experimental study. *Constr. Build. Mater.* **2021**, *293*, 123525. [CrossRef]
40. Samb, N.; Chaallal, O.; El-Saikaly, G. Multilayer versus monolayer externally bonded CFRP sheets for shear strengthening of concrete T-Beams. *J. Compos. Constr.* **2021**, *25*, 04021025. [CrossRef]
41. Mukhtar, F.M.; Shehadah, M.E. Shear behavior of flexural CFRP-strengthened RC beams with crack-induced delamination: Experimental investigation and strength model. *Compos. Struct.* **2021**, *268*, 113894. [CrossRef]
42. Mansour, W. Numerical analysis of the shear behavior of FRP-strengthened continuous RC beams having web openings. *Eng. Struct.* **2021**, *227*, 111451. [CrossRef]
43. Gemi, L.; Alsdudi, M.; Aksoylu, C.; Yazman, S.; Ozkilic, Y.O.; Arslan, M.H. Optimum amount of CFRP for strengthening shear deficient reinforced concrete beams. *Steel Compos. Struct.* **2022**, *43*, 735–757.
44. Siddika, A.; Al Mamun, M.A.; Alyousef, R.; Amran, Y.M. Strengthening of reinforced concrete beams by using fiber-reinforced polymer composites: A review. *J. Build. Eng.* **2019**, *25*, 100798. [CrossRef]
45. Capozucca, R.; Magagnini, E.; Vecchietti, M.V.; Khatir, S. RC beams damaged by cracking and strengthened with NSM CFRP/GFRP rods. *Frat. Ed Integrità Strutt.* **2021**, *15*, 386–401. [CrossRef]
46. Panigrahi, A.K.; Biswal, K.; Barik, M. Strengthening of shear deficient RC T-beams with externally bonded GFRP sheets. *Constr. Build. Mater.* **2014**, *57*, 81–91. [CrossRef]
47. Boumaaza, M.; Bezazi, A.; Bouchelaghem, H.; Benzennache, N.; Amziane, S.; Scarpa, F. Behavior of pre-cracked deep beams with composite materials repairs. *Struct. Eng. Mech.* **2017**, *63*, 575–583.
48. Aksoylu, C. Experimental analysis of shear deficient reinforced concrete beams strengthened by glass fiber strip composites and mechanical stitches. *Steel Compos. Struct. Int. J.* **2021**, *40*, 267–285.
49. Rahman, Z. Strength and ductility behaviour of FRC beams strengthened with externally bonded GFRP laminates. *Mater. Today: Proc.* **2021**, *37*, 2542–2546.
50. Kumari, A.; Nayak, A. Strengthening of shear deficient RC deep beams using GFRP sheets and mechanical anchors. *Can. J. Civ. Eng.* **2021**, *48*, 1–15. [CrossRef]
51. Ali, H.; Assih, J.; Li, A. Flexural capacity of continuous reinforced concrete beams strengthened or repaired by CFRP/GFRP sheets. *Int. J. Adhes. Adhes.* **2021**, *104*, 102759. [CrossRef]
52. Abbass, M.; Medhlom, M.; Ali, I. Strength Capacity Cracks Propagations Deflection and Tensile Enhancement of Reinforced Concrete Beams Warped by Glass Fiber Reinforced Polymer Strips. *Int. J. Eng.* **2021**, *34*, 1094–1104.
53. Miruthun, G.; Vivek, D.; Remya, P.; Elango, K.; Saravanakumar, R.; Venkatraman, S. Experimental investigation on strengthening of reinforced concrete beams using GFRP laminates. *Mater. Today: Proc.* **2021**, *37*, 27448. [CrossRef]
54. Al-Shalif, S.A.; Akın, A.; Aksoylu, C.; Arslan, M.H. *Strengthening of Shear-Critical Reinforced Concrete T-Beams with Anchored and Non-Anchored GFRP Fabrics Applications*; Elsevier: Amsterdam, The Netherlands, 2022; pp. 809–827.
55. Peng, J.; Tang, H.; Zhang, J. Structural behavior of corroded reinforced concrete beams strengthened with steel plate. *J. Perform. Constr. Facil.* **2017**, *31*, 04017013. [CrossRef]
56. Kazem, H.; Rizkalla, S.; Kobayashi, A. Shear strengthening of steel plates using small-diameter CFRP strands. *Compos. Struct.* **2018**, *184*, 78–91. [CrossRef]
57. Alam, M.A.; Onik, S.A.; Mustapha, K.N.B. Crack based bond strength model of externally bonded steel plate and CFRP laminate to predict debonding failure of shear strengthened RC beams. *J. Build. Eng.* **2020**, *27*, 100943. [CrossRef]
58. Aykac, S.; Özbek, E. Strengthening of reinforced concrete T-beams with steel plates. *Tek. Dergi* **2011**, *22*.
59. Acar, D. Çelik Levha ve Karbon Kumaşlarla Güçlendirilmiş Betonarme Kirişlerin Davranış ve Dayanımı. Master Thesis, Gazi University, Ankara, Turkey, 2014.
60. Aykaç, B.; Acar, D. Betonarme Kirişlerin Dıştan Yapıştırılmış Karbon Kumaş Ve Çelik Levhalardan Oluşan Kompozit Malzemeyle Güçlendirilmesi. *J. Fac. Eng. Archit. Gazi Univ.* **2014**, *29*, 1. [CrossRef]
61. Abdul-Razzaq, K.S.; Ali, H.I.; Abdul-Kareem, M.M. A new strengthening technique for deep beam openings using steel plates. *Int. J. Appl. Eng. Res.* **2017**, *12*, 15935–15947.
62. Demir, A.; Ercan, E.; Demir, D.D. Strengthening of reinforced concrete beams using external steel members. *Steel Compos. Struct.* **2018**, *27*, 453–464.
63. Osman, B.H.; Wu, E.; Ji, B.; Abdulhameed, S.S. Repair of Pre-cracked Reinforced Concrete (RC) Beams with Openings Strengthened Using FRP Sheets Under Sustained Load. *Int. J. Concr. Struct. Mater.* **2017**, *11*, 171–183. [CrossRef]
64. Xu, C.-X.; Peng, S.; Deng, J.; Wan, C. Study on seismic behavior of encased steel jacket-strengthened earthquake-damaged composite steel-concrete columns. *J. Build. Eng.* **2018**, *17*, 154–166. [CrossRef]
65. Xu, C.C.X.; Sheng, P.S.; Wan, C.C. Experimental and Theoretical Research on Shear Strength of Seismic-Damaged SRC Frame Columns Strengthened with Enveloped Steel Jackets. *Adv. Civ. Eng.* **2019**, *2019*, 6401730. [CrossRef]

66. Alshlash, S.; Aksoylu, C.; Erkan, I.H.; Arslan, M.H. Kesme Kapasitesi Yetersiz On Hasarli Betonarme Kirislerin "Dikis Demirleri" Ile Onarim/Guclendirilmesi". In Proceedings of the IV International Scientific and Vocational Studies Congress—Engineering, Ankara, Turkey, 12–15 December 2019.

67. Hamoush, S.; Ahmad, S. Concrete crack repair by stitches. *Mater. Struct.* **1997**, *30*, 418–423. [CrossRef]

68. Altın, S.; Anıl, Ö.; Gökten, Y. *Betonarme Kirişlerin Kesmeye Karşı Güçlendirilmesinde bir Kelepçe Uygulaması*; Gazi Üniversitesi Mühendislik Mimarlık Fakültesi Dergisi: Ankara, Turkey, 2004; Volume 19, pp. 415–422.

69. Rizal, A.R.; Wibowo, A.; Wijatmiko, I.; Remayanti, C. Effect of Repairing with Retrofit Method (Concrete Jacketing) Using Bamboo Reinforcement on Flexural Capacity of Reinforced Concrete Beam with Initial Damage Variation. *Res. J. Adv. Eng. Sci.* **2019**, *4*, 340–344.

70. Chalioris, C.E.; Kytinou, V.K.; Voutetaki, M.E.; Papadopoulos, N.A. Repair of heavily damaged RC beams failing in shear using U-shaped mortar jackets. *Buildings* **2019**, *9*, 146. [CrossRef]

71. Aldhafairi, F.; Hassan, A.; Abd-El-Hafez, L.M.; Abouelezz, A.E.Y. Different techniques of steel jacketing for retrofitting of different types of concrete beams after elevated temperature exposure. *Structures* **2020**, *28*, 713–725. [CrossRef]

72. Di Trapani, F.; Malavisi, M.; Marano, G.C.; Sberna, A.P.; Greco, R. Optimal seismic retrofitting of reinforced concrete buildings by steel-jacketing using a genetic algorithm-based framework. *Eng. Struct.* **2020**, *219*, 110864. [CrossRef]

73. Yuan, F.; Chen, M.; Pan, J. Flexural strengthening of reinforced concrete beams with high-strength steel wire and engineered cementitious composites. *Constr. Build. Mater.* **2020**, *254*, 119284. [CrossRef]

74. Murthy, A.R.; Ganesh, P.; Sakthi Priya, G.N. Flexural Behaviour of Severely Damaged RC Beams Strengthened with Ultra-High Strength Concrete. In *Recent Advances in Structural Engineering*; Rao, A.R.M., Ramanjaneyulu, K., Eds.; Springer: Singapore, 2019; Volume 2, pp. 699–707.

75. Hassan, A.; Baraghith, A.T.; Atta, A.M.; El-Shafiey, T.F. Retrofitting of shear-damaged RC T-beams using U-shaped SHCC jacket. *Eng. Struct.* **2021**, *245*, 112892. [CrossRef]

76. Ganesh, P.; Ramachandra Murthy, A. Analytical model to predict the fatigue life of damaged RC beam strengthened with GGBS based UHPC. *Structures* **2021**, *33*, 2559–2569. [CrossRef]

77. Chandrakar, J.; Singh, A.K. Retrofitting of Rcc Structral Members Using Concrete Jacketing. *Int. J. Recent Dev. Civ. Environ. Eng.* **2017**, *2*. ISSN 25814117.

78. Rodrigues, H.; Pradhan, P.M.; Furtado, A.; Rocha, P.; Vila-Pouca, N. Structural Repair and Strengthening of RC Elements with Concrete Jacketing. In *Strengthening and Retrofitting of Existing Structures*; Costa, A., Arêde, A., Varum, H., Eds.; Springer: Singapore, 2018; pp. 181–198.

79. TS500, Enstitü, T.S. TS 500 Betonarme Yapıların Tasarım ve Yapım Kuralları. Ankara, Türkiye. 2000. Available online: https://www.resmigazete.gov.tr/eskiler/2000/07/20000712M1-23.pdf (accessed on 25 August 2022).

80. Özkılıç, Y.O.; Aksoylu, C.; Arslan, M.H. Experimental and numerical investigations of steel fiber reinforced concrete dapped-end purlins. *J. Build. Eng.* **2021**, *36*, 102119. [CrossRef]

81. Aksoylu, C.; Özkılıç, Y.O.; Arslan, M.H. Damages on prefabricated concrete dapped-end purlins due to snow loads and a novel reinforcement detail. *Eng. Struct.* **2020**, *225*, 111225. [CrossRef]

 **buildings**

Article

A Comparative Probabilistic Seismic Hazard Analysis for Eastern Turkey (Bitlis) Based on Updated Hazard Map and Its Effect on Regular RC Structures

Ercan Işık [1] and Ehsan Harirchian [2,*]

1   Department of Civil Engineering, Bitlis Eren University, Bitlis 13100, Turkey
2   Institute of Structural Mechanics (ISM), Bauhaus-Universität Weimar, 99423 Weimar, Germany
*   Correspondence: ehsan.harirchian@uni-weimar.de

Abstract: Determining the earthquake hazard of any settlement is one of the primary studies for reducing earthquake damage. Therefore, earthquake hazard maps used for this purpose must be renewed over time. Turkey Earthquake Hazard Map has been used instead of Turkey Earthquake Zones Map since 2019. A probabilistic seismic hazard was performed by using these last two maps and different attenuation relationships for Bitlis Province (Eastern Turkey) were located in the Lake Van Basin, which has a high seismic risk. The earthquake parameters were determined by considering all districts and neighborhoods in the province. Probabilistic seismic hazard analyses were carried out for these settlements using seismic sources and four different attenuation relationships. The obtained values are compared with the design spectrum stated in the last two earthquake maps. Significant differences exist between the design spectrum obtained according to the different exceedance probabilities. In this study, adaptive pushover analyses of sample-reinforced concrete buildings were performed using the design ground motion level. Structural analyses were carried out using three different design spectra, as given in the last two seismic design codes and the mean spectrum obtained from attenuation relationships. Different design spectra significantly change the target displacements predicted for the performance levels of the buildings.

Keywords: Eastern Turkey; seismic risk; adaptive pushover; design spectra; Bitlis

Citation: Işık, E.; Harirchian, E. A Comparative Probabilistic Seismic Hazard Analysis for Eastern Turkey (Bitlis) Based on Updated Hazard Map and Its Effect on Regular RC Structures. *Buildings* **2022**, *12*, 1573. https://doi.org/10.3390/buildings12101573

Academic Editor: Marco Di Ludovico

Received: 2 September 2022
Accepted: 27 September 2022
Published: 30 September 2022

Publisher's Note: MDPI stays neutral with regard to jurisdictional claims in published maps and institutional affiliations.

1. Introduction

The main priority for reducing earthquake damage depends on the reliable determination of the seismic hazard. Risk is the combination of the probability or frequency of a defined threat and the magnitude of the consequences of the occurrence. In other words, the level of risk is proportional to the hazard's intensity (or magnitude) and the vulnerability of the affected elements [1,2]. After each earthquake, significant loss and damage to life and properties emphasize the necessity to demonstrate the seismic hazard reliably. Therefore, determining the seismic risk of any region is an integral part of modern pre-earthquake disaster management [3–7]. This is also important for the seismic sensitivity assessment and retrofit decision making of structures [8,9].

The assessment of seismic risk is recognized as an early and emerging new discipline that was introduced as a logical continuation of seismic hazard research that Luis Esteva (1967, 1968) [10,11] and Allin Cornell [12] carried out. As stated in an elementary definition of this discipline, outlined by the EERI Committee on Seismic Risk in 1984, "seismic risk is the probability that social and economic consequences of earthquakes will equal or exceed specified values at a site, at various sites or in an area during a specified exposure time" [13]. Opinions in the literature on initial earthquake risk assessment differ greatly. For example, Luis Esteva (1967, 1968) and Allin Cornell (1968) were the first to initiate seismic risk analysis in 1968. On the other hand, Whitman et al. pointed to several previous assessments of earthquake loss estimations, such as the NOAA1 study for San Francisco [14], and state that earthquake loss studies follow this for more than thirty US regions [15]. On the other hand, long before these studies, John Freeman's Earthquake Damage and Earthquake Insurance [16], accepted as an earthquake damage prediction today, was published [17]. Afterwards, earthquake damage prediction was mostly regarded as part of the insurance sector until the publication of Cornell's work in 1968 [2]. After that, seismic hazard and risk analyses for different parts of the world were out using different methods, such as the Philippines [18], Bangladesh [19], Iran [20], Korean Peninsula [21], Pakistan [22,23], Croatia [5,24,25], Brazil [26], Italy [27], Argentina [28], Bosnia-Herzegovina [29], Malaysia and Singapore [30] and Turkey [31,32]. In addition, Turkey has smaller-scale studies for different provinces in the Eastern Anatolia region, such as Bingöl [33] and Van [34–36].

Seismicity is based on geological, tectonic and statistical data. Macro seismic data regarding the earthquake origin time, location, epicenter, source parameters and magnitude are the most important parameters in determining the seismic hazard of any region. Furthermore, the seismicity of a region is an indicator of a future earthquake in that region [37–39]. Therefore, the data from destructive earthquakes significantly contribute to determining seismic hazard zones more realistically and developing fundamental principles for the design of earthquake-resistant structures. Thus, Turkey's demand for renewal in seismic hazard maps and principles for designing earthquake-resistant structures, especially after the 2011 Van earthquakes, emerged. Thanks to the studies performed, both seismic hazard maps and seismic design codes were updated in 2018 and started to be used in 2019 [31,40–42].

In this study, Bitlis province was selected as it is located in Lake Van Basin in Eastern Turkey. Lake Van Basin is one of Turkey's current and intensive seismic activity regions. Specifically, the earthquakes, whose epicenter was Van province located in this basin, and the losses that come after the earthquakes have, once again, revealed the seismic risk of the basin. Figure 1 displays the districts of the Bitlis province and its geographical location.

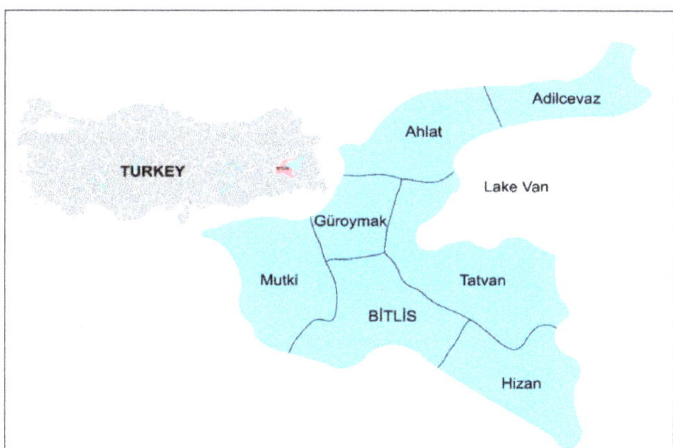

Figure 1. Location of Bitlis and its districts.

Bitlis is a historical city surrounded by mountains, located on the strait passages connecting Eastern Anatolia to South-Eastern Anatolia, located between 41°33′–43° and 37°54′–38°58′. Bitlis is in a position worth examining due to the seismicity in Bitlis and especially its close surroundings and the earthquakes that occurred in the past. This study conducted a probabilistic seismic hazard analysis for Bitlis province, considering the seismic sources and attenuation relationships. The results are compared to the design spectra in the last two seismic design codes. In addition, structural analyses were performed by using the mean design spectrum obtained from the used attenuation relationships and the last two design spectra. The study is important regarding the site-specific seismic hazard analysis and comparing the last two earthquake hazard values for Bitlis. Furthermore, different design spectra were tried to reveal at what level the earthquake hazard changes affect the building performance. Therefore, in light of current data and studies, the region's seismicity, the earthquake hazard and their behaviour under the effect of earthquakes should be reviewed.

2. Tectonics and Seismicity of Bitlis

It is a known fact that local geological soil conditions directly affect and alter seismic activity characteristics and may damage existing structures on these soils [43,44]. In the center of Bitlis and its vicinity, the metamorphic rocks of Bitlis massif, upper cretaceous mélange in the Ahlat-Adilcevaz area, Eocene aged Ahlat conglomerate, Miocene aged Adilcevaz Limestone, polio-quaternary volcanism and alluvium outcrop are present. Rock assemblages are in the Van Lake Basin formed in the Paleozoic Era–present time period and alluvial sediment outcrop. Generally, the metamorphic rocks of the Bitlik Massif in are the south of the basin, volcanic and volcanoclastic rocks that are products of young Nemrut and Süphan are in the west and north areas, volcanic rocks and ophiolite components of the Yüksekova Complex are in the east of the basin, young-present streams and lacustrine sediments and carbonates outcrop [45–51]. The Bitlis Massif contains ophiolites of the old ocean floor and rock assemblages containing different metamorphic facies [52].

The Eastern Anatolia Region, located on the Alpine-Himalayan seismic belt, one of the most important seismic belts in the world, is seismically active. This area where these two faults in the nature of intra-continental transform faults limit the Anatolian plate and cross fault systems developed among them are a region with the highest density of active faults in Turkey. Eastern Anatolia is under continental shortening and thickening effects due to the continental collision between the Arabian and Eurasian plates [53–57]. In addition, the volcanism in the region was developed by this collision [49,58]. This active continental collision forces the Anatolian Plate to move counterclockwise to the west, along the North Anatolian Fault Zone (NAFZ) and East Anatolian Fault Zone (EAFZ), two major strike-slip fault zones (Figure 2) [59,60]. Sinistral-slip NAFZ and dextral-slip EAFZ join the Karliova Triple Junction (KTJ) located in Eastern Anatolia [61]. Many fault lines in Turkey have developed due to this active continental collision [59,62–66]. Both Bitlis-Zagros Suture Belt and Karliova Triple Junction are close to Bitlis province.

Lake Van, which is a product of the tectonic pressure led by the collision of the Arabian and Eurasian Plates [59,67,68] and partially remained in the province of Bitlis, is in a tectonic structure that has undergone intense deformation in Eastern Anatolia [69,70]. With its volume of 607 km^3 and a maximum depth of 451 m, Lake Van is the fourth largest lake in the world among inland lakes in terms of water content, after the Caspian Sea, the Aral Sea and Lake Issyk-Kul [71]. Therefore, earthquake activity is very high around Lake Van [72,73]. In addition, a major and destructive earthquake that may occur in the Lake Van Basin can closely affect the Bitlis city center and districts in this basin. Table 1 shows some major and destructive earthquakes around Bitlis in the instrumental period. The M $\geq$ 5.0 earthquakes in and around Bitlis between 1900 and 2017 are shown in Figure 3.

Table 1. Some significant earthquakes in and around Bitlis [78,79].

Date	Location	M	Earthquake Losses
1903	Malazgirt (Muş)	6.7	600 casualties and 450 building damage
1941	Van	5.9	192 casualties and 600 building damage
1946	Varto (Muş)	5.9	839 casualties and 3000 building damage
1949	Karlıova(Bingöl)	6.7	450 casualties and 3500 building damage
1966	Varto (Muş)	6.9	2396 casualties and 20,007 building damage
1971	Bingöl	6.7	878 casualties and 9111 building damage
1975	Lice (Diyarbakır)	6.6	2385 casualties and 8149 building damage
1976	Muradiye (Van)	7.5	3840 casualties and 9232 building damage
2003	Bingöl	6.4	176 casualties and 6000 building damage
2004	Ağrı	5.1	17 casualties and 1000 building damage
2011	Tabanlı (Van)	7.2	644 casualties and 17,005 building damage
2011	Edremit (Van)	5.6	40 casualties

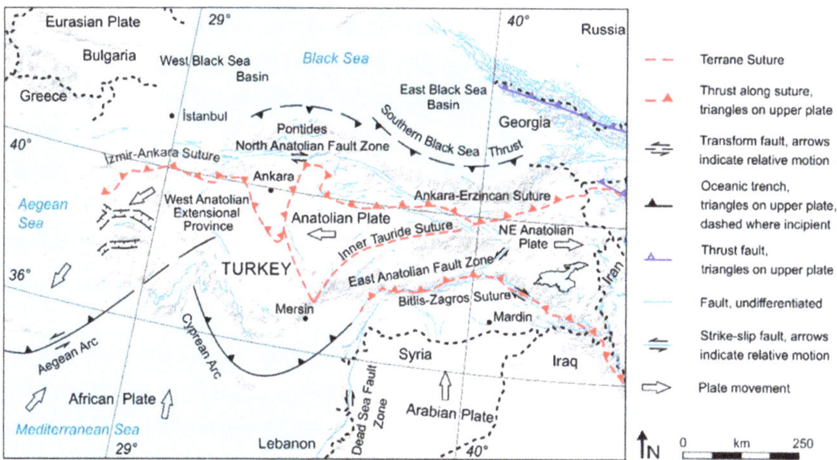

Figure 2. Simplified neotectonic map of Turkey and the surroundings [74–77].

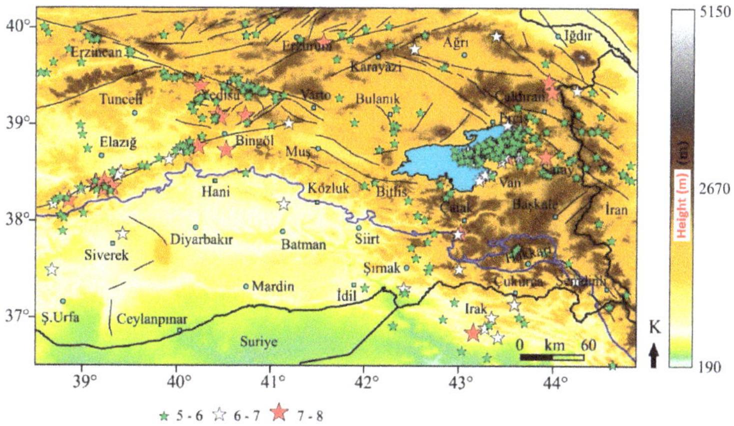

Figure 3. M ≥ 5.0 earthquakes surroundings Bitlis between 1900 and 2017 [80].

3. Current Seismic Parameters of Bitlis

The fact that December 27, 1939, Erzincan, December 20, 1942, Niksar–Erbaa, June 20, 1943, Adapazari-Hendek, November 26, 1943, Tosya-Ladik and February 1, 1944, Bolu-Gerede earthquakes occurred at close time intervals and led to huge economic losses and casualties triggered the efforts to reduce earthquake losses in Turkey [81,82]. Turkey's first official earthquake zonation map was prepared in 1945 following these earthquakes [83,84]. The historical development in these maps is shown in Table 2.

Table 2. The published official maps of earthquake zones in Turkey.

Year	Name	Method
1945	Map of earthquake zones	Based on damage data
1947	Map of earthquake zones	Based on damage data
1963	Turkey Earthquake Zones Map	Based on deterministic approach
1972	Turkey Earthquake Zones Map	Based on deterministic approach
1996	Turkey Earthquake Zones Map	Based on probabilistic approach
2018	Turkish Earthquake Hazard Map	Based on probabilistic approach

In the seismic hazard models used in the creation of the earthquake hazard map published in 1996, the errors resulting from the use of the attenuation relationship obtained from Western US measurements were ignored due to lack of data in the earthquake catalogue, uncertainties in the geographical boundaries of earthquake source faults and lack of local data [85]. The Seismic Zoning Map of Turkey, which entered into force in 1996, was renewed by the Disaster & Emergency Management Authority, Presidential of Earthquake Department and published in 2018 and became effective as of January 1, 2019. The new map was prepared in cooperation with the public and universities through the project titled Updating Turkey Earthquake Hazard Maps, which was supported by the AFAD National Earthquake Research Programme (UDAP). The new map was prepared with much more detailed data, considering the most up-to-date earthquake source parameters, earthquake catalogues and next-generation mathematical models. Unlike the previous map, the new map includes peak ground acceleration values rather than earthquake zones and the concept was removed [86–88].

In the studies of determining the source zone forming the basis for creating the earthquake hazard map that entered into force in 2019, a total of 105 seismic sources was identified by taking into account the active fault database of Turkey [89] and the earthquake catalogue [90]. The highest earthquake magnitudes of these seismic sources were determined after statistical analysis of instrumental and historical earthquake catalogues. In addition, the ground motion databases of Turkey, Greece, Italy, and California, which have similar seismotectonic structures, were compiled, and a broad, strong ground motion database was created. Moreover, four ground motion prediction equations that best represent this database were used in probabilistic seismic hazard analysis with a different emphasis [91]. An updated earthquake hazard map of Turkey is given in Figure 4.

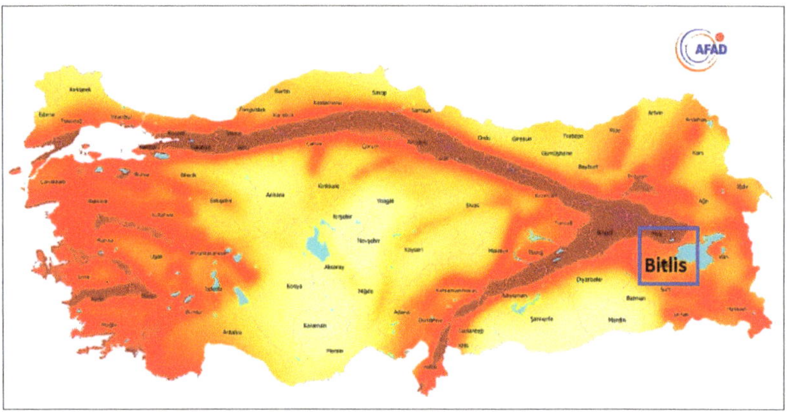

Figure 4. Updated Turkish Earthquake Hazard Map.

Turkish Earthquake Hazard Maps have started to be used with the Turkish Building Earthquake Code (TBEC-2018). Thanks to these maps, earthquake and earthquake-building parameters of any geographical location can be determined. These values can be obtained practically using the Turkey Earthquake Hazard Maps Interactive Web Application (TEHMIWA). Ground motion levels for four different exceedance probabilities were identified in the TBEC-2018 (Table 3).

The peak ground acceleration (PGA) and peak ground velocity (PGV) obtained for different probabilities of exceedance of all neighborhoods in Bitlis province are shown in Table 4.

Table 3. Earthquake ground motion levels [40].

Earthquake Level	Repetition Period (Year)	Probability of Exceedance (in 50 Years)	Description
DD-1	2475	2%	Largest earthquake ground motion
DD-2	475	10%	Standard design earthquake ground motion
DD-3	72	50%	Frequent earthquake ground motion
DD-4	43	68%	Service earthquake movement

Table 4. PGA and PGV values for all neighbourhoods in the city center of Bitlis.

Neighbourhood	Peak Ground Acceleration (g) Probability of Exceedance in 50 Years				Peak Ground Velocity (cm/s)-PGV Probability of Exceedance in 50 Years			
	2%	10%	50%	68%	2%	10%	50%	68%
Atatürk	0.490	0.260	0.107	0.077	28.306	15.149	6.547	4.878
Beş Minare	0.493	0.261	0.107	0.078	28.503	15.276	6.620	4.936
Gazi Bey	0.490	0.260	0.106	0.077	28.249	15.104	6.525	4.861
Hersan	0.489	0.260	0.106	0.076	28.132	15.016	6.475	4.821
Hüsrev Paşa	0.491	0.260	0.107	0.077	28.403	15.211	6.585	4.909
İnönü	0.490	0.260	0.106	0.077	28.213	15.078	6.509	4.848
Muştakbaba	0.490	0.260	0.106	0.077	28.214	15.080	6.505	4.843
Saray	0.490	0.260	0.106	0.077	28.176	15.050	6.487	4.828
Sekiz Ağustos	0.490	0.259	0.106	0.077	28.148	15.025	6.484	4.830
Şemsi Bitlis	0.490	0.260	0.106	0.077	28.197	15.066	6.501	4.841
Taş	0.490	0.260	0.107	0.077	28.308	15.153	6.545	4.874
Yükseliş	0.491	0.261	0.107	0.077	28.397	15.222	6.573	4.892
Zeydan	0.490	0.260	0.106	0.077	28.260	15.116	6.523	4.856

There are seven districts (Adilcevaz, Ahlat, Güroymak, Hizan, Mutki and Tatvan) in Bitlis province, including the central district. The southern side of the Nemrut volcanic mount in the province of Bitlis is located 10 km from the Tatvan district and about 24 km from Bitlis central district. Mount Süphan, which is approximately 85 km away from Bitlis city center and part of which is located within the boundaries of Adilcevaz district, is the highest mountain of volcanic origin (4058 m) in Turkey after the Mount of Greater Agri (Ararat). Table 5 compares PGA and PGV values measured by different earthquake ground motion levels for seven districts of Bitlis province.

Table 5. PGA and PGV values for different probabilities of exceedance for Bitlis districts.

District	Peak Ground Acceleration (g)-PGA Probability of Exceedance in 50 Years				Peak Ground Velocity (cm/s)-PGV Probability of Exceedance in 50 Years			
	2%	10%	50%	68%	2%	10%	50%	68%
Adilcevaz	0.578	0.303	0.121	0.086	37.108	18.625	7.399	5.345
Ahlat	1.038	0.570	0.203	0.128	62.602	32.921	11.090	7.075
Güroymak	0.549	0.296	0.118	0085	32.817	17.744	7.485	5.405
Hizan	0.522	0.281	0.110	0.078	28.693	14.955	6.239	4.588
Merkez	0.490	0.260	0.106	0.078	28.193	15.063	6.500	4.841
Mutki	0.522	0.280	0.109	0.078	30.507	16.212	6.770	4.964
Tatvan	0.502	0.265	0.109	0.079	29.017	15.512	6.731	5.042

Turkish Earthquake Hazard Map Interactive Web Application (TEHMIWA) has become available for the computation of earthquake parameters used in structural analyses for any geographic location since the beginning of 2019 [86,88]. The seismic hazard maps obtained for Bitlis and its districts for the earthquake ground motion level (DD-1) that is a 2% probability of exceedance (repetition period 2475 years) in 50 years is given in

Figure 5A, for 10% is given in Figure 5B, for 50% is given in Figure 5C and for 68% is given in Figure 5D by using TEHMIWA.

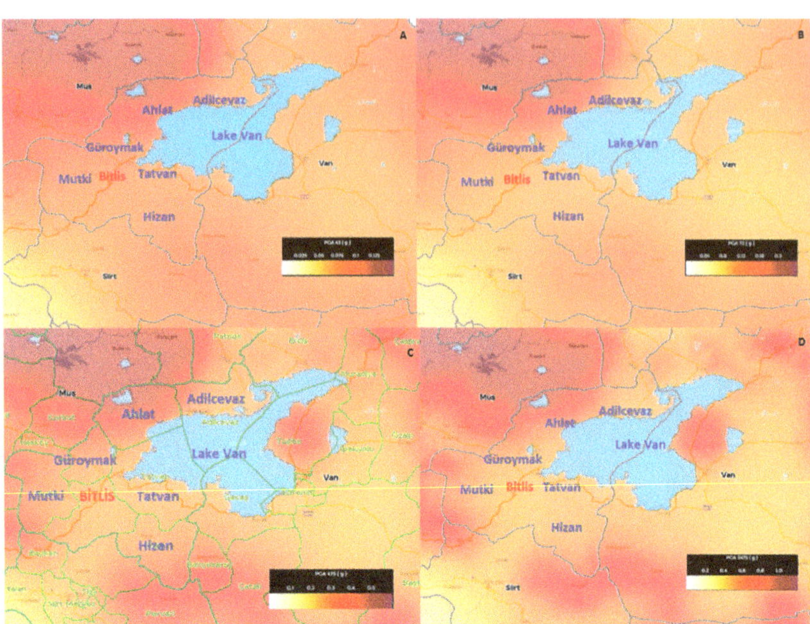

Figure 5. Seismic hazard maps for different probabilities of exceedance; (**A**) 2%, (**B**) 10%, (**C**) 50%, (**D**) 68%.

In order to make comparisons for design spectra, the ZB class was chosen as the local ground condition from the local soil class given in TBEC-2018. The features of this soil class are given in Table 6.

Table 6. Local soil class type ZB [40].

Local Soil Class	Soil Type	Upper Average at 30 m		
		$(V_S)_{30}$ [m/s]	(N60)30 [Pulse/30 cm]	(cu)30 [kPa]
ZB	Slightly weathered, medium tough rocks	760–1500	—	—

Figures 6 and 7 compare horizontal and vertical elastic design spectra obtained when the DD-2 ground motion level of Bitlis districts and local soil profile belongs to the ZB soil type. The vertical elastic design spectrum first started to be used with TBEC-2018.

According to TSDC-2007 and TBEC-2018, the spectral acceleration coefficients and dominant ground periods of the design earthquake (DD-2) with a 10% probability of exceedance in 50 years are shown in Table 7. DD-2 ground motion level was chosen because it is included in the last two seismic design codes.

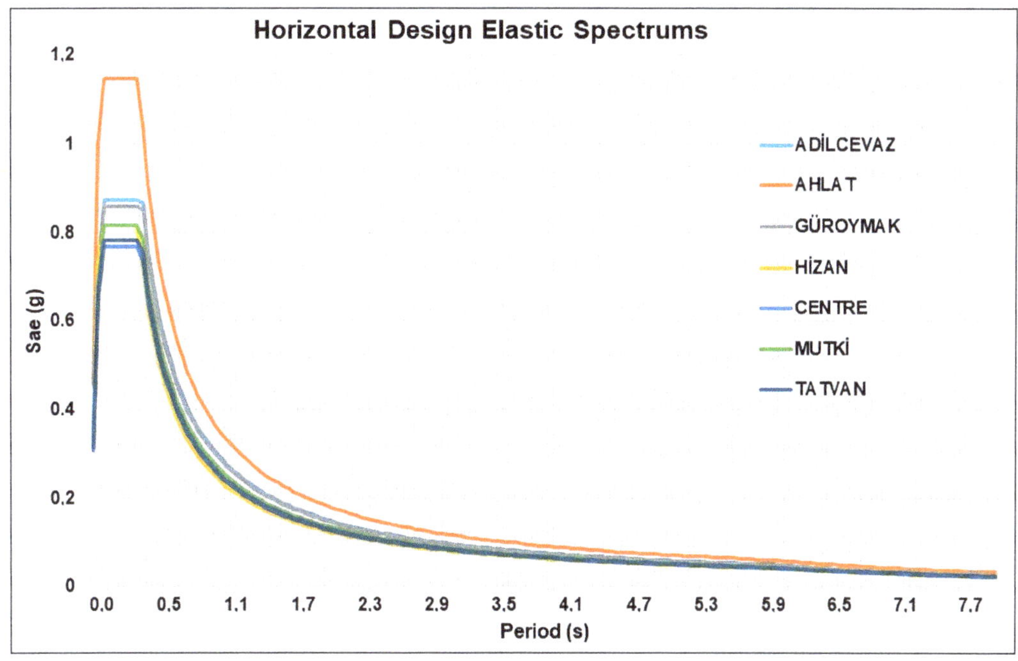

Figure 6. Comparison of horizontal elastic design spectra (DD-2/ZB).

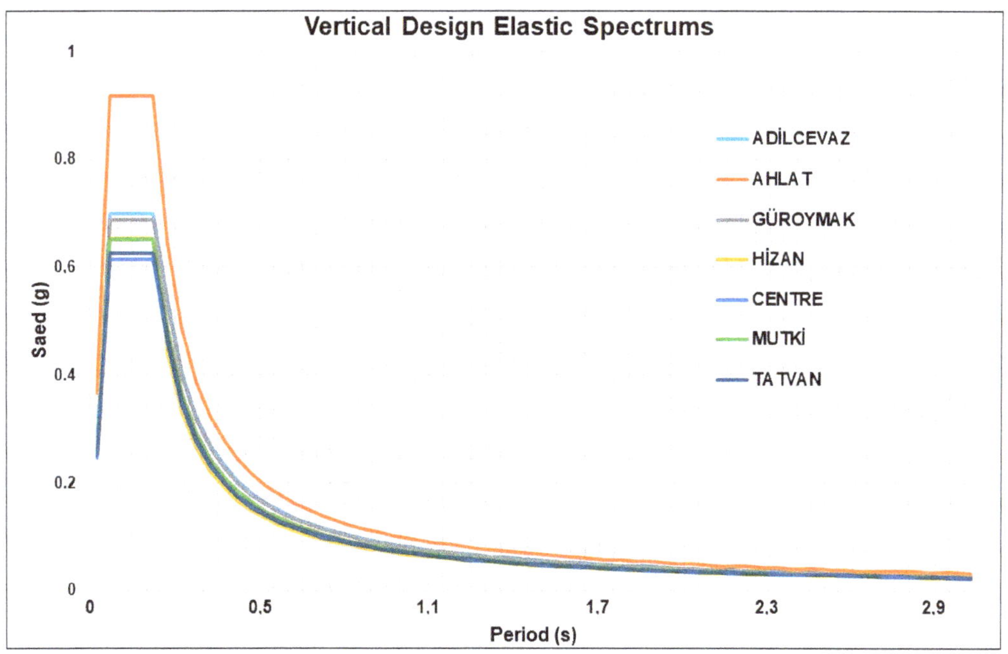

Figure 7. Comparison of vertical elastic design spectra (DD-2/ZB).

Table 7. The comparison of spectral acceleration coefficients with ground type ZB.

DD-2 District	Spectral Acceleration Coefficients All Ground Types TSDC-2007		ZB TBEC-2018		Horizontal ZB TSDC-2007		ZB TBEC-2018		Vertical ZB TSDC-2007	ZB TBEC-2018	
	S_{DS}	$0.40S_{Ds}$	S_{DS}	$0.40S_{Ds}$	T_A	T_B	T_A	T_B	T_{AD} T_{BD}	T_{AD}	T_{BD}
Adilcevaz	1	0.4	0.638	0.255	0.15	0.40	0.050	0.249	There is no vertical spectrum curve in this code	0.017	0.083
Ahlat	1	0.4	1.227	0.491	0.15	0.40	0.043	0.214		0.014	0.071
Güroymak	0.75	0.3	0.634	0.254	0.15	0.40	0.050	0.250		0.017	0.083
Hizan	1	0.4	0.597	0.239	0.15	0.40	0.045	0.223		0.015	0.074
Merkez	1	0.4	0.552	0.221	0.15	0.40	0.050	0.249		0.017	0.083
Mutki	1	0.4	0.593	0.237	0.15	0.40	0.049	0.247		0.016	0.082
Tatvan	0.75	0.3	0.559	0.224	0.15	0.40	0.050	0.249		0.017	0.083

4. Seismic Hazard Analyses

The threat posed by earthquakes on human activities in many parts of the world is a sufficient reason for carefully considering earthquakes in design structures and facilities. Seismic hazard analysis is the first step in earthquake risk assessment. Seismic hazard analysis involves quantitatively estimating ground-shaking hazards in a particular area. The main purpose of the seismic hazard analysis is to measure the parameters related to seismic ground motion (acceleration, velocity, displacement) for calculating the seismic loading conditions that the ground and engineering structures will be exposed to in the future [92,93]. In this study, Probabilistic Seismic Hazard Analysis (PSHA) for Bitlis city center was made using EZ-FRISK v7.43 software developed by Robin McGuire. In the calculations, two data sets were used to select earthquake sources. First, a study was conducted by considering regional data in the EZ-FRISK program database valid for Greece, Turkey, Lebanon, Syria and Israel. In this context, the areal earthquake sources of GT Area 26 (North Anatolian Fault), GT Area 31 (Bitlis Thrust Belt-East), GT Area 32 (Bitlis Thrust Belt-West) and GT Area 33 (Van-North) were taken into account in the analysis (Figure 8). Secondly, Bitlis's fault groups and surroundings were defined as areal sources. Since there are many fault and fault groups in the region and the fault parameters cannot be defined, the necessity of defining areal sources as earthquake sources has emerged. The study defined fault groups as Kavakbaşı, Bitlis Thrust-North, Bitlis Thrust, Van East, Suphan, Ahlat, and Malazgirt zones (Figure 9).

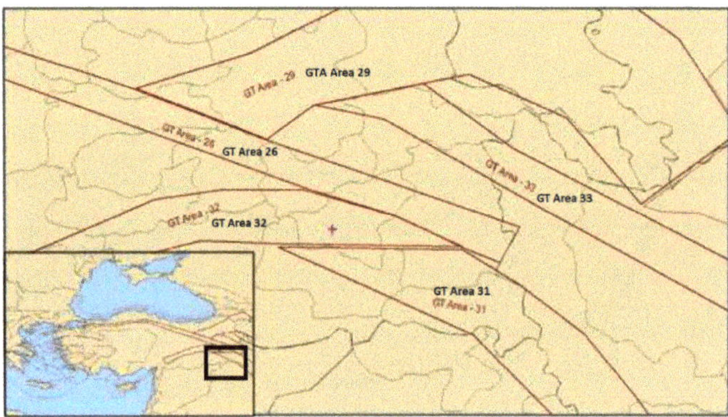

Figure 8. Areal earthquake sources in the database of EZ-FRISK software for Bitlis.

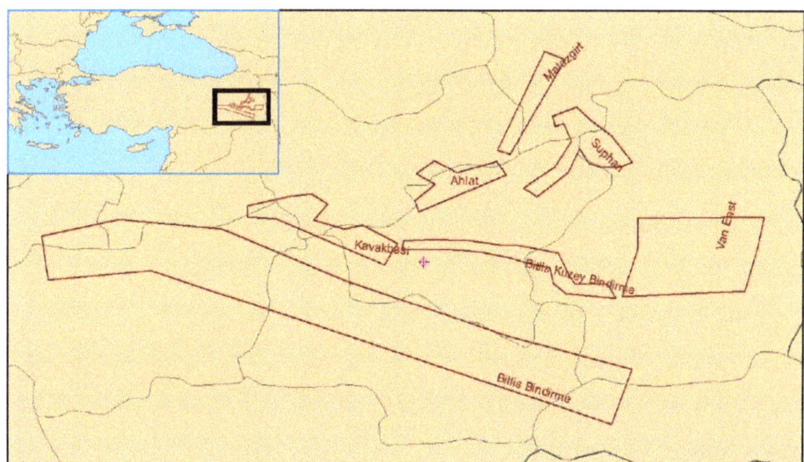

Figure 9. Definition of fault groups as areal earthquake source in Bitlis and its vicinity.

These areal earthquake sources are used to measure the change in spectral accelerations at periods for the earthquakes with a 50%, 10% and 2% probability of exceedance in 50 years. The strong ground motion acceleration records for the Eastern Anatolia region are very limited. The Abrahamson-Silva (1997) [94] and Campbell (2003) NGA Ground-Motion Relations [95], which is valid for shallow earthquakes around the world, the Graizer-Kalkan (2009) [96], which is valid in active tectonic zones around the world and Idriss (2008) [97] attenuation relations, which is developed for strike-slip shallow earthquakes, were used in the study. The software program obtained peak ground acceleration values as a function of the return periods. Uniform probability response spectra were obtained for the selected return periods. Figure 10 shows the response spectra for the Ahlat district with the highest risk of earthquake hazard, which has a 50% probability of exceedance in 50 years and a 72-year return period. A comparison of response spectra with a probability of exceedance in 50 years and a recurrence period of 475 years for Ahlat is shown in Figure 11.

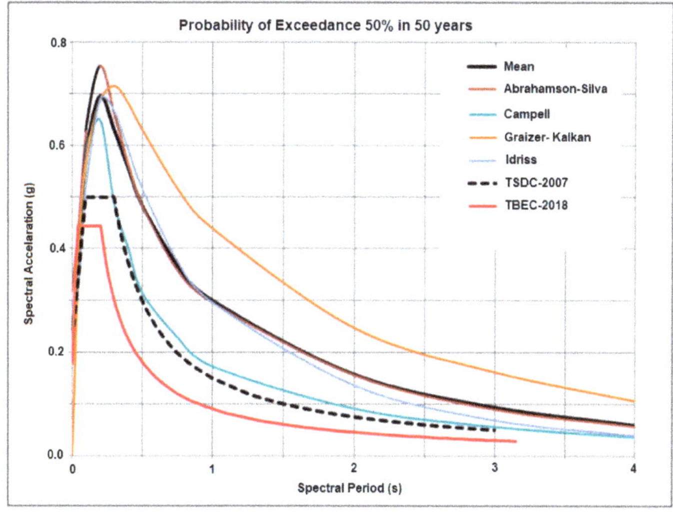

Figure 10. Comparison of response spectra with a return period of 72 years for Ahlat district.

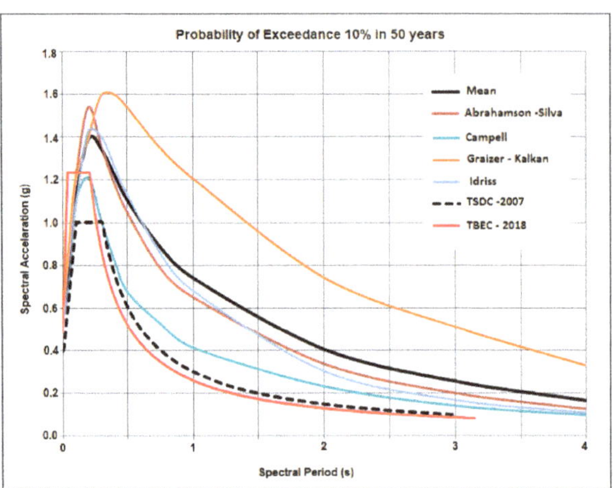

Figure 11. Comparison of response spectra with a return period of 475 years for Ahlat district.

A comparison of response spectra with a probability of exceedance in 50 years and a recurrence period of 2475 years for Ahlat is shown in Figure 12.

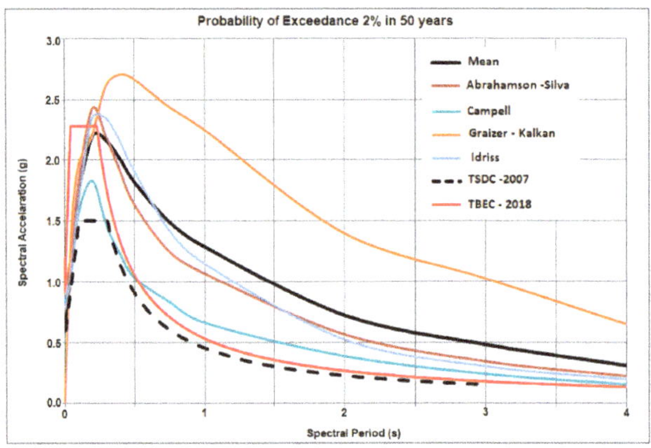

Figure 12. Comparison of response spectra with a return period of 2475 years for Ahlat district.

5. Structural Analysis

Structural analyses were carried out using academic licensed Seismostruct software.

Static pushover analysis has been widely used to determine the seismic behavior of structures. Various pushover analysis methods have been developed, including modal pushover, adaptive pushover, and cyclic pushover, where some of the weaknesses of the traditional pushover method are eliminated [98]. In this study, the static adaptive pushover analysis method was used. In this method, the effect of the frequency content and deformation of the ground motion on the structure's dynamic behavior is considered to determine the structure's capacity under horizontal loads. Furthermore, in this method, analysis was carried out taking into account the mode shapes and participation factors obtained from the eigenvalue analyses performed at each step. As a result, load distributions and strain profiles can be obtained for the structure with the help of the method. In conventional pushover analysis, the input functionality and load control types are similar to static adaptive pushover analysis [99–105]. This procedure can be expressed under four main headings: (i) definition of nominal load vector and inertia mass, (ii) computation of load factor, (iii) calculation of normalized scaling vector, and (iv) update of loading displacement vector [106]. The flow chart of the adaptive pushover analyses is given in Figure 13.

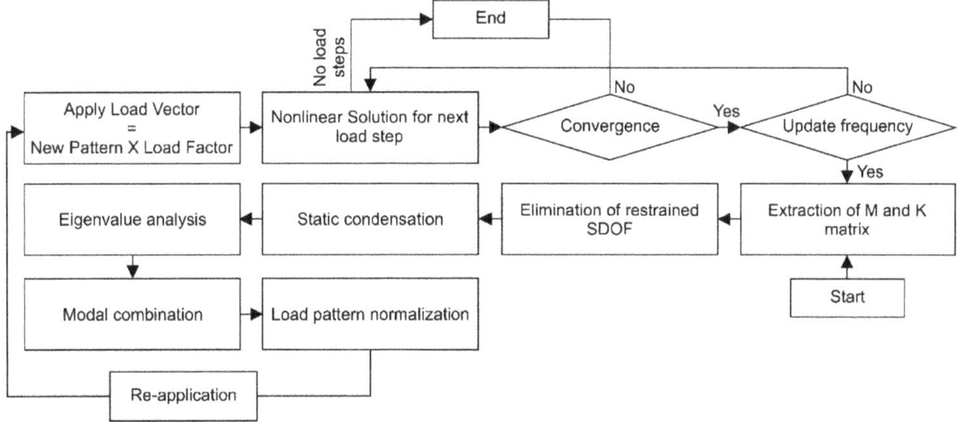

Figure 13. Flow chart of the adaptive pushover method (modified from [107]).

A seven-story RC building with the same structural characteristics was chosen as an example to reveal the structural analysis result differences for the settlements on the same fault zone. The analyses were performed in only one direction, since the RC building was chosen symmetrically in both directions. Five equal spans of 5 m length are considered in both the X and Y directions. A seven-story RC building with a total length of 2500 cm in both X and Y directions was chosen as the structural model. The blueprint of the selected RC building is given in Figure 14.

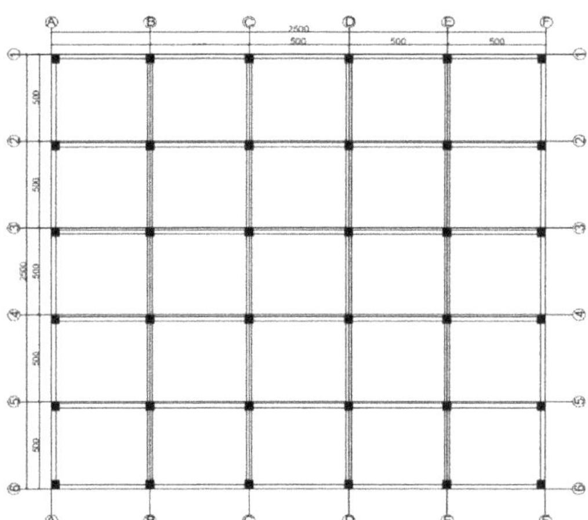

Figure 14. The blueprint of sample RC building.

Permanent and incremental loads were applied to the structure of the software. Incremental load values were selected as displacement. Permanent load values of 5.00 kN were taken into consideration. The target displacement was selected as 0.42 m. All these values were the same in all structural models used in this study. The three-dimensional model obtained in the software for the structure and the loads that were applied are given in Figure 14. The Menegotto-Pinto steel model (stl_mp) for all reinforcement in structural elements and Mander et al. nonlinear concrete model (con_ma) for concrete were selected as material model.

Each story had an equal height and was taken as 3 m. The material class used for all load-bearing elements of the structure was selected as C25-S420. All columns were selected as 0.40 × 0.50 m and beams were selected as 0.25 × 0.60 m. The transverse reinforcements used in both elements were selected as φ10/10. The reinforcements used in the columns were selected as 4φ20 at the corners and 4φ16 on the top, bottom and left–right sides. The reinforcements used in the beams were selected as 4φ16 on the lower side, 5φ14 on the upper side and 2φ12 on the side. The damping ratio was taken as % 5 in all structural models. The ZB class was chosen as the ground class. The importance of structure was taken into consideration in Class II. The slabs were selected as rigid diaphragms. The 2D and 3D structural models are shown in Figure 15.

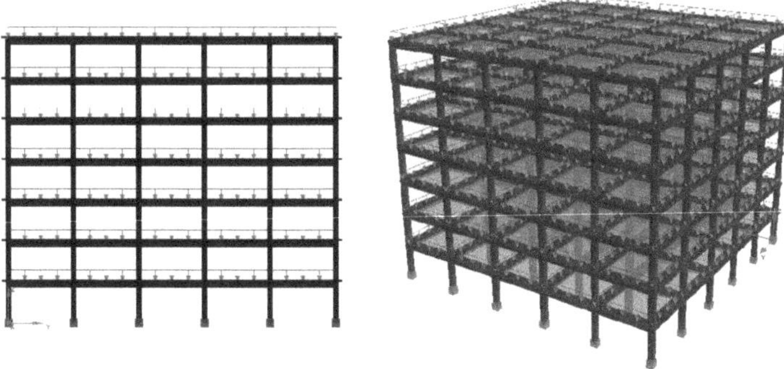

Figure 15. 2D and 3D models of the sample RC building.

Structural characteristics taken into account while creating the sample reinforced concrete building model are shown in Table 8.

Table 8. The structural characteristics of sample RC building.

Parameter			Value
Concrete grade			C25
Reinforcement grade			S420
Story height			3.0 m
Beams			250 × 600 mm
Height of floor			120 mm
Cover thickness			25 mm
Columns			400 × 500 mm
	Corners		4Φ20
Longitudinal Reinforcement	Top bottom side		4Φ16
	Left right side		4Φ16
Transverse reinforcement			Φ10/100
Steel material Model			Menegotto-Pinto steel model (stl_mp)
Concrete material model			Mander et al. nonlinear concrete model (con_ma)
Constraint type			Rigid diaphragm
Permanent Load			5 kN/m
Target Displacement			0.42 m
Ground Type			ZB
Importance Class			II
Damping ratio			5%

The force-based plastic-hinged frame members (infrmFBPH) are selected for structural elements in the sample RC building model. These elements model the spread inelasticity based on force and only limit the plasticity to a finite length. In total, 100 fiber elements are defined for the selected sections. This value is sufficient for such sections. Plastic-hinge length (Lp/L) was chosen as 16.67%.

The sample RC building was analyzed using the three different horizontal design spectrum curves, such as mean, TSDC-2007 and TBEC-2018, obtained for Ahlat. As a result of the analysis, the base shear forces were calculated for each spectrum. The displacement values were obtained for three different points on the idealized curve. The first value refers to displacement at the moment of yield, the second value refers to the intermediate (d_{int}) displacement and the third value refers to the target displacement. Elastic stiffness (K_elas) and effective stiffness (K_eff) values were calculated separately for all models. Three different performance criteria were obtained for damage estimation. These are considered as near collapse (NC), significant damage (SD) and damage limitation (DL). All these values are calculated separately for different design spectra. The comparison of all values obtained in the X direction as a result of structural analyses is shown in Table 9.

Table 9. Comparison of the result values obtained for X direction.

Spectrum Type	Base Shear (kN)	Displacement (m)	K_elas (kN/m)	K-eff (kN/m)	DL (m)	SD (m)	NC (m)
Mean	8981.73	0.1019 0.2091 0.4181	202,200.17	88,118.27	0.2162	0.2773	0.4807
TSDC-2007	8973.51	0.1039 0.2098 0.4191	202,200.17	88,065.00	0.1018	0.1304	0.2263
TBEC-2018	8980.65	0.1021 0.2093 0.4182	202,200.17	87,944.03	0.1468	0.1884	0.3226

The stiffness value of any structural reinforced-concrete element differs from the estimated stiffness value under the impact of earthquake. Therefore, the concept of effective cross-sectional stiffness has emerged in the analysis and design of RC structural members. The stiffness of the cracked sections of RC structural systems is taken into account to determine its performance under earthquake loads. The effective stiffness of the cracked sections is obtained using the predicted stiffness reduction coefficients of the elastic stiffness value [108–110]. In this study, elastic stiffness (K_elas) and effective stiffness (K_eff) values were directly obtained by using the stiffness reduction coefficients estimated in the software used.

6. Conclusions

During the update of earthquake hazard maps in 2018, the province of Bitlis and its districts located in the Eastern Anatolia Region were considered a region with a high earthquake risk in Turkey. PGA and PGV values were obtained for different probabilities of exceedance. According to the values obtained within the scope of this study, PGA values in 50 years for the province were found as follows: 0.49–1.04 g for 2% probability of exceedance; 0.26–0.57 g for 10% probability of exceedance; 0.010–0.20 g for 50% probability of exceedance; and 0.08–0.13 g for 68% probability of exceedance, respectively. The study obtained horizontal and vertical elastic design spectra for each district by choosing the same local soil class. The order of magnitude of PGA values has also remained valid for design spectrum. Computation of design spectrum on a point basis indicates that the earthquake behaviour of structures can be calculated more realistically.

This study is important regarding the joint implementation of the Turkey Building Earthquake Code that entered into force in 2019 and the Turkey Earthquake Hazard Maps presented with this code. Changes were observed in the result values obtained for all neighborhoods and districts in Bitlis province. It was concluded that the reason for these differences is due to factors, such as site-specific seismicity characteristics, fault groups and their characteristics, the distance of the selected geographical locations to the fault/fault groups and earthquake history of the region. The results indicate that obtaining design spectra by considering the site-specific earthquake hazard stipulated in the new earthquake code is remarkable. Furthermore, earthquake data will give applicable and practical results thanks to the transition from macro-zoning to micro-zoning.

Since the structural properties were kept constant in the sample RC building model considered in all settlements, the base shear force, elastic and effective stiffness values and period values were approximately equal. However, the differentiation in the design spectrum significantly changed the target displacements predicted for the expected performance levels of the structure. This reveals once again that the design spectrum significantly affects the target displacements expected from the buildings and, thus, the building performance level under impact of earthquake. While the greatest displacement values were obtained for the design spectrum obtained by considering the average of the attenuation relations, the lowest values were obtained using the design spectrum stipulated in the previous code. By comparing the stipulated values in the last two codes, it is concluded that the requests for displacement requests in the last regulation were greater.

Author Contributions: Conceptualization, E.I. and E.H.; methodology, E.I. and E.H.; software, E.I.; validation, E.I. and E.H.; formal analysis, E.H.; investigation E.I.; resources, E.I.; data curation, E.I. and E.H., writing—original draft preparation, E.I. and E.H.; writing—review and editing, E.H. and E.I.; visualization, E.H.; supervision, E.H. and E.I.; project administration, E.I.; funding acquisition, E.H. All authors have read and agreed to the published version of the manuscript.

Funding: This research received no external funding.

Institutional Review Board Statement: Not applicable.

Informed Consent Statement: Not applicable.

Data Availability Statement: Most data are included in the manuscript.

Acknowledgments: We acknowledge the support of the German Research Foundation (DFG) and the Bau-haus-Universität Weimar within the Open-Access Publishing Programme.

References

1. Okuyama, Y.; Chang, S.E. *Modeling Spatial and Economic Impacts of Disasters*; Springer Science & Business Media: Berlin, Germany, 2004.
2. Coburn, A.; Spence, R. *Earthquake Protection*, 2nd ed.; John Wiley & Sons: Chichester, UK, 2002.
3. Isik, E.; Isik, M.F.; Bulbul, M.A. Web based evaluation of earthquake damages for reinforced concrete buildings. *Earthq. Struct.* **2017**, *13*, 387–396.
4. Strukar, K.; Sipos, T.K.; Jelec, M.; Hadzima-Nyarko, M. Efficient damage assessment for selected earthquake records based on spectral matching. *Earthq. Struct.* **2019**, *17*, 271–282.
5. Pavić, G.; Hadzima-Nyarko, M.; Bulajić, B. A contribution to a UHS-based seismic risk assessment in Croatia—A Case Study for the City of Osijek. *Sustainability* **2020**, *12*, 1796. [CrossRef]
6. Büyüksaraç, A.; Işık, E.; Bektaş, Ö. A comparative evaluation of earthquake code change on seismic parameter and structural analysis; a case of Turkey. *Arab. J. Sci. Eng.* **2022**, 1–21. [CrossRef]
7. Avcil, F.; Işık, E.; Bilgin, H.; Özmen, H.B. Tbdy-2018'de verilen tasarım spektrumlarının anıtsal yığma yapı sismik davranışına etkisi. *Adıyaman Üniversitesi Mühendislik Bilim. Derg.* **2022**, *9*, 165–177.
8. Shabani, A.; Alinejad, A.; Teymouri, M.; Costa, A.N.; Shabani, M.; Kioumarsi, M. Seismic vulnerability assessment and strengthening of heritage timber buildings: A review. *Buildings* **2021**, *11*, 661. [CrossRef]
9. Xu, M.; Zhang, P.; Cui, C.; Zhao, J. An ontology-based holistic and probabilistic framework for seismic risk assessment of buildings. *Buildings* **2022**, *12*, 1391. [CrossRef]
10. Esteva, L. Criterios para la construcción de espectros para diseño sísmico. In Proceedings of the XII Jornadas Sudamericanas de Ingeniería Estructural y III Simposio Panamericano de Estructuras, Caracas, Venezuela, 2–8 July 1967; Boletín del Instituto de Materiales y Modelos Estructurales, Universidad Central de Venezuela: Caracas, Venezuela, 1967.
11. Esteva, L. Bases Para la Formulacion de Decisiones de Diseño Sísmico. Ph.D. Thesis, Universidad Autonoma Nacional de México, Mexico City, Mexico, 1968.
12. Cornell, C.A. Engineering seismic risk analysis. *Bull. Seismol. Soc. Am.* **1968**, *58*, 1583–1606. [CrossRef]

13. Lang, D. Earthquake Damage and Loss Assessment—Predicting the Unpredictable. Ph.D. Thesis, University of Bergen, Bergen, Norway, 2012.
14. Algermissen, S.T.; Rinehart, W.A.; Dewey, J.; Steinbrugge, K.V.; Degenkolb, H.J.; Cluff, L.S.; McClure, F.E.; Gordon, R.F.; Scott, S.; Lagorio, H.J. *A Study of Earthquake Losses in the San Francisco Bay Area: Data and Analysis*; Office of Emergency Preparedness and National Oceanic and Atmospheric Administration: Washington, DC, USA, 1972.
15. National Institute of Building Sciences. *Assessment of State of-the-Art Earthquake Loss Estimation Methodologies, FEMA249*; Federal Emergency, Management Agency: Washington, DC, USA, 1994.
16. Freeman, J.R. *Earthquake Damage and Earthquake Insurance: Studies of a Rational Basis for Earthquake Insurance, also Studies of Engineering Data for Earthquake Resisting Construction*; McGraw-Hill Book Co.: New York, NY, USA, 1932.
17. Kircher, C.A.; Reitherman, R.K.; Whitman, R.V.; Arnold, C. Estimation of earthquake losses to buildings. *Earthq. Spectra* **1997**, *13*, 703–720. [CrossRef]
18. Peñarubia, H.C.; Johnson, K.L.; Styron, R.H.; Bacolcol, T.C.; Sevilla, W.I.G.; Perez, J.S.; Bonita, J.D.; Narag, I.C.; Solidum, R.U., Jr.; Pagani, M.M.; et al. Probabilistic seismic hazard analysis model for the Philippines. *Earthq. Spectra* **2020**, *36*, 44–68. [CrossRef]
19. Rahman, M.Z.; Siddiqua, S.; Kamal, A.M. *Seismic* source modeling and probabilistic seismic hazard analysis for Bangladesh. *Nat. Hazards* **2020**, *103*, 2489–2532. [CrossRef]
20. Mahsuli, M.; Rahimi, H.; Bakhshi, A. Probabilistic seismic hazard analysis of Iran using reliability methods. *Bull. Earthq. Eng.* **2019**, *17*, 1117–1143. [CrossRef]
21. Alam, J.; Kim, D.; Choi, B. Seismic probabilistic risk assessment of weir structures considering the earthquake hazard in the Korean Peninsula. *Earthq. Struct.* **2017**, *13*, 421–427.
22. Khan, S.; Waseem, M.; Khan, M.A.; Ahmed, W. Updated earthquake catalogue for seismic hazard analysis in Pakistan. *J. Seismol.* **2018**, *22*, 841–861. [CrossRef]
23. Abid, M.; Isleem, H.F.; Shahzada, K.; Khan, A.U.; Kamal Shah, M.; Saeed, S.; Aslam, F. Seismic hazard assessment of Shigo Kas Hydro-Power Project (Khyber Pakhtunkhwa, Pakistan). *Buildings* **2021**, *11*, 349. [CrossRef]
24. Šipoš, T.K.; Hadzima-Nyarko, M. Seismic risk of Croatian cities based on building's vulnerability. *Tehnički Vjesnik* **2018**, *25*, 1088–1094.
25. Hadzima-Nyarko, M.; Kalman Sipos, T. Insights from existing earthquake loss assessment research in Croatia. *Earthq. Struct.* **2017**, *13*, 365–375.
26. Almeida, A.A.D.; Assumpção, M.; Bommer, J.J.; Drouet, S.; Riccomini, C.; Prates, C.L. Probabilistic seismic hazard analysis for a nuclear power plant site in southeast Brazil. *J. Seismol.* **2019**, *23*, 1–23. [CrossRef]
27. Ebrahimian, H.; Jalayer, F.; Forte, G.; Convertito, V.; Licata, V.; d'Onofrio, A.; Santo, A.; Silvestri, F.; Manfredi, G. Site-specific probabilistic seismic hazard analysis for the western area of Naples, Italy. *Bull. Earthq. Eng.* **2019**, *17*, 4743–4796. [CrossRef]
28. Gregori, S.D.; Christiansen, R. Seismic hazard analysis for central-western Argentina. *Geod. Geodyn.* **2018**, *9*, 25–33. [CrossRef]
29. Ademović, N.; Hadzima-Nyarko, M.; Zagora, N. Seismic vulnerability assessment of masonry buildings in Banja Luka and Sarajevo (Bosnia and Herzegovina) using the macroseismic model. *Bull. Earthq. Eng.* **2020**, *18*, 3897–3933. [CrossRef]
30. Looi, D.T.; Tsang, H.H.; Hee, M.C.; Lam, N.T. Seismic hazard and response spectrum modelling for Malaysia and Singapore. *Earthq. Struct* **2018**, *15*, 67–79.
31. Akkar, S.; Kale, Ö.; Yakut, A.; Ceken, U. Ground-motion characterization for the probabilistic seismic hazard assessment in Turkey. *Bull. Earthq. Eng.* **2018**, *16*, 3439–3463. [CrossRef]
32. Nas, M.; Lyubushin, A.; Softa, M.; Bayrak, Y. Comparative PGA-driven probabilistic seismic hazard assessment (PSHA) of Turkey with a Bayesian perspective. *J. Seismol.* **2020**, *24*, 1109–1129. [CrossRef]
33. Balun, B.; Nemutlu, O.F.; Benli, A.; Sari, A. Estimation of probabilistic hazard for Bingol province, Turkey. *Earthq. Struct* **2020**, *18*, 223–231.
34. Kutanis, M.; Ulutaş, H.; Işik, E. PSHA of Van province for performance assessment using spectrally matched strong ground motion records. *J. Earth Syst. Sci.* **2018**, *127*, 99. [CrossRef]
35. Büyüksaraç, A.; Işık, E.; Harirchian, E. A case study for determination of seismic risk priorities in Van (Eastern Turkey). *Earthq. Struct.* **2021**, *20*, 445–455.
36. Selcuk, L.; Selcuk, A.S.; Beyaz, T. Probabilistic seismic hazard assessment for Lake Van basin, Turkey. *Nat. Hazards* **2010**, *54*, 949–965. [CrossRef]
37. Kramer, S.L. Seismic hazard analysis. In *Geotechnical Earthquake Engineering*; Springer Science & Business Media: Berlin, Germany, 1996; pp. 106–142.
38. Ozmen, B.; Can, H. Deterministic seismic hazard assessment for Ankara, Turkey. *J. Fac. Eng. Arch. Gazi Uni.* **2016**, *31*, 9–18.
39. Işık, E. Comparative investigation of seismic and structural parameters of earthquakes (M $\geq$ 6) after 1900 in Turkey. *Arab. J. Geosci.* **2022**, *15*, 971. [CrossRef]
40. *TBEC-2018, Turkish Building Earthquake Code, T.C. Resmi Gazete*; Disaster and Emergency Management Presidency of Turkey: Ankara, Turkey, 2018.
41. AFAD. 2021. Available online: https://tdth.afad.gov.tr (accessed on 15 November 2021).
42. Çeken, U.; Dalyan, İ.; Kılıç, N.; Köksal, T.S.; Tekin, B.M. Türkiye Deprem Tehlike Haritaları İnteraktif Web Uygulaması. 4. In Proceedings of the International Earthquake Engineering and Seismology Conference, Bucharest, Romania, 14–17 June 2017.

43. Borcherdt, R.D. A theoretical model for site coefficients in building code provisions. In Proceedings of the 13th World Conference on Earthquake Engineering, Vancouver, BC, Canada, 1–6 August 2004; pp. 1–6.
44. Işik, E.; Büyüksaraç, A.; Aydin, M.C. Effects of local soil conditions on earthquake damages. In *Journal of Current Construction Issues. Civil Engineering Present Problems, Innovative Solutions—Sustainable Development in Construction*; Górecki, J., Ed.; BGJ Consulting: Bydgoszcz, Poland, 2016; pp. 191–198.
45. Ketin, İ. Van Gölü ile İran sınırı arasındaki bölgede yapılan jeoloji gözlemlerinin sonuçları hakkında kısa bir açıklama. *Türkiye Jeol. Kurumu Bülteni* **1977**, *20*, 79–85.
46. Ternek, Z. Van Gölü Güney Doğu Bölgesinin jeolojisi. *Türkiye Jeol. Bülteni* **1953**, *4*, 1–32.
47. Goncuoglu, M.C.; Turhan, N. Geology of the Bitlis metamorphic belt. In *Geology of the Taurus Belt. International Symposium*; MTA: Ankara, Turkey, 1984; pp. 237–244.
48. Helvaci, C.; Griffin, W.L. *Rb-Sr Geochronology of the Bitlis Massif, Avnik (Bingöl) Area, SE Turkey*; Special Publications, Geological Society: London, UK, 1984; Volume 17, pp. 403–413.
49. Yılmaz, Y.; Güner, Y.; Şaroğlu, F. Geology of the Quaternary volcanic centres of the East Anatolia. *J. Volcanol. Geotherm. Res.* **1998**, *85*, 173–210. [CrossRef]
50. Ustaömer, P.A.; Ustaömer, T.; Collins, A.S.; Robertson, A.H. Cadomian (Ediacaran–Cambrian) arc magmatism in the Bitlis Massif, SE Turkey: Magmatism along the developing northern margin of Gondwana. *Tectonophysics* **2009**, *473*, 99–112. [CrossRef]
51. Işık, E. Seismic Performance Analysis of Bitlis City. Ph.D. Thesis, Institute of Natural Science, Sakarya University, Sakarya, Turkey, 2010.
52. Yılmaz, Y.; Dîlek, Y.; Işik, H. Gevaş (Van) ofiyolitinin jeolojisi ve sinkinematik bir makaslama zonu. *Türkiye Jeol. Kurumu Bülteni* **1981**, *24*, 37–45.
53. Şengör, A.M.C.; Kidd, W.S.F. Post-collisional tectonics of the Turkish-Iranian plateau and a comparison with Tibet. *Tectonophysics* **1979**, *55*, 361–376. [CrossRef]
54. Şengör, A.M.C.; Yilmaz, Y. Tethyan evolution of Turkey: A plate tectonic approach. *Tectonophysics* **1981**, *75*, 181–241. [CrossRef]
55. Dewey, J.F.; Hempton, M.R.; Kidd, W.S.F.; Saroglu, F.A.; Şengör, A.M.C. *Shortening of Continental Lithosphere: The Neotectonics of Eastern Anatolia—A Young Collision Zone*; Special Publications, Geological Society: London, UK, 1986; Volume 19, pp. 1–36.
56. Utkucu, M.; Durmus, H.; Yalçin, H.; Budakoglu, E.; Isik, E. Coulomb static stress changes before and after the 23 October 2011 Van, eastern Turkey, earthquake (MW = 7.1): Implications for the earthquake hazard mitigation. *Nat. Hazards Earth Syst. Sci.* **2013**, *13*, 1889. [CrossRef]
57. Alkan, H.; Büyüksaraç, A.; Bektaş, Ö.; Işık, E. Coulomb stress change before and after 24.01. 2020 Sivrice (Elazığ) Earthquake (Mw = 6.8) on the East Anatolian Fault Zone. *Arab. J. Geosci.* **2021**, *14*, 2648. [CrossRef]
58. Koçyiğit, A.; Yilmaz, A.; Adamia, S.; Kuloshvili, S. Neotectonics of East Anatolian Plateau (Turkey) and Lesser Caucasus: Implication for transition from thrusting to strike-slip faulting. *Geodin. Acta* **2001**, *14*, 177–195. [CrossRef]
59. McKenzie, D. Active tectonics of the Mediterranean region. *Geophys. J. Int.* **1972**, *30*, 109–185. [CrossRef]
60. Burke, K.; Şengör, A.M.C. Tectonic Escape in the Evolution of the Continental Crust. In *Reflection Seismology: The Continental Crust*; Barazangi, M., Brown, L., Eds.; American Geophysical Union: Washington, DC, USA, 1986; pp. 41–53.
61. Gök, R.; Mahdi, H.; Al-Shukri, H.; Rodgers, A.J. Crustal structure of Iraq from receiver functions and surface wave dispersion: Implications for understanding the deformation history of the Arabian–Eurasian collision. *Geophys. J. Int.* **2008**, *172*, 1179–1187. [CrossRef]
62. Ketin, İ. Über die tektonisch-mechanischen Folgerungen aus den großen anatolischen Erdbeben des letzten Dezenniums. *Geol. Rundsch.* **1948**, *36*, 77–83. [CrossRef]
63. Dewey, J.F.; Şengör, A.M.C. Aegean and surrounding regions: Complex multiplate and continuum tectonics in a convergent zone. *Geol. Soc. Am. Bull.* **1979**, *90*, 84–92. [CrossRef]
64. Turkelli, N.; Sandvol, E.; Zor, E.; Gok, R.; Bekler, T.; Al-Lazki, A.; Barazangi, M. Seismogenic zones in eastern Turkey. *Geophys. Res. Lett.* **2003**, *30*. [CrossRef]
65. Horasan, G.; Boztepe-Güney, A. Observation and analysis of low-frequency crustal earthquakes in Lake Van and its vicinity, eastern Turkey. *J. Seismol.* **2007**, *11*, 1–13. [CrossRef]
66. Ateş, Y.; Yakupoğlu, T. Assessment of lacustrine/fluvial clays as liners for waste disposal (Lake Van Basin, Turkey). *Environ. Earth Sci.* **2012**, *67*, 653–663. [CrossRef]
67. Öztürk, B.; Balkıs, N.; Güven, K.C.; Aksu, A.; Görgün, M.; Ünlü, S.; Hanilci, N. Investigations on the sediment of Lake Van, II. heavy metals, sulfur, hydrogen sulfide and thiosulfuric acid S-(2-amino ethyl ester) contents. *J. Black Sea/Medit. Environ.* **2005**, *11*, 125–138.
68. Utkucu, M. 23 October 2011 Van, Eastern Anatolia, earthquake (M w 7.1) and seismotectonics of Lake Van area. *J. Seismol.* **2013**, *17*, 783–805. [CrossRef]
69. Işık, E. Bitlis ili'nin depremselliği. *Erciyes Üniversitesi Bilim. Enstitüsü Bilim. Dergisi* **2013**, *29*, 267–273.
70. Utkucu, M.; Budakoğlu, E.; Yalçin, H.; Durmuş, H.; Gülen, L.; Işık, E. Seismotectonic characteristics of the 23 October 2011 Van (Eastern Anatolia) earthquake (Mw = 7.1). *Bull. Earth Sci. Appl. Res. Cent. Hacet. Univ.* **2014**, *35*, 141–168.
71. Degens, E.T.; Wong, H.K.; Kempe, S.; Kurtman, F.J.G.R. A geological study of Lake Van, eastern Turkey. *Geol. Rundsch.* **1985**, *73*, 701–734. [CrossRef]

72. Toker, M.; Krastel, S.; Demirel-Schlueter, F.; Demirbağ, E.; Imren, C. Volcano-seismicity of Lake Van (Eastern Turkey), a comparative analysis of seismic reflection and three component velocity seismogram data and new insights into volcanic lake seismicity. In Proceedings of the International Earthquake Symposium, Kocaeli, Turkey, 22–26 October 2007; pp. 103–109.

73. Toker, M.; Sengor, A.C.; Schluter, F.D.; Demirbag, E.; Cukur, D.; Imren, C. The structural elements and tectonics of the Lake Van basin (Eastern Anatolia) from multi-channel seismic reflection profiles. *J. Afr. Earth Sci.* **2017**, *129*, 165–178. [CrossRef]

74. Okay, A.I.; Tüysüz, O. *Tethyan Sutures of Northern Turkey*; Special Publications, Geological Society: London, UK, 1999; Volume 156, pp. 475–515.

75. USGS. *Porphyry Copper Assessment of the Tethys Region of Western and Southern Asia*; Scientific Investigations Report 2010-5090-V; U.S. Geological Survey: Reston, VA, USA, 2010.

76. Ekinci, Y.L.; Yiğitbaş, E. Interpretation of gravity anomalies to delineate some structural features of Biga and Gelibolu peninsulas, and their surroundings (north-west Turkey). *Geodin. Acta* **2015**, *27*, 300–319. [CrossRef]

77. Işık, E.; Büyüksaraç, A.; Ekinci, Y.L.; Aydın, M.C.; Harirchian, E. The effect of site-specific design spectrum on earthquake-building parameters: A case study from the Marmara Region (NW Turkey). *Appl. Sci.* **2020**, *10*, 7247. [CrossRef]

78. Anonymous. Historical Earthquakes. 2021. Available online: https://deprem.afad.gov.tr (accessed on 15 May 2022).

79. Anonymous. Historical Earthquakes. 2021. Available online: http://www.koeri.boun.edu.tr (accessed on 15 May 2022).

80. Ekinci, R.; Büyüksaraç, A.; Ekinci, Y.L.; Işık, E. Bitlis ilinin doğal afet çeşitliliğinin değerlendirilmesi. *Doğal Afetler Ve Çevre Derg.* **2020**, *6*, 1–11.

81. Özmen, B. Türkiye deprem bölgeleri haritalarının tarihsel gelişimi. *Türkiye Jeol. Bülteni* **2012**, *55*, 43–55.

82. Işık, E.; Ekinci, Y.L.; Sayıl, N.; Büyüksaraç, A.; Aydın, M.C. Time-dependent model for earthquake occurrence and effects of design spectra on structural performance: A case study from the North Anatolian Fault Zone, Turkey. *Turk. J. Earth Sci.* **2021**, *30*, 215–234. [CrossRef]

83. Pampal, S.; Özmen, B. Development of earthquake zoning maps of Turkey. In Proceedings of the Sixth National Conference on Earthquake Engineering, Istanbul, Turkey, 16–20 October 2007.

84. Işık, E. A comparative study on the structural performance of an RC building based on updated seismic design codes: Case of Turkey. *Challenge* **2021**, *7*, 123–134. [CrossRef]

85. Gülkan, P.; Koçyiğit, A.; Yücemen, M.S.; Doyuran, V.; Başöz, N. *En son verilere göre hazırlanan Türkiye deprem bölgeleri haritası. Report No: METU/EERC, 93-101*; Department of Civil Engineering, Middle East Technical University: Ankara, Türkiye, 1993.

86. Aksoylu, C.; Arslan, M.H. 2007 ve 2019 deprem yönetmeliklerinde betonarme binalar için yer alan farklı deprem kuvveti hesaplama yöntemlerinin karşılaştırmalı olarak irdelenmesi. *Int. J. Eng. Res. Dev.* **2021**, *13*, 359–374.

87. Nemutlu, Ö.F.; Balun, B.; Benli, A.; Sarı, A. Bingöl ve Elazığ illeri özelinde 2007 ve 2018 Türk deprem yönetmeliklerine göre ivme spektrumlarının değişiminin incelenmesi. *Dicle Üniversitesi Mühendislik Fakültesi Mühendislik Derg.* **2020**, *11*, 1341–1356.

88. AFAD. Available online: https://www.afad.gov.tr/ (accessed on 18 May 2022).

89. Emre, Ö.; Duman, T.Y.; Özalp, S.; Şaroğlu, F.; Olgun, Ş.; Elmacı, H.; Çan, T. Active fault database of Turkey. *Bull. Earthq. Eng.* **2018**, *16*, 3229–3275. [CrossRef]

90. Kadirioğlu, F.T.; Kartal, R.F.; Kılıç, T.; Kalafat, D.; Duman, T.Y.; Azak, T.E.; Özalp, S.; Emre, Ö. An improved earthquake catalogue (M ≥ 4.0) for Turkey and its near vicinity (1900–2012). *Bull. Earthq. Eng.* **2018**, *16*, 3317–3338. [CrossRef]

91. Akkar, S.; Sandıkkaya, M.A.; Bommer, J.J. Empirical ground-motion models for point-and extended-source crustal earthquake scenarios in Europe and the Middle East. *Bull. Earthq. Eng.* **2014**, *12*, 359–387. [CrossRef]

92. Moehle, J.; Deierlein, G.G. A framework methodology for performance-based earthquake engineering. In Proceedings of the 13th World Conference on Earthquake Engineering, Vancouver, BC, Canada, 1–6 August 2004; Volume 679.

93. Abrahamson, N.A.; Bommer, J.J. Probability and uncertainty in seismic hazard analysis. *Earthq. Spectra* **2005**, *21*, 603–607. [CrossRef]

94. Abrahamson, N.A.; Silva, W.J. Empirical response spectral attenuation relations for shallow crustal earthquakes. *Seismol. Res. Lett.* **1997**, *68*, 94–127. [CrossRef]

95. Campbell, K.W.; Bozorgnia, Y. Updated near-source ground-motion (attenuation) relations for the horizontal and vertical components of peak ground acceleration and acceleration response spectra. *Bull. Seismol. Soc. Am.* **2003**, *93*, 314–331. [CrossRef]

96. Graizer, V.; Kalkan, E. Ground motion attenuation model for peak horizontal acceleration from shallow crustal earthquakes. *Earthq. Spectra* **2007**, *23*, 585–613. [CrossRef]

97. Idriss, I.M. An NGA empirical model for estimating the horizontal spectral values generated by shallow crustal earthquakes. *Earthq. Spectra* **2008**, *24*, 217–242. [CrossRef]

98. Shafigh, A.; Ahmadi, H.R.; Bayat, M. Seismic investigation of cyclic pushover method for regular reinforced concrete bridge. *Struct. Eng. Mech.* **2021**, *78*, 41–52.

99. Antoniou, S.; Pinho, R. Advantages and limitations of adaptive and non-adaptive force-based pushover procedures. *J. Earthq. Eng.* **2004**, *8*, 497–522. [CrossRef]

100. Ferracuti, B.; Pinho, R.; Savoia, M.; Francia, R. Verification of displacement-based adaptive pushover through multi-ground motion incremental dynamic analyses. *Eng. Struct.* **2009**, *31*, 1789–1799. [CrossRef]

101. Pinho, R.; Casarotti, C.; Antoniou, S. A comparison of single-run pushover analysis techniques for seismic assessment of bridges. *Earthq. Eng. Struct. Dyn.* **2007**, *36*, 1347–1362. [CrossRef]

102. Casarotti, C.; Pinho, R. An adaptive capacity spectrum method for assessment of bridges subjected to earthquake action. *Bull. Earthq. Eng.* **2007**, *5*, 377–390. [CrossRef]
103. Seismosoft. SeismoStruct 2021—A Computer Program for Static and Dynamic Nonlinear Analysis of Framed Structures. 2021. Available online: http://www.seismosoft.com (accessed on 10 June 2022).
104. Antoniou, S.; Pinho, R. Development and verification of a displacement-based adaptive pushover procedure. *J. Earthq. Eng.* **2004**, *8*, 643–661. [CrossRef]
105. Navideh, M.; Hamid, R.A.; Hamed, M. A comparative study on conventional push-over analysis method and incremental dynamic analysis (IDA) approach. *Sci. Res. Essays* **2012**, *7*, 751–773. [CrossRef]
106. Pinho, R.; Antoniou, S. A displacement-based adaptive pushover algorithm for assessment of vertically irregular frames. In Proceedings of the Fourth European Workshop on the Seismic Behaviour of Irregular and Complex, Structures, Thessaloniki, Greece, 26–27 August 2005.
107. Papanikolaou, V.K.; Elnashai, A.S. Evaluation of conventional and adaptive pushover analysis I: Methodology. *J. Earthq. Eng.* **2005**, *9*, 923–941. [CrossRef]
108. Caglar, N.; Demir, A.; Ozturk, H.; Akkaya, A. A simple formulation for effective flexural stiffness of circular reinforced concrete columns. *Eng. Appl. Artif. Intel.* **2015**, *38*, 79–87. [CrossRef]
109. Wilding, B.V.; Beyer, K. The effective stiffness of modern unreinforced masonry walls. *Earthq. Eng. Struct. Dyn.* **2018**, *47*, 1683–1705. [CrossRef]
110. Ugalde, D.; Lopez-Garcia, D.; Parra, P.F. Fragility-based analysis of the influence of effective stiffness of reinforced concrete members in shear wall buildings. *Bull. Earthq. Eng.* **2020**, *18*, 2061–2082. [CrossRef]

 *buildings*

Article

Investigation of the Earthquake Performance Adequacy of Low-Rise RC Structures Designed According to the Simplified Design Rules in TBEC-2019

Nur Seda Yel [1], Musa Hakan Arslan [1], Ceyhun Aksoylu [1,*], İbrahim Hakkı Erkan [1], Hatice Derya Arslan [2] and Ercan Işık [3]

[1] Department of Civil Engineering, Konya Technical University, Konya 42130, Turkey
[2] Department of Architecture, Necmettin Erbakan University, Konya 42090, Turkey
[3] Department of Civil Engineering, Bitlis Eren University, Bitlis 13100, Turkey
* Correspondence: caksoylu@ktun.edu.tr

Abstract: In this study, earthquake performance of the structures was tested which were modeled according to the minimum criteria of simplified analysis approach proposed in TBEC-2019. For this purpose, 144 reinforced-concrete building models were designed according to parameters such as earthquake design class, building height (number of storey), number of spans, soil type and three different simplified formulas suggested in the code. The level of structural performance of buildings models was determined by the linear (L) and nonlinear performance analysis (NL) methods that given in TBEC-2019. The base shear force, top displacements and over-strength factor (Ω) of each structural model were obtained, and performance analysis was performed by comparatively. As a result of the structural analyses, it was seen that some of the buildings model designed according to minimum column sectional criteria given in simplified methods could not meet the suggested seismic performance level. While the number of structural models that provide the controlled damage (CD) level in the L analysis method is 44 (30.55%), it is 107 (74.3%) in the NL analysis method. The insufficient performance was obtained in both L and NL methods in models which have over-strength values below 3. It has been observed that multi-criteria of building performance are not met with the weakening of local soil conditions. It was also seen that the L method chosen in the performance analysis gave more conservative results with this study.

Keywords: simplified design rules; earthquake; linear method; nonlinear method; performance analysis

Citation: Yel, N.S.; Arslan, M.H.; Aksoylu, C.; Erkan, İ.H.; Arslan, H.D.; Işık, E. Investigation of the Earthquake Performance Adequacy of Low-Rise RC Structures Designed According to the Simplified Design Rules in TBEC-2019. *Buildings* **2022**, *12*, 1722. https://doi.org/10.3390/buildings12101722

Academic Editor: Enrico Tubaldi

Received: 23 August 2022
Accepted: 14 October 2022
Published: 18 October 2022

Publisher's Note: MDPI stays neutral with regard to jurisdictional claims in published maps and institutional affiliations.

1. Introduction

Turkey is located in a region with high seismicity due to its geographical location. Many "sub-standard buildings" in Turkey have not received sufficient or, in some cases, any engineering services and have been built without professional supervision. In sub-standard buildings, there are critical problems such as faulty construction techniques or quality issues relating to use of inappropriate building materials. Earthquakes that caused significant loss of life and property were experienced in Turkey in the 20th century and the first quarter of the 21st century [1–12]. The 1939 Erzincan earthquake (M_w 7.2) and the 1999 Marmara earthquake (M_w 7.4) are the two most destructive earthquakes that occurred in the 20th century [13]. In the first quarter of the 21st century, many earthquakes caused significant losses, such as 2002 Sultandağı Afyon (M_w 6.5), 2003 Bingöl (M_w 6.4), 2011 Van (M_w 7.2), 2020 Elazığ (M_w 6.5) and 2020 İzmir (M_w 6.6) [14,15]. In particular, the 1000 km long North Anatolian Fault line and the 400 km long East Anatolian Fault line within the borders of Turkey, surround the country in east–west and southeast–northeast axes. Severe earthquakes occur at certain repetitions on these fault lines. In addition, a significant part of the existing building stock in Turkey is reinforced concrete, and does not have sufficient earthquake performance due to the reasons mentioned above, which have been mentioned frequently in the literature [16–20]. Continuous improvement studies have been carried out to reduce the destructive effects of earthquakes on structures in Turkey.

Many seismic design codes have been put into effect in Turkey since the 1940s. The codes have been updated on average every eight years since 1940, and 10 codes have entered into force in the last 80 years [21,22]. In 2007, the "Regulation on Buildings to be Constructed in Earthquake Zones (TEC-2007)" [23] came into force, so the nonlinear calculation method in existing reinforced concrete buildings and the Turkish Buildings Earthquake Code in 2019 (hereafter, TBEC-2019) [24] became the most up-to-date. Changes in codes over time and earthquake load calculation methods are shown schematically in Figure 1 and Table 1. These codes contain quite comprehensive and innovative information as of the date of their publication.

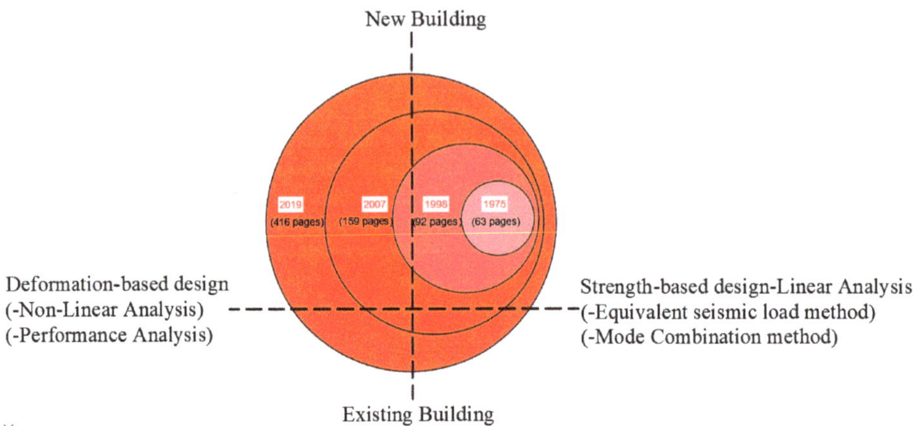

Figure 1. Earthquake codes and analysis methods.

Table 1. Earthquake Code Analysis Methods.

	TEC-1975	**TEC-1998**	**TEC-2007**	**TBEC-2019**
Force-Based Seismic Analysis of New RC Buildings	Eq.Static Anly.	Eq.Static Anly. /Modal Anly/ Time History Anly.	Eq.Static Anly./ Modal Anly/ Time History Anly.	Eq.Static Anly./ Modal Anly/ Time History Anly.
Deformation Based Seismic Analysis of New RC Buildings	—	—	—	NL Resp.Hist. Anly/ NL Static Pushover Anly.
Seismic Analysis of Existing RC Buildings	—	—	NL Static Pushover Anly.	NL Resp.Hist. Anly/ NL Static Pushover Anly.

Finally, in the current code, the TBEC-2019, radical changes have been made compared to the previous code (TEC-2007). New earthquake hazard maps have been updated in the current code. In addition, map spectral acceleration coefficients (S_S, S_1), spectral acceleration coefficient (S_{DS}), earthquake design classes (EDC, in Turkish DTS), local ground effect coefficients (F_S, F_1), and building height classes (BHL in Turkish BYS) started to be used for design and calculations. New concepts such as building usage classes (BUC in Turkish BKS), building performance levels, design according to strength (DGT), evaluation and design according to deformation (in Turkish ŞGDT), over-strength factor (Ω in Turkish D) and effective section stiffness are defined in the current code.

It is seen that studies on linear (L) and nonlinear (NL) behavior of buildings related to the TBEC-2019 are sufficient when the literature is examined. Studies are generally in the form of a comparison between the last two seismic design codes. Base shear forces, building displacements, periods and acceleration spectrum were investigated with the L comparisons made with NL analyses. Moreover, studies were carried out to determine the performance of buildings. L using the equivalent earthquake load method or mode superposition method as analysis by Aksoylu and Arslan [25], Özmen and Sayın [26], Korkmaz [27], Başaran and Hicyilmaz [28], Doğan [29], Aksoylu and Arslan [30,31], Bayrakcı and Baran [32], Başaran [33], Işık and Velioğlu [34], conducted research for frame and frame + shear walled buildings. In addition, Karaca et al. [35] compared five reinforced concrete buildings located in the city centre of Niğde and designed according to the TBEC-2019 in the context of structural design according to the last two codes. As a result of the structural analyses, it has been observed that more concrete will need to be used in a building to be designed according to the 2018 earthquake code, but there is a tendency to decrease the amount of reinforcement in general. Ünsal et al. [36] investigated the effect of building height on base shear force and top displacement according to the TEC-2007 and TBEC-2019 codes. They observed that if the base shear force values obtained according to the last two codes are close, the maximum displacement obtained based on the TBEC-2018 is much larger than the maximum displacement obtained based on the TEC-2007. Studies have also been carried out on static pushover analysis for nonlinear analysis [37] and time history analysis. Güllü [38], Çoban and Çeribaşı [39], Aydemir and Jakayev [40] carried out studies using time history analysis. Furthermore, Kasap [41], Özer and Yüksel [42], Işık [43], Çolakoğlu [44], used static pushover analysis, while Dalyan and Şahin [45], investigated building performances by using modal pushover analysis. Kap et al. (2019) [46] determined the capacities of the load-bearing elements in an existing school building, which was exposed to the 1999 Marmara and Düzce earthquakes, by performing an earthquake performance analysis according to the TBEC-2019. Aksoylu et al. [47] investigated reinforced concrete frame-type buildings of different elevations using ETABS, according to linear equivalent seismic load method for the TEC-2007, TBEC-2019 and ASCE 7–16. As a result of the study, the closest results for the three codes occurred on the softest soils. Kürkçü [48] designed a 20-storey reinforced concrete structure according to the TBEC-2019 and determined the earthquake performance with the calculation method in the time history. In the study conducted by Akçora [49], a 30-storey RC high-rise building was examined according to the TBEC-2019. As a result of analyses, it was determined that the plastic deformation and rotation values obtained for the sections met the limits given in the TBEC-2019. Capa [50] determined the earthquake performances of the three, five and seven storey buildings, which he took into account together with the linear and nonlinear calculation methods given in the TBEC-2019, and compared the obtained results with each other. Severcan and Sinani [51] determined the earthquake performance of an existing eight storey RC building using static pushover analysis according to the TEC-2007 and Eurocode-8 (2004). As a result of the evaluation, it has been seen that the performance levels of the elements are close to each other, especially in the vertical bearing elements, but the TEC-2007 is on the safer side compared to Eurocode-8. Ulutaş (2019) [52], compared the 2007 and 2018 earthquake codes in terms of sectional damage limits. As a result of the examination, it was concluded that the TBEC-2019 is safer in terms of earthquake safety than the TEC-2007 earthquake code. Sarı [53] determined the earthquake performance of an existing residential building according to the TBEC-2019 and TEC-2007 by static pushover analysis. According to both codes, the type of residence examined determined that the building met the target performance level. Ayaz [54] investigated the nonlinear behavior of a building strengthened according to the TEC-2007 and TBEC-2019. It was determined that the building met the requirements of the TBEC-2019 by conducting performance analyses according to DD-1 and DD-3 earthquake ground motion levels. Seşetyan et al. [55] carried out seismic hazard analysis for the Marmara Region according to new data. Sianka et al. [56] conducted seismic analysis for the Marmara Region, and they determined a good agreement with the updated Turkish

Earthquake Hazard Map. It is seen that these studies are generally focused on mid-rise or high-rise buildings.

If the RC building stock is divided into low-, mid- and high-rise according to the height of the structural system (Figure 2), some buildings should be made low-rise, especially due to architectural constraints and seismic zone. In low-rise buildings, having a regular load-carrier system may be preferred due to simpler design details. However, the design of low-rise buildings according to earthquake effects is easier than medium and high-rise buildings. It is also easier for these buildings to show the theoretically expected performance in a real earthquake. For this reason, there are very limited studies on low-rise RC buildings. Although easier to design, Erberik [57] stated that the recent devastating earthquakes in Turkey have shown that low-rise buildings are very sensitive to seismic effects. Particularly, the low-rise building design should be at a level that can easily meet the seismic actions (especially shear forces and displacements demanded by the earthquake). Therefore, consideration of seismic effects is as critical in the design of low-rise buildings as in the design of mid- and high-rise buildings. In addition, skyscrapers (which are taller than 70 m in a high seismic hazard zone, 91 m in a moderate seismic hazard zone and 105 m in a low seismic hazard zone), that occupy a very small place in the building stock but need to receive very important engineering services, are generally buildings with 30 storeys and above and are considered as part of the high-rise category. The design criteria for skyscrapers are completely different. Three-stage earthquake calculations are required for such buildings in the TBEC-2019. Performance analysis is conducted for DD-2 in the first stage (preliminary design stage), DD-3 and DD-4 in the second stage (evaluation of immediate occupancy), and DD-1 in the third stage (evaluation of near collapse or collapse prevention). The performance target expected from the structure at each stage is different. The first two stages are force based; the last stage is deformation-based analysis. In the last stage, 11 different ground motion records are used for time history analyses. Time history analysis is a step-by-step analysis of the dynamic response of a structure to a specified loading that may vary with time.

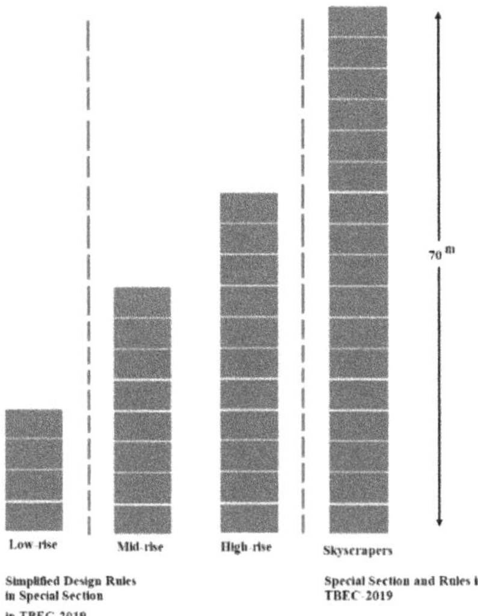

Figure 2. Schematic representation of buildings according to their heights.

One of the most remarkable innovations with the TBEC-2019 is the presence of a new section in which the analysis procedure for the design of simple low-rise structures, which will be carried out without detailed earthquake load calculations, is explained. These simplifying approaches, new to the TBEC-2019, are described by İlki et al. [58,59]. In the document prepared by the National Institute of Building Sciences for Development of Simplified Seismic Design Procedures in 2010 (NIBSDSSDP, 2010) [60] "The project's engineers are aware of the complexity of the existing regulation rules. Especially it was initiated to respond to concerns that earthquake resistant structure design for buildings reduces the effectiveness and reliability of the buildings" [59]. Additionally, in this document (NIBSDSSDP, 2010) it is recommended to continue work on the development of simplified design rules for earthquake resistance. Apart from that, ASCE 7–10 (2010) [61], and FEMA 450–2003 [62], separate simple shear walls or frame systems into simplified alternative structural design criteria [58,59]. A simplified calculation of the earthquake base shear force is given in these codes. In addition, after the base shear force is distributed to the storeys, the distribution of these shear forces to the load-carrier member is made by considering the relative lateral stiffness of the elements separately for flexible and rigid diaphragm behaviors. The document ISO/TC 71/SC5–2010 [63], "Guidelines for Simplified Seismic Assessment and Rehabilitation of Concrete Buildings" has given alternative simple approaches to find load-bearing element capacities and has made a suggestion for the simple calculation of lateral displacement ratios of buildings. The document also defines limits for lateral displacement ratios for different types of load-carrier systems. With these developments, simplified design rules for regular and straightforward cast-in-place RC buildings have been addressed in the TBEC-2019. Studies in the related literature are quite limited. To the authors' knowledge, only Balun et al. [64] has taken into account the simplifying design method in the TBEC-2019; the base shear forces of a structure for two different DTS were compared using the ETABS v18 program. Local soil classes and building height are determined as a variable within the boundaries of this section. As a result of the analysis, it was stated that the earthquake design class, local soil class and the number of storeys of the building made a difference between the simplified earthquake load calculation and the standard earthquake calculation. As a result, it is stated that only the shear forces obtained from the simplified calculation are on the safe side by obtaining higher values than the sole shear forces obtained by the standard method.

As can be seen from the comprehensive literature review above, although this issue concerns Turkish building stock, it is an area that has not been evaluated much. The question of whether the complex procedure in the earthquake codes is necessary in the design of simple structures can always be asked. Based on this problem, a simplified design process for low-rise buildings and minimum codes for calculations were proposed in the TBEC-2019, unlike other codes in the world. The main motivation for this study was this question: will the earthquake performance be sufficient for the buildings that were modeled according to the minimum criteria of the simplified design rules for low-rise RC buildings? Therefore, in this study, the authors wanted to investigate whether the earthquake performance of buildings designed with a simple method would be sufficient, especially since there is a very detailed calculation procedure in determining the earthquake performance. In particular, the relationship between the over-strength coefficient (Ω, in Turkish D) of the buildings and structural performance was examined, and how the over-strength coefficients change when low-rise buildings are designed according to minimum conditions was also investigated.

The determination of the behavior of RC structures can sometimes be difficult, because the earthquake load and behavior of RC are quite complex, especially under earthquake loading. For this reason, there are many assumptions in earthquake codes to facilitate these calculations, even though these assumptions sometimes contain important errors in mid- and high-rise buildings, having complex geometry. They can give predictable results in simple planned and relatively low-rise buildings. The TBEC-2019 has proposed a series of analysis procedures in which RC sections can be determined without making detailed earthquake calculations, for RC structures with simple geometry and limited building height. However, there is no study in the literature that indicates whether these provisions in the TBEC-2019 ensure that the structure has sufficient strength and ductility in terms of earthquake performance. Therefore, this study contains originality in terms of the subject it deals with. In this study, the simplifying rules given in the TBEC-2019 were examined through created structural models according to a set of parameter groups [65]. For this purpose, 144 RC building models with different parameters such as DTS, building height (H_N), number of spans, and soil type were used with the simplified formulas in the relevant part of the code, using ProtaStructure [66] (a software that gives the same results as SAP2000 v23.3.1 (2012) [67]. Then, the level of the structural performance of these designed buildings was determined by linear (L) and nonlinear analysis (NL). At the end of this study, the base shear force of the structures for earthquake calculation was performed, and the structural performance and over-strength were obtained by comparing with the capacity of the structures. As a result of a detailed performance analysis, the performance levels of the structural models for L and NL were determined, according to the damage levels of the structural elements. As a result, the findings were evaluated in the context of structural-earthquake engineering and suggestions were made.

2. Design Procedure of TBEC-2019/Section 17

Earthquake-resistant structure design is a very complicated and multi-unknown problem. Design engineers prefer application-oriented calculations that can be terminated with simple computations rather than a multiplicity of calculations [68]. Notably, the "Simplified Design Rules for Regular Cast-in-situ Reinforced Concrete Buildings" section in the TBEC-2019 includes simple formulas and approaches that engineers can use in building design. In the relevant section of the code, it is foreseen that simplified design rules can be used in the design of buildings with a simple and symmetrical floor plan, without irregularities in the plan, and vertically, that adversely affect the earthquake performance, and with limited building height. The simplified design rules facilitate the structural design stages for engineers, as they offer a more practical solution for RC buildings, which are designed quite simply and constitute an essential part of the building stock in Turkey.

In the TBEC-2019, the structural design is created according to DTS, S_{DS}, BOC, and BHL in the relevant section. DTSs are determined based on S_{DS} and BHL. According to this, the DTS values of the buildings are calculated for the appropriate location from the web application of the Turkey Seismic Hazard Maps [69]. A building's DTS value takes one of the numbers 1, 2, 3 and 4 from the highest to the lowest according to S_{DS} and the building importance coefficient values [70]. In addition, the building has the status of being 1a, 2a, 3a, and 4a depending on the building occupancy class: BOC = 1, 2 and 3. Here, the index "a" indicates the importance of the structure. Therefore, while classes DTS 1 and 1a represent an ordinary and significant building located in a critical earthquake zone, 4 and 4a are structures built in the area with the lowest earthquake risk. BHL takes BHL 1 to BHL 8, depending on the building's total height (H_N) and DTS. BHL 1 represents the highest and BHL 8 the lowest height structures in this classification.

The TBEC-2019/Section 17 is valid only for RC buildings with a BOC value of 3, and the structures in which the simplified calculation can be applied are dimensionally limited. In order to apply the design rules in the TBEC-2019, the limit values of the D, BOC, BHL, and building floor plan model that the building model must have, as well as the general rules, dimensioning of the building elements, and the lower limit values of the reinforcement of the elements are explained in this section. The relevant section in the TBEC-2019 requires that the buildings in which the design rules will be used have a floor plan close to rectangular. In addition, the section requires that there be no discontinuity or off-axis shift in the axes of the load-carrier system and that there should be at least two spans in each direction of the load-carrier system. In addition, specific approaches for dimensioning vertical load-bearing members in the simplified design of cast-in-situ RC buildings are given according to the DTS value. Building heights are limited for workplaces and residences as BHL $\geq$ 6 for DTS = 1 and 2 and BHL $\geq$ 7 for DTS = 3 and 4. In addition, framed buildings with high acceptable ductility levels and frame + shear wall buildings are expressed in terms of DTS. The cross-sectional areas of the columns are limited according to the choice of the load-bearing system, considering the axial compressive stresses and sufficient shear strength (Figure 2). It has been stated that the $(g + q)$ value, which is the sum of the average distributed dead load and average distributed live load values to be used in determining the cross-sectional areas of the vertical bearing members, cannot be taken less than 15 kN/m^2, and the $(g + 0.3q)$ value cannot be taken less than 13 kN/m^2. However, the distributed dead load values should also include the weight of the lateral beam, vertical load-bearing elements (column and shear wall), and non-bearing elements (infill walls) along with the slab loads. The live load acting on the roof of the building should also include the reduced snow load with a load factor of 0.3. The smallest square column cross-section should be 30 cm $\times$ 30 cm. It is stated that the ratio of the long side to the short side of the columns should not be more than 2. The minimum width should be 30 cm for beam cross-sections, and the minimum height should be 50 cm.

The TBEC-2019 proposes three different formulas for sizing columns (Figure 2). For buildings with high ductility levels whose structural system consists of frames, the A_{ci} value in the first and second formulas represents the cross-sectional area of the column taken into account, and for the column whose ΣA_{oi} value is considered, the sum of the area shares accumulated along with all floors. The value of $\Sigma(I_i/H_i^2)$ in the third formula represents the sum of the values in the direction taken into account on the ground floor (column cross-section moment of inertia / the square of the storey height), and the value of ΣA_{pi} represents the sum of the storey areas of the building. The simplified design rules of the TBEC-2019 for reinforced concrete buildings are summarized in Figure 3.

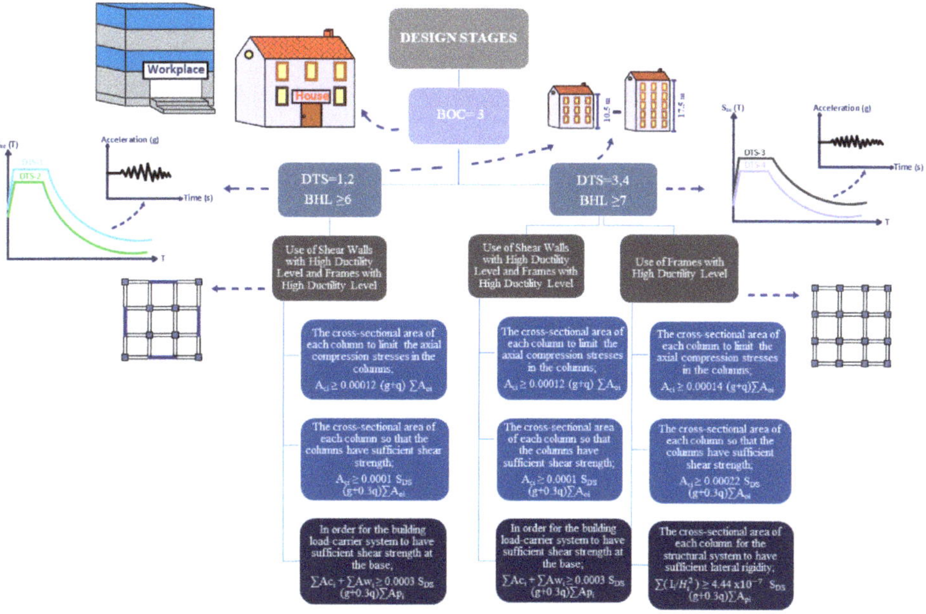

Figure 3. Simplified calculation procedure in TBEC-2019.

According to the relevant section of the TBEC-2019, it is stated that if the column is square, the longitudinal reinforcement ratio (ρ_t) should not be less than 1%, and the transverse reinforcement ratio ($\rho_{sh} = A_{sh}/s_b$) in the middle of the column should not be less than 0.165%. These ratios can be at least 1.5% and 0.25% in rectangular cross-section columns, respectively. In addition, in beam support sections where columns or shear walls join beams, the beam upper and lower longitudinal reinforcement ratios ($\rho = A_s/b_w.d$) vary from 0.6% and 0.4%, respectively, to the transverse reinforcement ratio in the beam middle region ($\rho_w = A_{sw}/b_w.s$) should not be less than 0.25%. The beam height should not be less than 1/11 of the spanning in buildings whose carrier system consists of frames and 1/12 of the spanning in buildings whose carrier system consists of shear walls and frames. It is also desirable that the beam height should not be more than 1/4 of the simple supported beamed span. If the slab-on-beam system is used in the building, the slab thickness should be at least 15 cm. The code states that in buildings to be designed with simplified design rules, a concrete grade with strength lower than 25 MPa or higher than 50 MPa and reinforcing steel other than S420 or B420c classes, cannot be used. In Figure 4, the geometric boundaries given in the TBEC-2019 are summarized through an example.

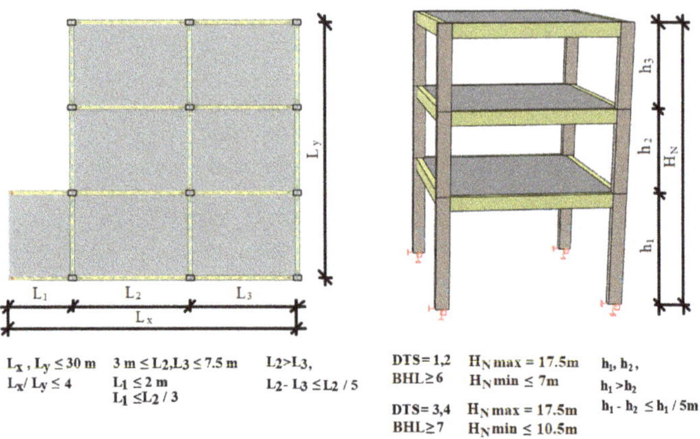

Figure 4. Building height limits according to building floor model plan limit values and earthquake design classes.

3. Numerical Analyses

Models with different parameters have been prepared to examine and control the behavior of the designed structures using the simple calculations in the related section of the TBEC-2019. A total of 144 building models were designed, depending on five different parameters: building models DTS, number of storeys, soil type, span length, and column cross-section calculation formula. Earthquake performance levels of the models were determined according to L and NL analysis methods in the Prota-Structure 2021 program. Prota-Structure is used for structural analyses. The analyses result of this program are the same as the results of SAP2000. It has been reported in the relevant sources [71,72] that it is precisely the same as SAP2000 v.23.3.1. While designing the building models, DTS = 3 and 4 were chosen. In the TBEC-2019, since the heights of the buildings are limited to be BHL $\geq$ 6 for D = 1 and 2 and BHL $\geq$ 7 for D = 3 and4, the building models have equal floor heights meeting the BHL $\geq$ 7 limit, and 3.5 m has been chosen, and three storeys (10.5 m), four storeys (14 m), five storeys (17.5 m). In the study, four different soil classes given in the TBEC-2019 were considered as ZA, ZB, ZC and ZD, from weak to vigorous, respectively. In selecting soil classes, ZE and ZF type soils, which are very weak soil classes with a high probability of liquefaction, were not considered because they require special designs.

The relevant section states that there should be at least two spans in both directions of the floor plan and that the spans should be at least 3 m and at most 7.5 m. Therefore, in the study, two different building floor plans were designed, three spans in both directions (6.66 m with equal span lengths) and four spans in both directions (5 m with equal span lengths), respectively. The regulation stipulated that the longest side of the floor plan should be 30 m at most. The building floor plan is square in both models in the study, and 20 × 20 m long was chosen. In the relevant section, it is requested that the carrier system typically consist of frames with high ductility levels or non-void walls with high ductility levels and frames with high ductility levels together. In the study, the structural system type of the models was chosen as the frame system with high ductility level, and the models were designed using three different column cross-section calculation formulas given in Figures 3 and 5, which are stipulated by the regulation for frames with high ductility level of the structural system type. The concrete grade was chosen in the models as C35 (characteristic compressive strength, f_{ck} = 35 MPa) and the reinforcement class as B420c (characteristic yield strength, f_{yk} = 420 MPa). The parameters taken into account in the design of the models are given in Table 2 in detail.

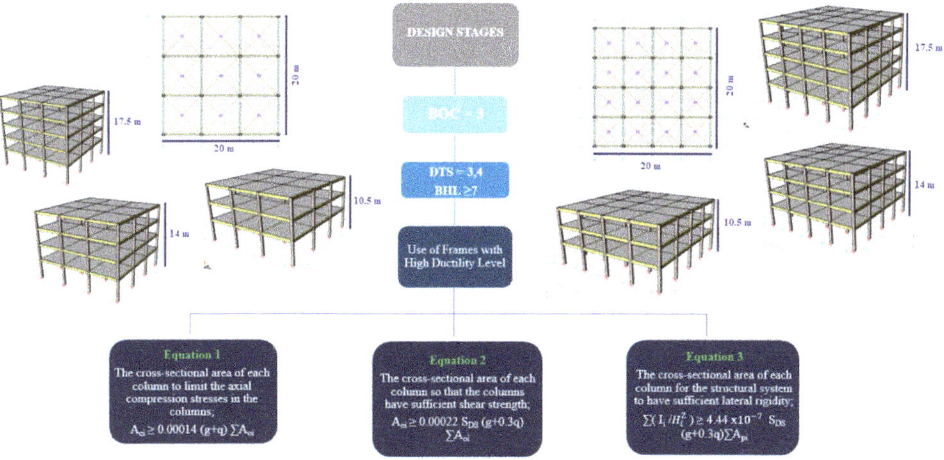

Figure 5. Flow chart of the formulas used in the building floor plan.

Table 2. Parameters used in building models design.

Plan Type/Span (m)	DTS	Soil Type	Number of Storeys and Building Height (H_N)	Equations Used in Calculation of Column Cross-Sectional Area
		ZA		**EQUATION 1** $A_{ci} \geq 0.00014\,(g + q)\,\sum A_{oi}$
/6.66 m	DTS-3	ZB ZC ZD	3-Storey 4-Storey	**EQUATION 2** $A_{ci} \geq 0.00022\,S_{DS}\,(g + 0.3q)\,\sum A_{oi}$
/5 m	DTS-4		5-Storey	**EQUATION 3** $\sum(I_i/H_i^2) \geq 4.44 \times 10^{-7}\,S_{DS}\,(g + 0.3q)\sum A_{pi}$

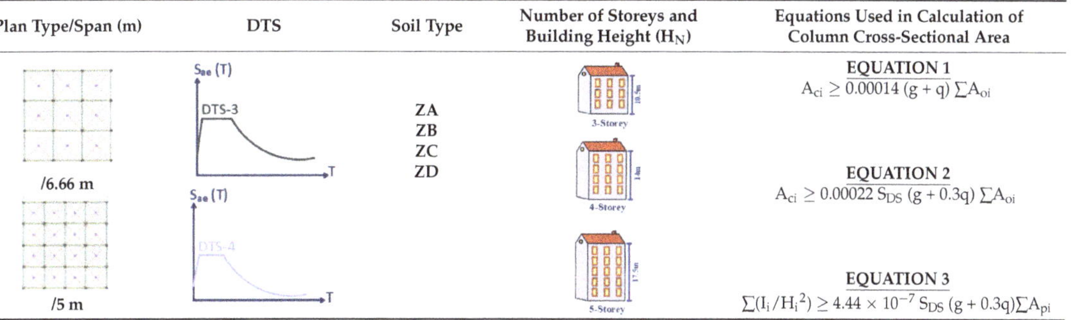

In Figure 5, there are visuals of two different floor plan models, designed as three-span (6.66 m) and four-span (5.00 m), designed as three storey, four storey, and five storey. The models' dead and live load values were chosen as g = 13 kN/m^2, q = 2 kN/m^2. Since the study aims to see how the building behaves for minimum values, columns, beams and slabs are dimensioned and designed according to the minimum values to meet the code. The width and height of the beams were chosen as 30 × 60 cm for three-span models and 30 × 50 cm for four-span models. Due to the requirement in the TBEC-2019 that the beam height will not be less than 1/11 of the span in buildings whose structural system consists of frames, the beam height was chosen as 61 cm in three-span models. The thickness of slabs with beams was chosen as 15 cm. Table 3 shows the beam reinforcement values calculated according to the code's relevant section reinforcement lower limits.

Table 3. Selection of beam reinforcements according to the general rules of TBEC-2019 (Section 17).

Model Type	Beam Longitudinal Reinforcements		
	Upper	Bottom	Stirrups
3 Spans	6Φ16	4Φ16	Φ8/13/10
4 Spans	6Φ14	4Φ14	Φ8/13/10

The same names are given to the columns with an equal sum of the area shares accumulated along all the slabs supported by the column, which is considered when calculating the across-sectional column areas. In addition, columns with an equal share of the accumulated area throughout all floors are shown with the same colour. For three-span building models and four-span building models, columns with an equal sum of accumulated area shares along all floors are shown in Figure 6.

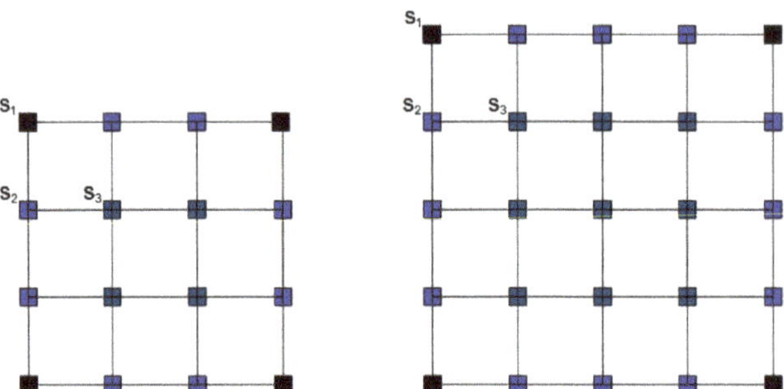

Figure 6. Location of columns S_1, S_2 and S_3 in plan, whose cross-sections are calculated for three-span and four-span building models.

The structural models designed in this study were named depending on the parameters. The criteria for naming the building models according to the parameters are explained in Figure 7.

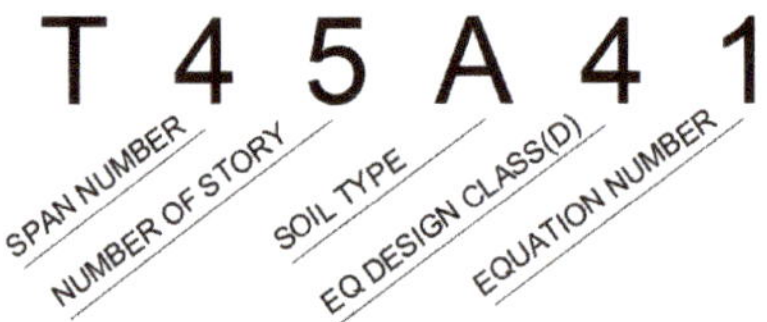

Figure 7. Codes of structural models.

In the relevant section of the TBEC-2019, the columns are dimensioned and reinforced using three formulas and reinforcement lower limits for the dimensioning of vertical load-bearing elements in the models selected as framed structural systems with high ductility levels. Since the formulas used in this study do not depend on the soil type, the column dimensions and reinforcement amounts do not change depending on the soil type. For this reason, 144 models depending on the parameters were analyzed in this study. However, only 36 of these models have different column cross-sections. Column dimensions and selected reinforcements of 36 models with different column cross-section values are shown in Table 4.

Table 4. Column dimensions and reinforcements details for each model.

Model Type	Column Sections S_1 (cm)	S_2 (cm)	S_3 (cm)	Column Longitudinal Reinforcement S_1	S_2	S_3	Column Lateral Reinforcement S_1	S_2	S_3
T45A41	30	36	51	8Φ14	6Φ14 + 2Φ16	8Φ16 + 4Φ18	Φ8/15/10	Φ8/16/10	Φ8/20/10
T45A42	30	30	31	8Φ14	8Φ14	8Φ14	Φ8/15/10	Φ8/15/10	Φ8/15/10
T45A43	36	36	36	6Φ14 + 2Φ16	6Φ14 + 2Φ16	6Φ14 + 2Φ16	Φ8/16/10	Φ8/16/10	Φ8/16/10
T35A41	34	48	68	8Φ14	12Φ16	16Φ18 + 4Φ14	Φ8/17/10	Φ8/20/10	Φ8/20/10
T35A42	30	30	41	8Φ14	8Φ14	2Φ14 + 6Φ18	Φ8/15/10	Φ8/15/10	Φ8/20/10
T35A43	41	41	41	2Φ14 + 6Φ18	2Φ14 + 6Φ18	2Φ14 + 6Φ18	Φ8/20/10	Φ8/20/10	Φ8/20/10
T45A31	30	36	51	8Φ14	6Φ14 + 2Φ16	8Φ16 + 4Φ18	Φ8/15/10	Φ8/16/10	Φ8/20/10
T45A32	30	31	43	8Φ14	8Φ14	8Φ18	Φ8/15/10	Φ8/15/10	Φ8/20/10
T45A33	43	43	43	8Φ18	8Φ18	8Φ18	Φ8/20/10	Φ8/20/10	Φ8/20/10
T35A31	34	48	68	8Φ14	10Φ14 + 4Φ16	16Φ18 + 4Φ14	Φ8/17/10	Φ8/20/10	Φ8/20/10
T35A32	30	41	58	8Φ14	2Φ14 + 6Φ18	4Φ14 + 14Φ16	Φ8/15/10	Φ8/20/10	Φ8/20/10
T35A33	48	48	48	10Φ14 + 4Φ16	10Φ14 + 4Φ16	10Φ14 + 4Φ16	Φ8/20/10	Φ8/20/10	Φ8/20/10
T44A41	30	32	46	8Φ14	8Φ14	6Φ16 + 4Φ18	Φ8/15/10	Φ8/16/10	Φ8/19/10
T44A42	30	30	30	8Φ14	8Φ14	8Φ14	Φ8/15/10	Φ8/15/10	Φ8/15/10
T44A43	35	35	35	8Φ14	8Φ14	8Φ14	Φ8/17/10	Φ8/17/10	Φ8/17/10
T34A41	31	43	61	8Φ14	8Φ18	10Φ16 + 4Φ24	Φ8/15/10	Φ8/20/10	Φ8/19/10
T34A42	30	30	36	8Φ14	8Φ14	6Φ14 + 2Φ16	Φ8/15/10	Φ8/15/10	Φ8/16/10
T34A43	39	39	39	8Φ16	8Φ16	8Φ16	Φ8/15/10	Φ8/15/10	Φ8/15/10
T44A31	30	32	46	8Φ14	8Φ14	4Φ14 + 8Φ16	Φ8/15/10	Φ8/16/10	Φ8/19/10
T44A32	30	30	39	8Φ14	8Φ14	8Φ16	Φ8/15/10	Φ8/15/10	Φ8/15/10
T44A33	41	41	41	2Φ14 + 6Φ18	2Φ14 + 6Φ18	2Φ14 + 6Φ18	Φ8/20/10	Φ8/20/10	Φ8/20/10
T34A31	30	43	61	8Φ14	8Φ18	10Φ16 + 4Φ24	Φ8/15/10	Φ8/20/10	Φ8/19/10
T34A32	30	36	52	8Φ14	6Φ14 + 2Φ16	8Φ18 + 4Φ16	Φ8/15/10	Φ8/16/10	Φ8/20/10
T34A33	46	46	46	4Φ14 + 8Φ16	4Φ14 + 8Φ16	4Φ14 + 8Φ16	Φ8/19/10	Φ8/19/10	Φ8/19/10
T43A41	30	30	40	8Φ14	8Φ14	8Φ16	Φ8/15/10	Φ8/15/10	Φ8/20/10
T43A42	30	30	30	8Φ14	8Φ14	8Φ14	Φ8/15/10	Φ8/15/10	Φ8/15/10
T43A43	32	32	32	8Φ14	8Φ14	8Φ14	Φ8/16/10	Φ8/16/10	Φ8/16/10
T33A41	30	37	53	8Φ14	8Φ16	14Φ16	Φ8/15/10	Φ8/16/10	Φ8/20/10
T33A42	30	30	32	8Φ14	8Φ14	8Φ14	Φ8/15/10	Φ8/15/10	Φ8/16/10
T33A43	36	36	36	6Φ14 + 2Φ16	6Φ14 + 2Φ16	6Φ14 + 2Φ16	Φ8/16/10	Φ8/16/10	Φ8/16/10
T43A31	30	30	40	8Φ14	8Φ14	8Φ16	Φ8/15/10	Φ8/15/10	Φ8/20/10
T43A32	30	30	33	8Φ14	8Φ14	8Φ14	Φ8/15/10	Φ8/15/10	Φ8/15/10
T43A33	38	38	38	8Φ16	8Φ16	8Φ16	Φ8/15/10	Φ8/15/10	Φ8/15/10
T33A31	30	37	53	8Φ14	8Φ16	4Φ16 + 8Φ18	Φ8/15/10	Φ8/16/10	Φ8/20/10
T33A32	30	31	45	8Φ14	8Φ14	4Φ16 + 8Φ14	Φ8/15/10	Φ8/15/10	Φ8/20/10
T33A33	43	43	43	8Φ18	8Φ18	8Φ18	Φ8/20/10	Φ8/20/10	Φ8/20/10

4. Numerical Analyses Results

Both L and NL analysis methods have been performed for structural analysis to determine the building performance. Earthquake ground motion level was chosen as DD-2, since simple buildings were analyzed according to the simplifying rules of the TBEC-2019. Structural analyses were performed only for the design earthquake (DD-2) with a 10% probability of exceedance in 50 years (the repetition period for which is 475 years). The code was adapted for the DD-2 earthquake level for residential buildings and stated in the relevant section that the controlled damage (CD, in Turkish KH) level was sufficient for this earthquake ground motion level. The building knowledge level has been chosen extensively. The multipliers of dead loads, live loads, and earthquake loads are considered 1.0, 0.3, 1.0 (G + Q+E) respectively. Snow load is neglected in the roof in all the structural models. Considering that the vertical earthquake effect is small, it is assumed that the structures are not exposed to vertical earthquakes. The base shear force–top displacement relationship and performance points were calculated for each model, respectively. These calculations were determined as shown in Figure 8.

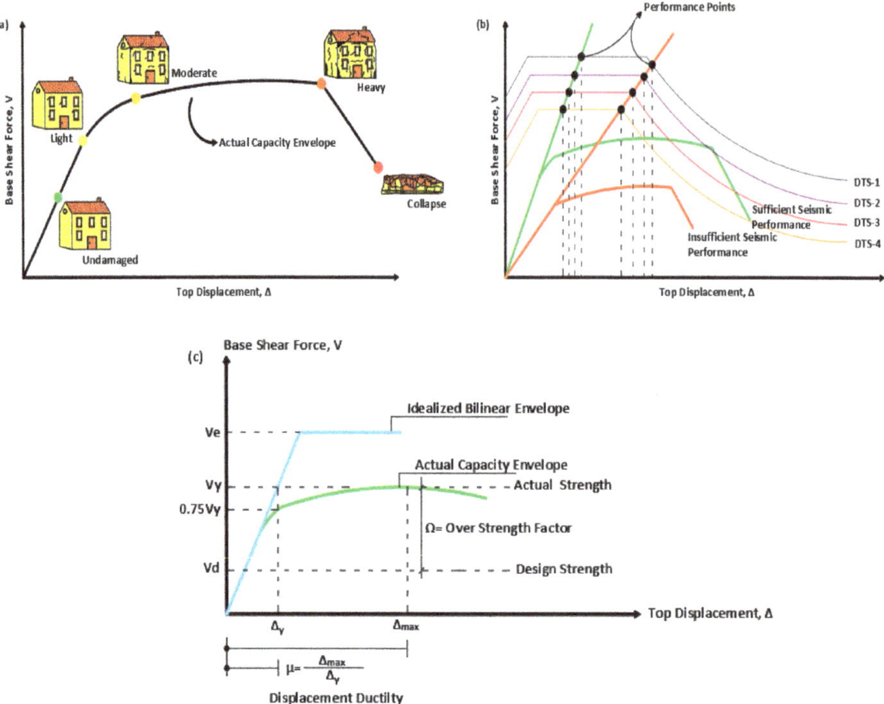

Figure 8. Lateral load–top displacement curve under earthquake and determination of performance point: (**a**) building performance levels; (**b**) comparison of capacity curves; (**c**) V_d/V_y.

The damage situation is determined according to the point where the structure is located on the capacity curve, drawn depending on the displacement of the structure under the effect of earthquake force in the graph (a) considering Figure 8. Graph (b) compares the capacity curves of the strong building and the weak building. Spectrum curves are elastic curves. The spectrum curves of the buildings do not intersect with the inelastic force–displacement graph. The force–displacement graphs obtained as a result of the pushover analysis are linearized, perpendiculars are drawn to the force–displacement graph from the point where they cut the elastic design spectra, and the displacement values that will occur in the structures under the influence of the force coming to the structures during the earthquake are seen. The point where the linearized force–displacement graph and the spectrum curve intersect is called the performance point. On the other hand, graph (c) shows how much the force (V_d) that may occur to the structure under the effect of an earthquake is below the actual strength (V_y) for which the structure is designed. The ratio of the actual strength of the structure to the design strength (V_y/V_d) is defined as the over-strength factor (Ω or D) [73,74]. It is foreseen as 3 in the TBEC-2019 for "buildings where reinforced concrete frames cover all the earthquake effects with high moment transmitting ductility level". The predicted value for Ω in the TBEC-2019 was evaluated by comparing the values obtained from the study.

4.1. Performance Parameters of Building Models

Existing buildings were analyzed according to both analysis methods L and NL. The behavior of the models, which vary depending on the number of storeys of a building, soil type, over-strength factor (Ω), column cross-section formula, and the number of spans in the building floor plan, are investigated. As a result of the analysis of buildings with different parameters, building shear capacities (V_y) and base shear forces (V_d) were determined, and Ω was calculated. As a result of the performance analyses of the existing buildings, the building performance levels were determined according to the damage conditions. In the TBEC-2019, for all buildings that are not classified as tall buildings, an ordinary performance target is set as CD under DD-2. In the results of the analysis, CD or collapse prevention (CP) was evaluated depending on the plastic rotation limits, and building performance levels were obtained. The relative storey drift values were also calculated from the ratio of the displacements of the buildings to their heights.

4.2. Comparisons of Parameter Results

The results obtained from the structural analysis for models and the effects of different parameters, such as Ω, number of storeys of a building, number of spans in the building floor plan, soil type, and the column cross-section formula used on the building performance, were examined. The performance analysis results of three storey, four storey and five storey models are given in Figure 9, respectively. The figures show the number of buildings with a CD (i.e., adequate performance) level depending on the respective soil type and the methods used. The columns show the total number of models (NM) designed for each soil type and how many of these models reached the CD level in L and NL. In terms of the number of building slabs, it is seen that the five storey models meet the performance target more than the three and four storey models. It is seen that the models belonging to ZA and ZB soil types are more likely to provide target performance among three different building storeys. It is seen that the target performance of the three storey models of the ZC and ZD soil types is relatively low. When the linear analysis results for the models of three different building floors are examined, it is seen that the amount of buildings providing the target performance is low, especially in the three storey models. As a result of nonlinear analysis in ZC and ZD soil types, it is seen that the number of models that provide target performance is relatively high for five storey models.

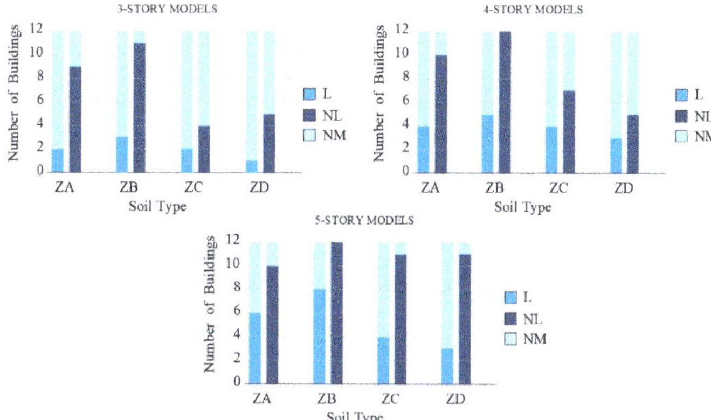

Figure 9. Distribution of buildings with sufficient performance according to the storey number/soil type parameters.

The analysis results of the models obtained according to Formulas 1, 2 and 3 in the TBEC-2019 are given in Figure 10. In the figures, the number of buildings with CD level is shown depending on the number of storeys of the respective building and the methods used. The columns show the total number of models designed for each column cross-section formula and how many of these models reach the CD level in L and NL. In the models designed using three different column cross-section formulas, it is seen that the number of building models that provide the target performance value is higher as a result of the L and NL analyses in the buildings calculated using the first and third formulas. When the analysis results of the models designed with the second formula are examined, it is seen that the second formula is not very sufficient. In the design with Formula 3, side and corner columns take higher values than side and corner columns. In Formulas 1 and 2, smaller cross-sections are obtained in columns calculated by using the second formula in the column cross-section formula. Since smaller column cross-section values were obtained in the models designed using the second formula, these models could not provide the target performance.

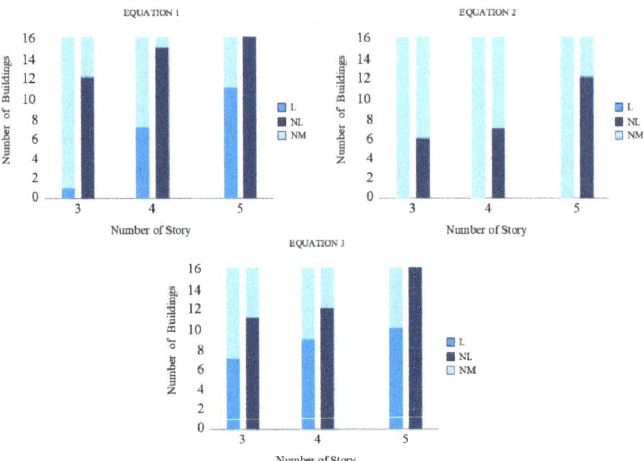

Figure 10. Distribution of buildings with sufficient performance according to the storey number / equation types.

Figure 11 shows the analysis results obtained by designing three different building storey numbers based on $S_{DS} < 0.33$ and $S_{DS} < 0.5$ values. The figures show the number of buildings with CD level depending on the number of slabs and the methods used for the relevant S_{DS} value. It is seen that the number of building models that provide the target performance is higher in the models designed for the $S_{DS} < 0.5$ value. Especially since the second and third formulas depend on the S_{DS} value, it is thought that the increase in the S_{DS} value contributes to the increase in the column cross-section. As the number of storey of the building increases, the number of models that provide the target performance within both S_{DS} values increases.

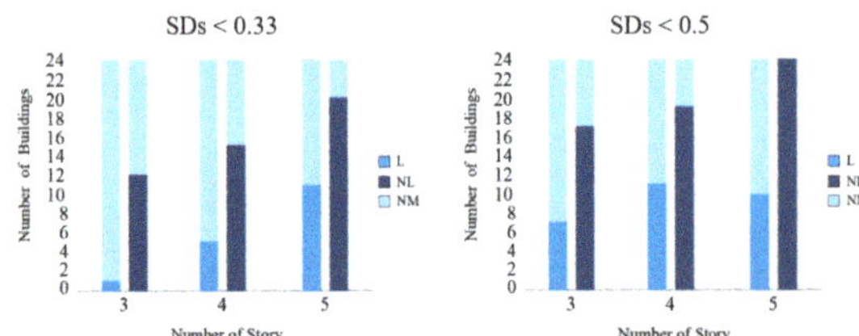

Figure 11. Distribution of buildings with sufficient performance according to the storey number/S_{DS}.

The analysis results for three storey, four storey and five storey building models with three spans (6.66 m) and four spans (5 m) are given in Figure 12. The figures show the number of buildings with CD levels depending on the number of spans involved and the methods used. The columns show the total number of models designed for each aperture number and how many of these models reached the CD level in the L and NL method. When the graphs are examined depending on the number of spans, it is seen that the number of buildings providing the target performance is higher in the four-span models. When investigate the number of storey in the two spans, it is seen that the number of models that provide the target performance is higher in five storey models.

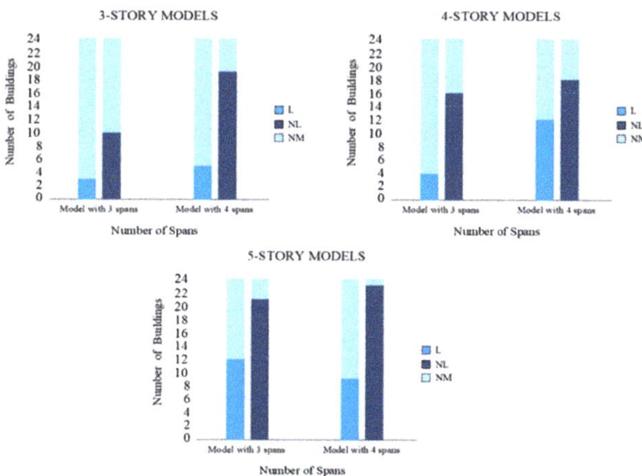

Figure 12. Distribution of buildings with sufficient performance according to the storey number/number of spans.

Figure 13 shows the analysis results of three storey, four storey and five storey building models calculated using the three different column cross-section formulas envisaged in the TBEC-2019. The figures show the number of buildings with a CD level based on the column cross-section formulas and the methods used for the number of storey in the relevant building. The columns show the total number of models designed for each column cross-section formula and how many of these models reached the CD level in the L and NL method. In general, it is seen that the models designed with the use of the first and third formulas in three different building storey are more likely to provide the target performance. In the models designed using the second formula, it is seen that no building model provides the target performance as a result of the linear analysis. It is seen that five storey models among three different building storeys provide the target performance more. It is seen that the number of models providing both analysis methods is high in four storey and five storey models designed using the first and third column cross-section formula. It has been observed that the column cross-section values calculated with three different formulas are larger in five storey models due to the larger area share. For this reason, it can be said that five storey models provide more target performance.

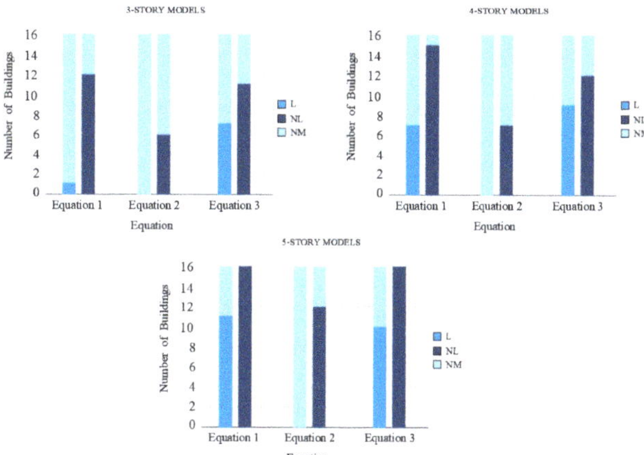

Figure 13. Distribution of buildings with sufficient performance according to the equation types/number of storeys.

Figure 14 shows the analysis results obtained by designing models with three different building floors according to three spans (6.66 m) and four spans (5 m). The figures show the number of buildings with a CD level, depending on the number of storeys and the methods used for the number of spans involved. The total number of models according to the number of building floors designed for each number of spans and how many of these models reached the CD level in the L and the NL are shown in the columns. It is seen that models with four spans provide more target performance than models with three spans in three different building storey. In the models with four spans, it is seen that the number of buildings providing the target performance is close to each other in the NL results within three different building storey. It is thought that the number of buildings providing target performance decreases as the span length increases. It is seen that the number of models that provide target performance decreases considerably with the increase in the span length, especially in three storey models.

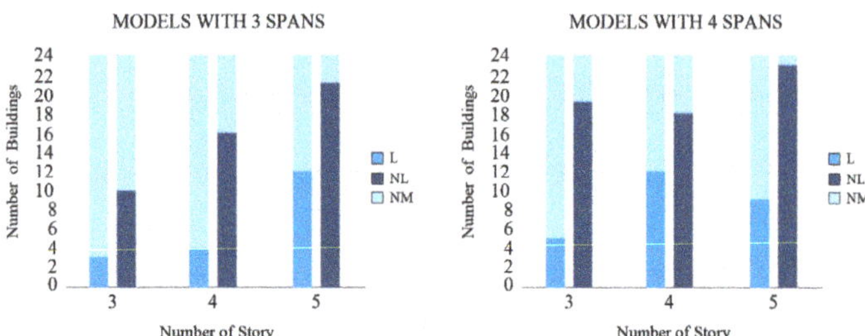

Figure 14. Distribution of buildings with sufficient performance according to the span number/number of storeys.

Figure 15 shows the analysis results for $S_{DS} \leq 0.33$ and $S_{DS} \leq 0.5$ for three storey, four storey and five storey models. The figures show the number of CD buildings based on the respective S_{DS} class and the methods used. The vertical axis shows the total number of models designed for each S_{DS} value, and how many of these models reached the CD level in the L and the NL method. In five storey models designed for $S_{DS} \leq 0.33$ and $S_{DS} \leq 0.5$ values, similar numbers of buildings provide target performance for both S_{DS} classes. For both S_{DS} values, it is seen that the number of building models that provide the target performance of the five storey models is higher than the number of buildings that provide the target performance of the three and four storey building models. When examining the three and four storey models, it is seen that the number of building models that provide the target performance of the models with the $S_{DS} \leq 0.5$ value is higher than the number of the building models that provide the target performance, especially in the linear analysis, the building models with the $S_{DS} \leq 0.33$ value.

Figure 16 shows the analysis results for four different soil types models. The figures show the number of buildings with a CD level, depending on the number of storey and the methods used for the respective soil types. The columns show the total number of models designed for different building floors for each floor class and how many of these models reached the CD level in the L and NL methods. Among the models with three different storeys, the target performance ratio is higher in ZA and ZB soil types. In models with ZC and ZD classes, it is seen that the amount of achieving the target performance for five storey buildings is high, while this amount is much lower in three and four storey models. In general, it is seen that the number of building models that provide target performance is higher in five storey models within four different soil types. It can be interpreted that the number of models providing target performance increases as the soil gets stronger.

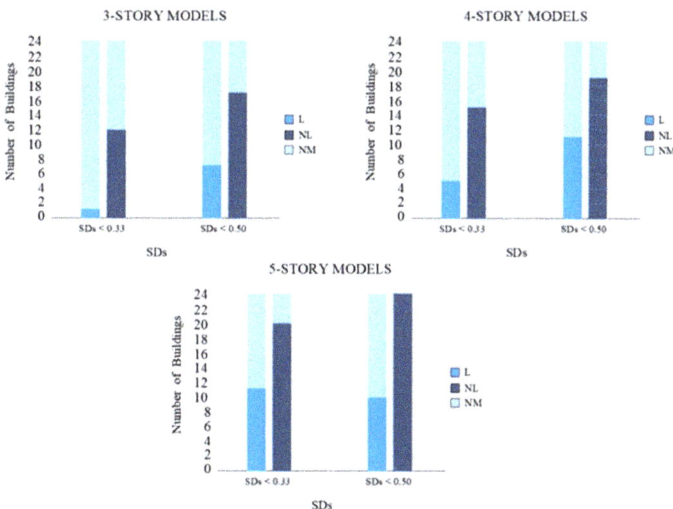

Figure 15. Distribution of buildings with sufficient performance according to the S$_{DS}$/storey number.

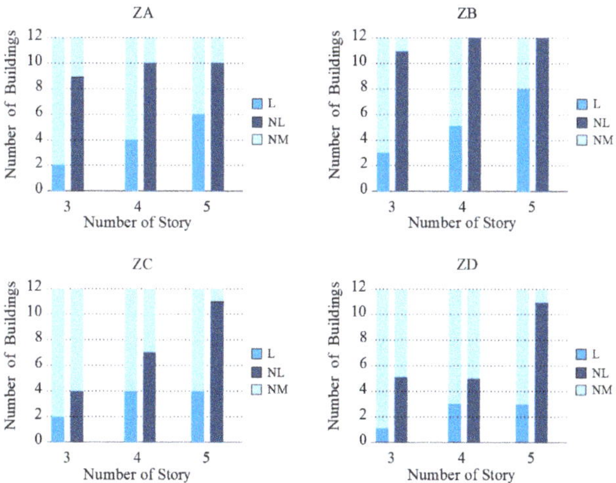

Figure 16. Distribution of buildings with sufficient performance according to the soil type/storey number.

Figure 17 shows the Ω coefficients obtained from the analysis of the models obtained by using three different column cross-section formulas predicted by the TBEC-2019. In the TBEC-2019, the number of storeys with Ω coefficients is foreseen as three for reinforced concrete buildings with carrier system-type frames. When the graphics are examined, it is seen that most of the models designed with the use of the first and third formulas have a coefficient of Ω of 3 and above 3. It is seen that approximately half of the models designed with the use of the second formula have a coefficient of Ω of 3 and above. It can be said that the reason for this situation may be that smaller column cross-sections are obtained as a result of using the second formula compared to other formulas.

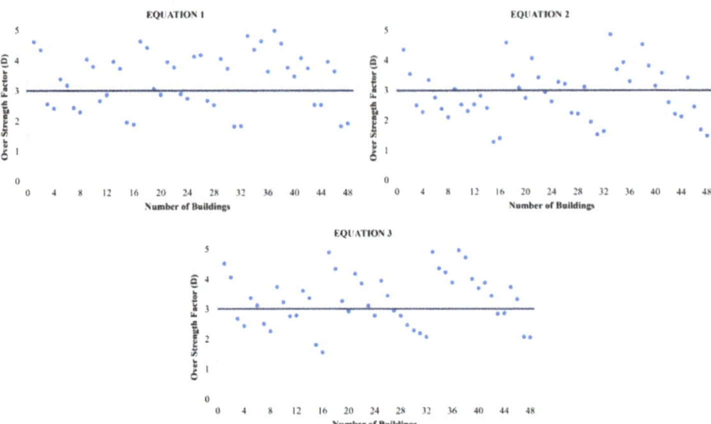

Figure 17. Distribution of the over-strength factor of the models of buildings designed different equations.

Figure 18 shows the relative storey drift values obtained according to the four different soil types and the number of building storeys. It is seen that the relative storey drift values for ZA and ZB soil types are also less for three different building storeys. For the ZD soil type, it is seen that this value is higher than other soil types. In the three and four storey models, it is seen that the relative storey drift values are quite high, especially in the ZC and ZD classes. In general, it is seen that the relative storey drift values of the five storey models are less. It is seen that the majority of the relative storey drift values of the models are generally 2‰–10‰.

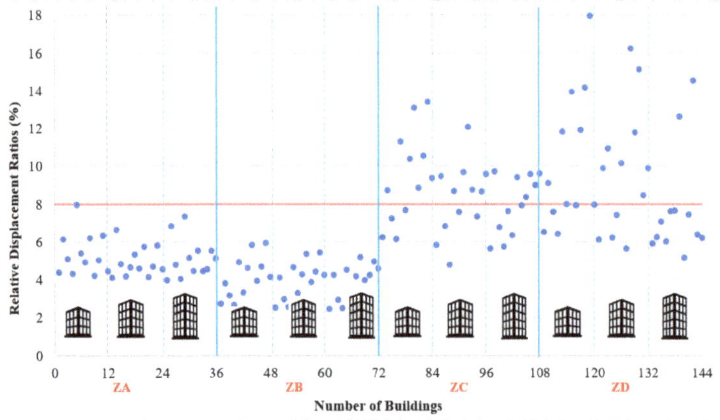

Figure 18. Distribution of the relative storey drift of all models according to the number of buildings and soil types.

In the TBEC-2019, if the infill walls are connected to the frame members without any flexible joints or gaps, the relative storey drift limit value is determined as 8‰.

In Figures 19 and 20, the values of Ω obtained according to the four different soil types and the number of building storeys of all designed models are shown. Ω (or D) is foreseen as 3 in reinforced concrete frame buildings in the TBEC-2019. In Figure 18, models that are below this value and cannot achieve the target performance (CD) due to the L analysis are scanned in red. In Figure 19, models with a coefficient of Ω below 3 and failing to provide CD due to NL analysis are scanned in red. When investigating the results of the models in general, it varies between 2.28–4.6 for ZA, 2.11–3.38 for ZB, 2.31–4 for ZC, and 1.29–3.96 for ZD in five storey models, respectively. For ZD, this ratio can give quite different results and remain below the desired value. For ZA, ZB and ZC models with 4 floors, the range is 2.5–4, and the desired value is generally provided. In models with ZD soil type, a few models are in the range of 2.5–4, and most of the models have an excess of strength below the desired value. In three storey models, it is seen that the excess strength coefficient of most of the models belonging to ZA, ZB and ZC soil types is generally 3 and above 3. For the ZD soil type, it is seen that there are models with an Ω above 3, as well as models with a range of 1.5–2, far below. In general, it is seen that the Ω value of the models belonging to the ZA and ZB floor classes in three and four storey models is both greater than the desired value and more than the five storey models. It is seen that the number of floors with Ω of the models belonging to the ZD soil type is relatively low.

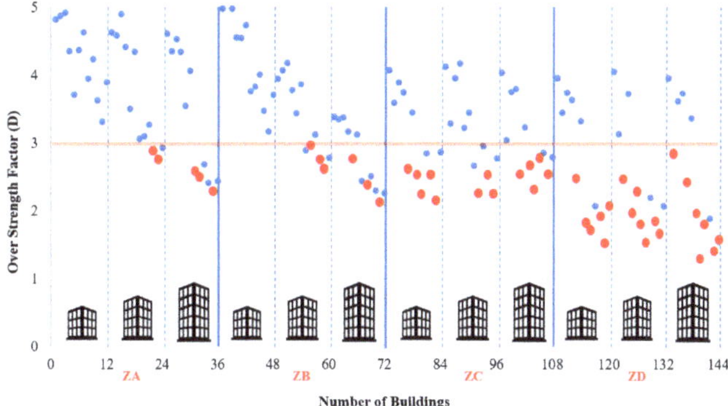

Figure 19. Distribution of the over-strength factor for all models according to the number of buildings and soil types (for linear analysis).

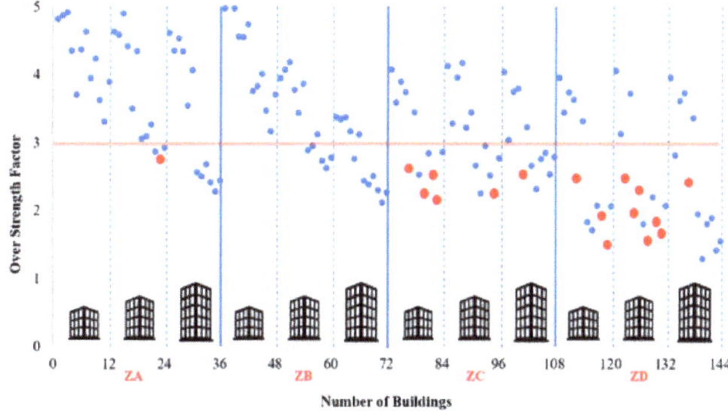

Figure 20. Distribution of the over-strength factor for all models according to the number of buildings and soil types (for nonlinear analysis).

5. Conclusions

In this study, a comprehensive numerical study was carried out with the performance analyses of 144 RC buildings, which were designed according to minimum column sectional criteria given in simplified methods of low-rise RC structures in the TBEC-2019. Models with different parameters (building models DTS, number of storeys, soil type, span length, and column cross-section) were prepared, to examine and control the behavior of the designed structures using the simple calculations in the related section of the TBEC-2019. The main findings of the study are summarized below;

- According to the analysis results, the number of models providing the CD performance level in the L analysis method is 44 (30.55%) and 107 (74.30%) in the NL analysis method. As a result, in the building designed according to minimum column sectional criteria given in simplified methods of low-rise RC structures, the performance criteria of the code cannot be met according to at least one method. It is noteworthy that such different results were obtained for the two analysis methods stipulated by the code.
- For different soil types, it is seen that the rate of achieving the target performance is higher in ZA and ZB soil types among models with three different building storeys. In the models with ZC and ZD classes, it is seen that the amount achieving the target performance for the five storey building model is high, while this amount is much lower in the three and four storey models.
- For the models designed using three different column cross-section formulas, it is seen that the number of building models that provide the target performance value is higher as a result of the L and NL analyses in the buildings calculated using the first and third formulas. When the analysis results of the models designed with the second formula are examined, it is seen that Formula 2 is not very sufficient due to the smaller column cross-section values obtained. For this reason, the relative storey drifts of the models were smaller in the models designed with the first and third formulas.
- For models designed with two different spans, when buildings with four spans (5 m) and three spans (6.66 m) are examined, it is seen that the building model that meets the performance target is higher in models with four spans. It can be said that the reason for this situation is that there are 25 columns in the four-span models, while 16 columns are included in the three-span models.
- For the models designed with two different S_{DS} values, it is seen that the models designed with the $S_{DS} < 0.5$ value reach the target performance more. For both S_{DS} values, it is seen that the number of building models that provide the target performance of the five storey models is higher than the number of buildings that provide the target performance of the three and four storey building models.
- For Ω, it is seen that the excess strength coefficient value for three and four storey models, especially for ZA and ZB soil types, generally reaches and exceeds the desired value of 3 in the TBEC-2019. In the five storey models, the coefficient of over-strength reached the value of 3 in general, but the results were not as high as in the three and four storey models. It is seen that the coefficient of over-strength is higher in the models designed with the first and third formulas. In addition, it is among the significant findings that the buildings with the over-strength coefficient below "3" will most likely not provide the performance level requested by the code.
- For the relative storey drift values, it is seen that the relative storey drift values of the five storey models are less than the three and four storey models. It is thought that in five storey models, the column cross-section values increase as the area share along all floors in the formula increases, and as a result, the relative storey drift is less than in three and four storey models. In terms of soil class, it is seen that the relative storey drift values are less for ZA and ZB soil types. It is seen that the relative storey drift values of the models designed with Formula 2 are relatively larger and they provide the code limit relative storey drift value less than the other formulas.
- For different building storeys, it is seen that the number of models that provide performance in five storey building models is higher than in three and four storey models. It is thought that this situation is due to the fact that the displacements of the three and four storey models are higher, although over-strength is large.
- Since the structures discussed in the study are new structures to be built, it is thought that the results obtained will change positively if the expected compressive strength of the concrete with 35 MPa characteristic compressive strength selected in the modelling is taken as the basis in the performance analysis.

- The simplified method proposed in the TBEC-2019 allows buildings that meet the relevant requirements to be designed very quickly without detailed seismic analysis. However, it is seen that the earthquake performance may not be sufficient in some of these structures designed according to minimum column sectional criteria.
- In this respect, the strength of the designed procedure explored in this study is simple and useful, and the weak side is that the desired earthquake performance of structures that are designed according to the simplified rules given in other parts of the TBEC-2019 could not be fully achieved in some buildings, especially with performance analysis where relatively complex analyses are required. The authors consider that while the simplifying rules are useful for engineers, the linear and nonlinear performance analysis section in the same code is partly complex for engineers. In this respect, structures designed according to minimum column sectional criteria, cannot meet a criteria of performance analysis which have complex rules, and this caused the results to be unsatisfactory.
- Differences in performance analysis approaches and acceptances in earthquake codes (such as Eurocode 8, etc.) will cause different results in modelling the same structures according to other codes.
- It is clear that the results will change if the structure systems are designed according to the detailed analysis procedure.

Author Contributions: Conceptualization, N.S.Y., M.H.A. and C.A.; methodology, N.S.Y., M.H.A., İ.H.E., H.D.A. and C.A.; software, N.S.Y., M.H.A., İ.H.E. and C.A.; validation, M.H.A., E.I., İ.H.E. and C.A.; formal analysis, H.D.A. and İ.H.E.; investigation, N.S.Y., M.H.A., C.A. and İ.H.E.; resources, N.S.Y., M.H.A. and H.D.A.; data curation, N.S.Y., M.H.A. and C.A., writing—original draft preparation, N.S.Y., M.H.A., C.A. and E.I.; writing—review and editing, M.H.A., İ.H.E. and E.I.; visualization, H.D.A.; supervision, M.H.A.; project administration, M.H.A.; funding acquisition, E.I. All authors have read and agreed to the published version of the manuscript.

Funding: This research received no external funding.

Institutional Review Board Statement: Not applicable.

Informed Consent Statement: Not applicable.

Data Availability Statement: Most data are included in the manuscript.

Acknowledgments: The authors appreciate the efforts of the Editorial Board of the Journal of Buildings. This research was produced from a master's thesis by the first author of the article. This article has been prepared by using some parts of the MSc thesis of "Evaluation of Simplified Design Rules for Regular Reinforced Concrete Building in The Context of Performance Analysis" (N. Seda YEL).

Conflicts of Interest: The authors declare no conflict of interest.

References

1. Alkan, H.; Büyüksaraç, A.; Bektaş, Ö.; Işık, E. Coulomb stress change before and after 24.01. 2020 Sivrice (Elazığ) Earthquake (Mw = 6.8) on the East Anatolian Fault Zone. *Arab. J. Geosci.* **2021**, *14*, 1–12. [CrossRef]
2. Bilgin, H.; Shkodrani, N.; Hysenlliu, M.; Ozmen, H.B.; Isik, E.; Harirchian, E. Damage and performance evaluation of masonry buildings constructed in 1970s during the 2019 Albania earthquakes. *Eng. Fail. Anal.* **2022**, *131*, 105824. [CrossRef]
3. Tekeli-Yesil, S.; Pfeiffer, C.; Tanner, M. The determinants of information seeking behaviour and paying attention to earthquake-related information. *Int. J. Disaster Risk Reduct.* **2020**, *49*, 101734. [CrossRef]
4. Işık, E.; Harirchian, E.; Büyüksaraç, A.; Ekinci, Y.L. Seismic and structural analyses of the eastern anatolian region (Turkey) using different probabilities of exceedance. *Appl. Sys. Innov.* **2021**, *4*, 89. [CrossRef]
5. Strukar, K.; Sipos, T.K.; Jelec, M.; Hadzima-Nyarko, M. Efficient damage assessment for selected earthquake records based on spectral matching. *Earthq. Struct.* **2019**, *17*, 271–282.
6. Harirchian, E.; Hosseini, S.E.A.; Jadhav, K.; Kumari, V.; Rasulzade, S.; Işık, E.; Lahmer, T. A review on application of soft computing techniques for the rapid visual safety evaluation and damage classification of existing buildings. *J. Build. Eng.* **2021**, *43*, 102536. [CrossRef]
7. Büyüksaraç, A.; Isik, E.; Harirchian, E. A case study for determination of seismic risk priorities in Van (Eastern Turkey). *Earthq. Struct.* **2021**, *20*, 445–455.

8. Bilgin, H.; Hadzima-Nyarko, M.; Isik, E.; Ozmen, H.B.; Harirchian, E. A comparative study on the seismic provisions of different codes for RC buildings. *Struct. Eng. Mech.* **2022**, *83*, 195–206.
9. Inel, M.; Meral, E. Seismic performance of RC buildings subjected to past earthquakes in Turkey. *Earthq. Struct.* **2016**, *11*, 483–503. [CrossRef]
10. Arslan, M.H.; Korkmaz, H.H. What is to be learned from damage and failure of reinforced concrete structures during recent earthquakes in Turkey? *Eng. Fail. Anal.* **2007**, *14*, 1–22. [CrossRef]
11. Arslan, M.H. An evaluation of effective design parameters on earthquake performance of RC buildings using neural networks. *Eng. Struct.* **2010**, *32*, 1888–1898. [CrossRef]
12. Kaltakci, M.Y.; Arslan, M.H.; Yilmaz, U.S.; Arslan, H.D. A new approach on the strengthening of primary school buildings in Turkey: An application of external shear wall. *Build. Environ.* **2008**, *43*, 983–990. [CrossRef]
13. Güner, B. Türkiye'deki deprem hasarlarina dönemsel bir yaklaşim; 3 dönem 3 deprem. *Doğu Coğrafya Derg.* **2020**, *25*, 139–152.
14. Doğan, T.P.; Kızılkula, T.; Mohammadi, M.; Erkan, İ.H.; Kabaş, H.T.; Arslan, M.H. A comparative study on the rapid seismic evaluation methods of reinforced concrete buildings. *Int. J. Disaster Risk Reduct.* **2021**, *56*, 102143. [CrossRef]
15. Dogan, G.; Ecemis, A.S.; Korkmaz, S.Z.; Arslan, M.H.; Korkmaz, H.H. Buildings damages after Elazığ, Turkey earthquake on 24 January 2020. *Nat. Hazards* **2021**, *109*, 161–200. [CrossRef]
16. Adalier, K.; Aydingun, O. Structural engineering aspects of the 27 June 1998 Adana–Ceyhan (Turkey) earthquake. *Eng. Struct.* **2001**, *23*, 343–355. [CrossRef]
17. Bruneau, M. Building damage from the Marmara, Turkey earthquake of 17 August 1999. *J. Seismol.* **2002**, *6*, 357–377. [CrossRef]
18. Sezen, H.; Whittaker, A.S.; Elwood, K.J.; Mosalam, K.M. Performance of reinforced concrete buildings during the 17 August 1999 Kocaeli, Turkey earthquake, and seismic design and construction practise in Turkey. *Eng. Struct.* **2003**, *25*, 103–114. [CrossRef]
19. Doğangün, A.; Yön, B.; Onat, O.; Öncü, M.E.; Sağıroğlu, S. Seismicity of East Anatolian of Turkey and failures of infill walls induced by major earthquakes. *J. Earthq. Tsunami* **2021**, *15*, 2150017. [CrossRef]
20. Celep, Z. *Betonarme Yapılar*; Beta Basım Yayın: İstanbul, Turkey, 2005.
21. Işık, E. A comparative study on the structural performance of an RC building based on updated seismic design codes: Case of Turkey. *Challenge* **2021**, *7*, 123–134. [CrossRef]
22. Özmen, B. Türkiye deprem bölgeleri haritalarının tarihsel gelişimi. *Türkiye Jeol. Bülteni* **2012**, *55*, 43–55.
23. *TEC-2007*; Turkish Earthquake Code. T.C. Resmi Gazete: Ankara, Turkey, 2007.
24. *TBEC-2019*; Turkish Building Earthquake Code. T.C. Resmi Gazete: Ankara, Turkey, 2018.
25. Aksoylu, C.; Arslan, M.H. 2007 ve 2019 Deprem yönetmeliklerinde betonarme binalar için yer alan farklı deprem kuvveti hesaplama yöntemlerinin karşılaştırmalı olarak irdelenmesi. *Int. J. Eng. Res. Dev.* **2021**, *13*, 359–374.
26. Özmen, A.; Sayin, E. Deprem etkisinde çok katlı betonarme bir binanın tdy-2007 ve tbdy-2018 deprem yönetmeliklerine göre eşdeğer deprem yüklerinin karşılaştırılması. *Osman. Korkut Ata Üniversitesi Fen Bilim. Enstitüsü Derg.* **2021**, *4*, 124–133. [CrossRef]
27. Korkmaz, M. Analysis of an educational building according to tec2007 and tec2018. *Sciennovation* **2020**, *1*, 43–50.
28. Başaran, V.; Hiçyilmaz, M. Betonarme çerçevelerde farklı deprem yer hareketi düzeyi etkilerinin incelenmesi. *J. Innov. Civ. Eng. Technol.* **2020**, *2*, 27–41.
29. Doğan, S. Farkli deprem etkilerine maruz kalan betonarme yapilar için deprem yükü azaltma katsayilarinin ilişkilerinin değerlendirilmesi. *TURAN Strat. Arast. Merk.* **2019**, *11*, 353–358.
30. Aksoylu, C.; Arslan, M.H. Çerçeve türü betonarme binaların periyod hesaplarının farklı ampirik bağıntılara göre irdelenmesi. *Bitlis Eren Üniversitesi Fen Bilim. Derg.* **2019**, *8*, 569–581. [CrossRef]
31. Aksoylu, C.; Arslan, M.H. Çerçeve+ perde türü betonarme binalarin periyod hesaplarinin tbdy-2019 yönetmeliğine göre ampirik olarak değerlendirilmesi. *Uludağ Üniversitesi Mühendislik Fakültesi Derg.* **2019**, *24*, 365–382.
32. Bayrakci, S.; Baran, T. Zemin dinamik davranışının eşdeğer lineer analiz yöntemi ile belirlenmesi. *Osman. Korkut Ata Üniversitesi Fen Bilim. Enstitüsü Derg.* **2018**, *1*, 10–15.
33. Başaran, V. Türkiye bina deprem yönetmeliğine (TBDY2019) göre Afyonkarahisar için deprem yüklerinin değerlendirilmesi. *Afyon Kocatepe Üniversitesi Fen Ve Mühendislik Bilim. Derg.* **2018**, *18*, 1028–1035.
34. Işık, E.; Velioğlu, E. A study on the consistency of different methods used in structural analysis. *AKU J. Sci. Eng.* **2017**, *17*, 1055–1065. [CrossRef]
35. Karaca, H.; Oral, M.; Erbil, M. Yapısal tasarım bağlamında 2007 ve 2018 deprem yönetmeliklerinin karşılaştırılması, Niğde örneği. *Niğde Ömer Halisdemir Üniversitesi Mühendislik Bilimleri Dergisi* **2020**, *9*, 898–903. [CrossRef]
36. Ünsal, İ.; Öncel, F.A.; Şahan, F. Tdy 2007 ve tbdy 2018 yönetmeliklerine göre yapi yüksekliğinin taban kesme kuvveti ve tepe deplasmani üzerindeki etkisinin incelenmesi. *Konya Mühendislik Bilimleri Dergisi* **2020**, *8*, 930–942.
37. Gunes, B.; Cosgun, T.; Sayin, B.; Mangir, A. Seismic performance of an existing low-rise RC building considering the addition of a new storey. *Rev. Construcción* **2019**, *18*, 459–475. [CrossRef]
38. Güllü, H. Tarihi yiğma yapili cendere köprüsünün deprem etkisinin incelenmesi. *Niğde Ömer Halisdemir Üniversitesi Mühendislik Bilim. Derg.* **2018**, *7*, 245–259. [CrossRef]
39. Çoban, S.; Çeribaşi, S. Gömülü çelik boru sistemlerinin zaman tanım alanında yapısal analizi. *Sakarya Uni. J. Sci.* **2018**, *22*, 581–592.
40. Jakayev, S.; Aydemir, M.E. Düzenli bir betonarme binada düşey deprem bileşeninin yapısal davranışa etkisi. *Afet Ve Risk Derg.* **2019**, *2*, 1–13. [CrossRef]

41. Kasap, T. Evaluation of earthquake behaviour of reinforced concrete frame buildings by nonlinear methods. *Sciennovation* **2021**, *2*, 27–35.
42. Özer, Ö.; Yüksel, S.B. Farkli betonarme bağ kirişi modellerinin tbdy (2018)'e göre yapi performansina etkisi. *Uludağ Üniversitesi Mühendislik Fakültesi Derg.* **2020**, *25*, 1169–1188.
43. Işık, E. Comparative investigation of seismic and structural parameters of earthquakes (M ≥ 6) after 1900 in Turkey. *Arab. J. Geosci.* **2022**, *15*, 1–21. [CrossRef]
44. Çolakoğlu, H.E. U şekilli betonarme perdelerin farklı yatay yük etkileri altında doğrusal olmayan davranışı. *Tek. Dergi* **2019**, *30*, 8887–8912. [CrossRef]
45. Dalyan, İ.; Şahin, B. Mevcut betonarme bir binanın 2007 ve 2018 deprem yönetmeliklerine göre deprem yükleri altındaki taşıyıcı sistem perfomansının değerlendirilmesi. *Türk Deprem Araştırma Derg.* **2019**, *1*, 134–147. [CrossRef]
46. Kap, T.; Özgan, E.; Uzunoğlu, M.M. Betonarme bir okul binasının 2018 deprem yönetmeliğine göre incelenmesi. *Düzce Üniversitesi Bilim Ve Teknol. Derg.* **2019**, *7*, 1140–1150. [CrossRef]
47. Aksoylu, C.; Mobark, A.; Arslan, M.H.; Erkan, İ.H. A comparative study on ASCE 7-16, TBEC-2018 and TEC-2007 for reinforced concrete buildings. *Rev. Construcción* **2020**, *19*, 282–305. [CrossRef]
48. Kürkçü, F. 20 Katlı Betonarme Bir Yapının Türkiye Bina Deprem Yönetmeliği'ne Göre Tasarımı Ve Deprem Performansının Belirlenmesi. Ph.D. Thesis, Yüksek Lisans Tezi, İstanbul Teknik Üniversitesi, Fen Bilimleri Enstitüsü, İstanbul, Türkiye, 2020.
49. Akçora, A.A. Betonarme Yüksek Binaların 2018 Yılı Türkiye Bina Deprem Yönetmeliğine Göre İncelenmesi: 30 Katlı Bina Örneği. Master's Thesis, Fen Bilimleri Enstitüsü, Yıldız Teknik Üniversitesi, İstanbul, Türkiye, 2020.
50. Çapa, Y.U. Kat Adetleri Farklı Betornarme Binaların Deprem Performanslarının İncelenmesi. Master's Thesis, Fatih Sultan Mehmet Vakıf Üniversitesi, Lisansüstü Eğitim Enstitüsü, İstanbul, Türkiye, 2020.
51. Severcan, M.H.; Sinani, B. Mevcut betonarme yapilarin deprem performansinin analizi. *Niğde Ömer Halisdemir Üniversitesi Mühendislik Bilim. Derg.* **2019**, *8*, 936–947.
52. Ulutaş, H. DBYBHY (2007) ve TBDY (2018) deprem yönetmeliklerinin kesit hasar sınırları açısından kıyaslanması. *Avrupa Bilim Ve Teknol. Derg.* **2019**, *17*, 351–359.
53. Sarı, O. Mevcut Konut Türü Bir Binanın Dbybhy–2007 Ve Tbdy–2018 Deprem Yönetmeliklerine Göre Deprem Performansının Değerlendirilmesi. Ph.D. Thesis, Yüksek Lisans–Tezi, Burdur Mehmet Akif Ersoy Üniversitesi Fen Bilimleri Enstitüsü, Burdur, Türkiye, 2020.
54. Ayaz, U. Mevcut Bir Betonarme Binanın 2007 Ve 2018 Türkiye Bina Deprem Yönetmeliklerine Göre Deprem Performansının Değer-Lendirilmesi Ve Güçlendirilmesi. Master's Thesis, Fen Bilimleri Enstitüsü, İzmit, Türkiye, 2020.
55. Şeşetyan, K.; Demircioğlu Tümsa, M.B.; Akinci, A. Evaluation of the seismic hazard in the Marmara region (Turkey) based on updated databases. *Geosciences* **2019**, *9*, 489. [CrossRef]
56. Sianko, I.; Ozdemir, Z.; Khoshkholghi, S.; Garcia, R.; Hajirasouliha, I.; Yazgan, U.; Pilakoutas, K. A practical probabilistic earthquake hazard analysis tool: Case study Marmara region. *Bull. Earthq. Eng.* **2020**, *18*, 2523–2555. [CrossRef]
57. Erberik, M.A. Fragility-based assessment of typical mid-rise and low-rise RC buildings in Turkey. *Eng Struct.* **2008**, *30*, 1360–1374. [CrossRef]
58. Ilki, A.; Comert, M.; Demir, C.; Orakcal, K.; Ulugtekin, D.; Tapan, M.; Kumbasar, N. Performance based rapid seismic assessment method (PERA) for reinforced concrete frame buildings. *Adv. Struct. Eng.* **2014**, *17*, 439–459. [CrossRef]
59. İlki, A.; Cömert, M.; Orakçal, K.; Gülkan, P. Alternative simplified seismic design for regular low-rise reinforced concrete buildings. In Proceedings of the Eighth National Conference on Earthquake Engineering, Istanbul, Turkey, 11–15 May 2015.
60. NIBSDSSDP. *Development of Simplified Seismic Design Procedures: Framework Report*; National Institute of Building Sciences for Development of Simplified Design Procedures: Anchorage, AK, USA, 2010.
61. *ASCE/SEI*; Minimum Design Loads for Buildings and Other Structures, ASCE/SEI 7-10. American Society of Civil Engineers: Reston, VA, USA, 2010.
62. FEMA. *NEHRP Recommended Provisions for Seismic Regulations for New Buildings and Other Structures*, FEMA 450; Federal Emergency Management Agency: Washington, DC, USA, 2003.
63. *ISO/DIS 28841*; İITS. Guidelines for Simplified Seismic Assessment and Rehabilitation of Concrete Buildings, Draft International Standard. ISO: Geneva, Switzerland, 2010.
64. Balun, B.; Nemutlu, Ö.F.; Sarı, A. TBDY 2018 basitleştirilmiş tasarım kurallarının taban kesme kuvvetine etkisinin incelenmesi. *Türk Doğa Ve Fen Derg.* **2020**, *9*, 173–181.
65. Yel, N.S. Düzenli Betonarme Binalar İçin Basitleştirilmiş Tasarım Kurallarının Performans Analizi Bağlamında Değerlendirilmesi. Master's Thesis., Konya Teknik Üniversitesi, Konya, Türkiye, 2021.
66. Integrated Analysis and Design Solution for Building Analysis. Prota Structure, 2022. Available online: https://www.protasoftware.com/documents/protastructure_2019_brochure.pdf (accessed on 20 May 2022).
67. *SAP2000*; User Guide. People's Communications Press: Beijing, China, 2012.
68. Crespi, P.; Zucca, M.; Valente, M. On the collapse evaluation of existing RC bridges exposed to corrosion under horizontal loads. *Eng. Fail. Anal.* **2020**, *116*, 104727. [CrossRef]
69. Turkish Earthquake Hazard Map Interactive Web Application. Available online: https://tdth.afad.gov.tr (accessed on 15 June 2022).

70. AFAD. Disaster and Emergency Management Presidency (AFAD). Available online: https://en.afad.gov.tr/2022 (accessed on 15 June 2022).
71. Yasin, M.F.; Başak, K.; Kubin, J.; Tan, M.T. Perdeli betonarme yapılar için doğrusal olmayan analiz metotları. In Proceedings of the Seventh National Conference on Earthquake Engineering, İstanbul, Turkey, 30 May–3 June 2011.
72. Fahjan, Y.; Kubin, J.; Tan, M. Nonlinear analysis Methods for Reinforced Concrete Buildings with Shear Walls. In Proceedings of the 14th European Conference on Earthquake Engineering, Ohrid, North Macedonia, 30 August–3 September 2010.
73. Arslan, M.H.; Erkan, İ.H. Betonarme Binaların Deprem Yükü Azaltma Katsayısı üzerine Yeni Bir Bakış. *Eng. Sci.* **2011**, *6*, 1486–1497.
74. Shendkar, M.R.; Kontoni, D.P.N.; Işık, E.; Mandal, S.; Maiti, P.R.; Harirchian, E. Influence of masonry infill on seismic design factors of reinforced-concrete buildings. *Shock. Vib.* **2022**, *2022*, 5521162. [CrossRef]

Article

Estimating the Concrete Ultimate Strength Using a Hybridized Neural Machine Learning

Ziwei Zhang

School of Civil Engineering, North China University of Technology, Beijing 100144, China; zhang_zi_wei111@sina.com

Abstract: Concrete is a highly regarded construction material due to many advantages such as versatility, durability, fire resistance, and strength. Hence, having a prediction of the compressive strength of concrete (CSC) can be highly beneficial. The new generation of machine learning models has provided capable solutions to concrete-related simulations. This paper deals with predicting the CSC using a novel metaheuristic search scheme, namely the slime mold algorithm (SMA). The SMA retrofits an artificial neural network (ANN) to predict the CSC by incorporating the effect of mixture ingredients and curing age. The optimal configuration of the algorithm trained the ANN by taking the information of 824 specimens. The measured root mean square error (RMSE = 7.3831) and the Pearson correlation coefficient (R = 0.8937) indicated the excellent capability of the SMA in the assigned task. The same accuracy indicators (i.e., the RMSE of 8.1321 and R = 0.8902) revealed the competency of the developed SMA-ANN in predicting the CSC for 206 stranger specimens. In addition, the used method outperformed two benchmark algorithms of Henry gas solubility optimization (HGSO) and Harris hawks optimization (HHO) in both training and testing phases. The findings of this research pointed out the applicability of the SMA-ANN as a new substitute to burdensome laboratory tests for CSC estimation. Moreover, the provided solution is compared to some previous studies, and it is shown that the SMA-ANN enjoys higher accuracy. Therefore, an explicit mathematical formula is developed from this model to provide a convenient CSC predictive formula.

Keywords: building material; concrete; compressive strength; neural network; slime mold algorithm

Citation: Zhang, Z. Estimating the Concrete Ultimate Strength Using a Hybridized Neural Machine Learning. *Buildings* **2023**, *13*, 1852. https://doi.org/10.3390/buildings13071852

Academic Editors: Tom Lahmer, Ehsan Harirchian and Viviana Novelli

Received: 12 May 2023
Revised: 18 June 2023
Accepted: 25 June 2023
Published: 21 July 2023

1. Introduction

In recent years, the world of engineering has witnessed significant developments that have enabled experts to solve problems with higher accuracy and convenience [1–3]. These developments include a wide range of civil engineering domains such as geotechnical [4,5] and water [6,7] analysis. Focusing on the structural aspects of this field, engineers have benefitted from various technologies and simulation tools for analyzing the behavior of structures (and their particular elements) [8–10] under different loading conditions [11–13].

Also, many laboratory tools and innovative approaches have been successfully employed for investigating the behavior of structural materials [14–16]. When it comes to construction materials, concrete is known as one of the most effective ones. Mechanical parameters, and particularly the compressive strength of concrete (CSC), play an appreciable role in determining the quality of this popular material [17,18]. Since the non-linear effect of different parameters should be incorporated for CSC analysis, machine learning models, and more particularly artificial neural networks (ANN), have been regarded for this aim. Some of the primary efforts in utilizing the ANNs for the CSC problem can be found in studies like [19,20]. Prasad et al. [21] used this model for predicting the CSC of self-compacting and high-performance concretes containing high-volume fly ash. Duan et al. [22] successfully modeled the CSC of recycled aggregate concrete using ANNs. Regarding the 99.55% correlation, as well as the mean absolute percentage error (MAPE) below 2%, they concluded the applicability of the ANN. Naderpour et al. [23] applied a similar methodology to environmentally friendly concrete. ANNs have also performed profitably for other properties of concrete like slump [24], creep and shrinkage [25], and strain [26].

In a more general sense, many scientific efforts have been dedicated to foreseeing a specific behavior, particularly prediction tasks [27–29]. As for the application of machine learning models for CSC modeling, many attempts can be found in the published literature [30–32]. Akande et al. [33] introduced support vector machine (SVM) as a stable approach for this objective. Also, the proposed SVM outperformed the ANN with respect to root mean square error (RMSE) values (23.14 vs. 27.15). Feng et al. [34] used an adaptive boosting algorithm (a combination of several learners) to predict the CSC. While the mean absolute percentage error (MAPE) for this model was around 6.8%, conventional benchmarks including ANN and SVM achieved MAPE of around 10 and 15%, respectively. Accordingly, the suggested model presented a promising approximation of the CSC. The authors also showed that specifying 80% of the data for pattern recognition leads to an acceptable accuracy. Moreover, scholars like Başyigit et al. [35] and Vakhshouri and Nejadi [36] proved the feasibility of fuzzy-based tools for handling the CSC estimation.

Moreover, optimization algorithms have served as prominent techniques for analyzing mechanical parameters of concrete such as CSC. Kandiri et al. [37] optimized an ANN by a multi-objective slap swarm algorithm for the prediction of CSC when the mixture contains ground granulated blast furnace slag. The authors compared the accuracy of the developed models with a well-known machine learning tool called the M5P model tree. The larger MAPE of the MP5 (12.05 vs. 7.25%) demonstrated the superiority of the optimal ANN. Naseri et al. [38] attained an optimal design of sustainable concrete by predicting the CSC using a potent metaheuristic algorithm called water cycle algorithm (WCA). In addition to the WCA, popular algorithms like ANN and SVM were also considered. The most sustainable mixtures were eventually detected among the 16 tested ones. Moreover, the cuckoo search algorithm (CSA) was used for a similar purpose by Boindala and Arunachalam [39]. As another usage of the WCA, Ashrafian et al. [40] coupled multivariate adaptive regression splines (MARS) with this technique to create a capable hybrid model. Similar to earlier efforts, the authors also proved the superiority of the developed model over a number of conventional tools like standard MARS and ANN. Grey wolf optimizer (GWO) is another well-known metaheuristic approach that was used by Golafshani et al. [41] for hybridizing the ANN and ANFIS toward simulating the CSC. In research by Zhang et al. [42], beetle antennae search (BAS) was employed to tune the parameters of a random forest model. The resultant hybrid method was applied to estimate the uniaxial CSC of concrete containing oil palm shell. Regarding the 96% correlation observed for the prediction phase, the developed BAS-based ensemble was introduced as an efficient approximator. Likewise, Akbarzadeh et al. [43] professed the outstanding accuracy of ANN tuned by electromagnetic field optimization (EFO) for predicting the CSC. The EFO algorithm outperformed several compatible techniques including sine cosine algorithm (SCA) and cuttlefish optimization algorithm (CFOA). This study presented the final solution in the form of a complex mathematical formula for convenient estimations of CSC. Further attempts concerning the use of metaheuristic algorithms for modeling different concrete characteristics can be found in studies [44–47].

As reported by the literature review, metaheuristic integrated approaches show high promise for studying CSC prediction. On the other hand, many old and reputable optimizers such as ICA [48], PSO [49], whale optimization algorithm [50], etc., have gained sufficient attention for this purpose. Therefore, due to the continuous development of metaheuristic algorithms, recent studies have mainly focused on evaluating newly devised techniques to keep the existing solutions updated. Some examples of these new metaheuristic algorithms that are coupled with ANN for CSC prediction are multi-tracker optimization algorithm (MTOA) [51], satin bowerbird optimizer (SBO) [52], equilibrium optimizer (EO) [53], beetle antennae search (BAS) [54], etc. Each of these algorithms follows a specific search scheme in order to find the optimum contribution between the CSC and concrete mixture components. Slime mold algorithm (SMA) [55] is another capable metaheuristic algorithm that has not been investigated in previous studies. This work presents various configurations of this algorithm that rules an ANN to attain an optimum non-linear prediction of the CSC. Moreover, two other metaheuristic algorithms of Harris hawks optimization (HHO) [56] and Henry gas solubility optimization (HGSO) [57] are employed as benchmark techniques to comparatively validate the performance of the SMA. The HHO and HGSO are also among newly developed techniques, and using them would add new insights to the body of knowledge regarding predicting mechanical properties of concrete, particularly the CSC. The final solution of this research is translated into a monolithic mathematical formula in order to provide a convenient predictive approach for the users.

2. Materials and Methods

2.1. Data and Statistics

It Yeh [19] collected the information (i.e., the conditions and results) of 1030 typical compressive strength tests (on the cylinder specimens with height 15 cm) to create the dataset used for capturing and reproducing the CSC behavior in this work. This dataset can be accessed at http://archive.ics.uci.edu/ml/datasets/Concrete+Compressive+Strength (accessed on 12 June 2021).

The amount of each ingredient of the specimen mixtures is considered as a separate independent factor for the corresponding CSC. Figure 1 shows the values of (a) cement, (b) blast furnace slag (BFS), (c) fly ash (FA1), (d) water, (e) superplasticizer (SP), (f) coarse aggregate (CA), (g) fine aggregate (FA2), and (h) age. The values of the mentioned factors are in the ranges of [102.00, 540.00], [0.00, 359.40], [0.00, 200.10], [121.75, 247.00], [0.00, 32.20], [801.00, 1145.00], [594.00, 992.60], and [1.00, 365.00], respectively. Meanwhile, the obtained CSCs are shown in Figure 1i with the minimum and maximum values of 2.33 to 82.60.

Table 1 provides some examples of the used data. As is seen, each row is labeled as either "Training" or "Testing". It determines the group that this data lies in. Based on random selection, as well as the division ratio of 80:20, a total of 824 and 206 samples form the training and testing datasets, respectively.

2.2. Methodology

Figure 2 shows the graphical methodology of the present study. The first section describes data provision, followed by the model development and prediction stage, and accuracy assessment. In the end, a comparison is carried out among the models to extract the formula from the outstanding one.

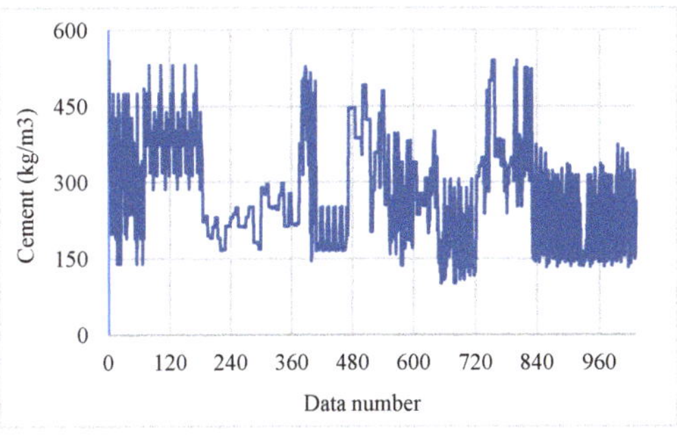

(a) cement

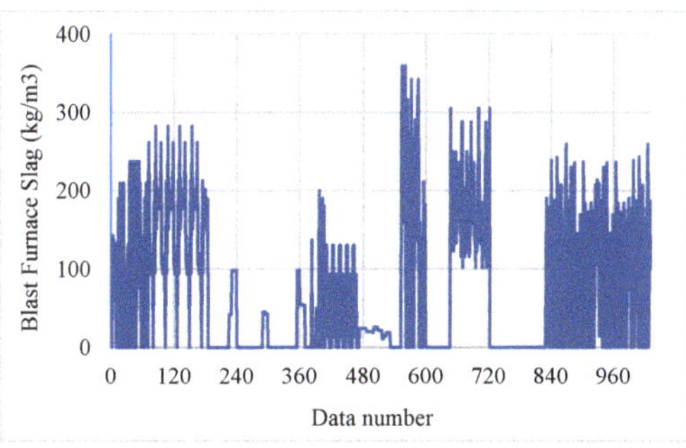

(b) BFS

Figure 1. *Cont.*

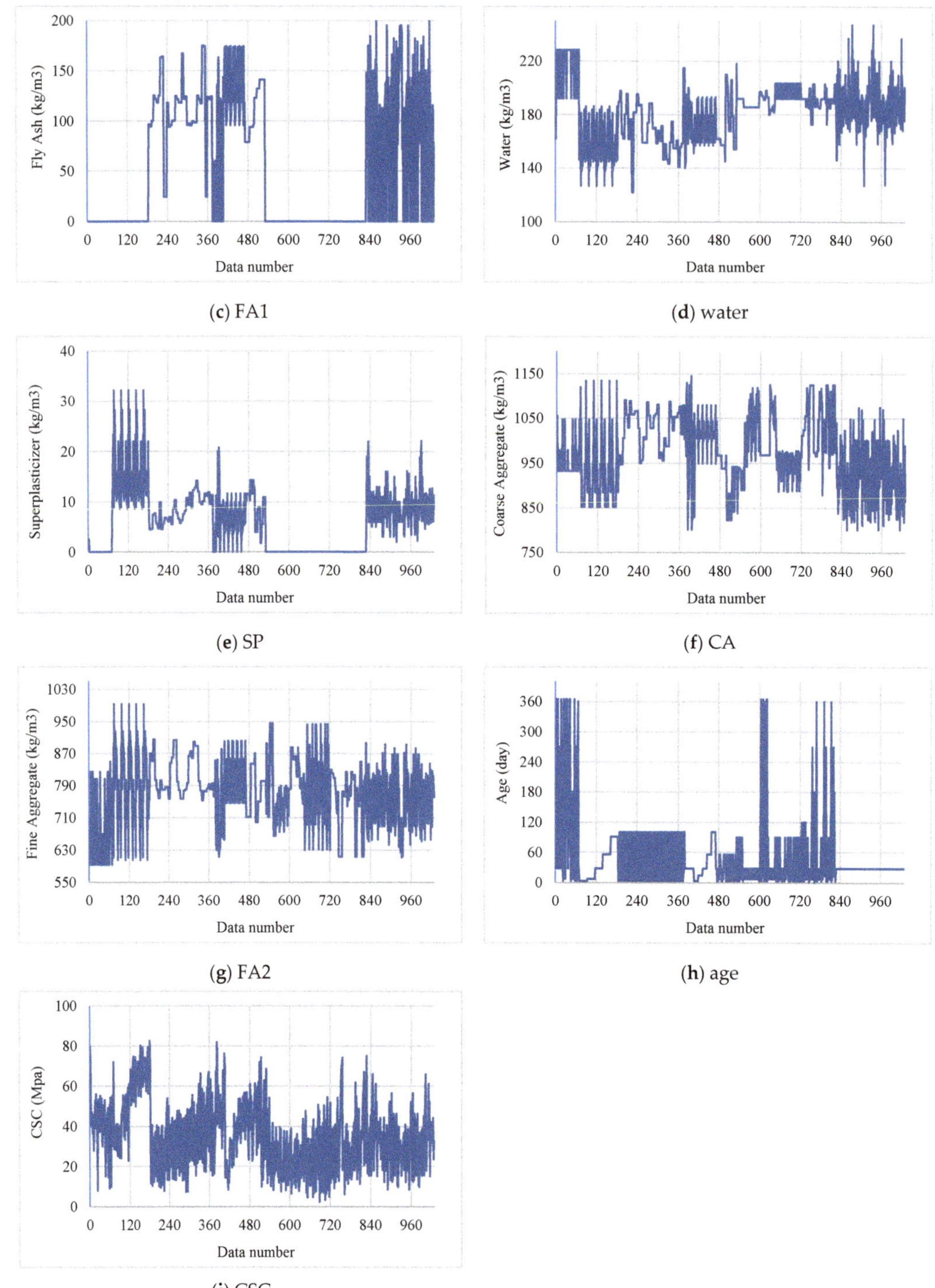

Figure 1. The variation in the CSC and independent factors.

Table 1. Information on some training and testing specimens.

Inputs									Target	Group
Cement (kg/m^3)	BFS (kg/m^3)	FA1 (kg/m^3)	Water (kg/m^3)	SP (kg/m^3)	CA (kg/m^3)	FA2 (kg/m^3)	Age (day)		CSC (MPa)	
149.00	236.00	0.00	176.00	13.00	847.00	893.00	28.00		32.96	Training
375.00	0.00	0.00	186.00	0.00	1038.00	758.00	7.00		26.06	Testing
213.76	98.06	24.52	181.74	6.65	1066.00	785.52	56.00		47.13	Testing
310.00	0.00	0.00	192.00	0.00	971.00	850.60	3.00		9.87	Training
290.35	0.00	96.18	168.08	9.41	961.18	865.00	100.00		48.97	Testing
277.00	117.00	91.00	191.00	7.00	946.00	666.00	28.00		43.57	Training
190.00	190.00	0.00	228.00	0.00	932.00	670.00	365.00		53.69	Training
446.00	24.00	79.00	162.00	11.64	967.00	712.00	3.00		25.02	Testing
236.00	157.00	0.00	192.00	0.00	972.60	749.10	28.00		32.88	Testing
214.90	53.80	121.89	155.63	9.61	1014.30	780.58	3.00		18.02	Training
330.50	169.60	0.00	194.90	8.10	811.00	802.30	28.00		56.62	Testing
181.38	0.00	167.01	169.59	7.56	1055.60	777.80	56.00		35.57	Training
475.00	118.80	0.00	181.10	8.90	852.10	781.50	91.00		74.19	Training
213.72	98.05	24.51	181.71	6.86	1065.80	785.38	100.00		53.90	Testing
218.23	54.64	123.78	140.75	11.91	1075.70	792.67	3.00		27.42	Training
300.00	0.00	0.00	184.00	0.00	1075.00	795.00	7.00		15.58	Testing
480.00	0.00	0.00	192.00	0.00	936.20	712.20	28.00		43.94	Testing
134.70	0.00	165.70	180.20	10.00	961.00	804.90	28.00		13.29	Training
397.00	0.00	0.00	185.00	0.00	1040.00	734.00	28.00		39.09	Training
218.23	54.64	123.78	140.75	11.91	1075.70	792.67	14.00		35.96	Testing

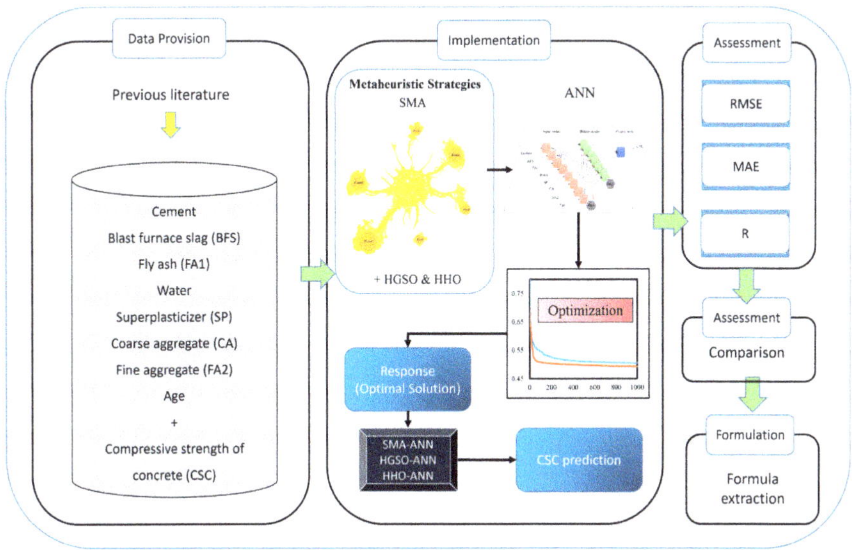

Figure 2. The methodology of the present study.

2.2.1. The SMA Algorithm

The name slime mold (SM) refers to Physarum polycephalum [58] that inhibits cool and humid areas. The main inspiration of the SMA algorithm is the dynamic nutritional behavior of the SM called Plasmodium. This stage includes three steps that an organic matter performs for seeking food, surrounding the discovered food, and secreting enzymes to digest it. Figure 3 shows the foraging morphology that forms an interconnected venous network using multiple food blocks at the same time.

Figure 3. Slime mold and hypothetical positions of food.

Inspired by the explained foraging behavior, Li, Chen, Wang, Heidari and Mirjalili [55] developed the SMA as a novel optimization approach. The objective of the SM is to find the optimal path to the largest concentration of nutrients [59]. Although the most promising food source is regarded by the SM, it needs to consider two important factors in foraging, namely speed and risk. Selecting the appropriate time for leaving the searched area (toward a new one) is another challenge for the SM. To figure it out, the algorithm uses heuristic or empirical rules. However, as explained, the algorithm can simultaneously exploit more than one source [60]. Overall, when several food sources with different qualities are at their disposal, an adaptive search strategy is executed to attain the best one.

Mathematically, the SMA comprises four major stages that are described below.

(a) Approaching food: Regarding the odor in the air, the SM approaches food based on the below equation:

$$\vec{X}(t+1) = \begin{cases} \vec{X_b}(t) + \vec{vb} \cdot \left(\vec{W} \cdot \vec{X_A}(t) - \vec{X_B}(t) \right), & r < p \\ \vec{vc} \cdot \vec{X}(t), & r \geq p \end{cases}, \tag{1}$$

where t signifies the current iteration, $\vec{vc}$ follows a linear decrease from 1 to 0, $\vec{vb}$ is a parameter ranging in $[-a, a]$ where $a = \text{arctanh}\left(-\left(\frac{t}{\max\_t} \right) + 1 \right)$, and $\vec{W}$ stands for the weight of the SM. Also, locations belonging to the SM, the individual with the largest odor concentration so far, and two randomly selected individuals are represented by $\vec{X}$, $\vec{X_b}$, $\vec{X_A}$ and $\vec{X_B}$, respectively. Moreover, given $S(i)$ as the fitness of $\vec{X}$ and DF as the best fitness obtained ever, the term p can be formulated as follows:

$$p = tanh|S(i) - DF| \quad i \in 1, 2, \ldots, n. \tag{2}$$

Equation (3) provides $\vec{W}$.

$$W(SmellIndex(i)) = \begin{cases} 1 + r \cdot \log\left(\frac{bF - S(i)}{bF - wF} + 1\right), & condition \\ 1 - r \cdot \log\left(\frac{bF - S(i)}{bF - wF} + 1\right), & others \end{cases}, \tag{3}$$

$$SmellIndex = sort(S), \tag{4}$$

where *condition* refers to ranking the first half of the population with respect to $S(i)$. The term r is a random number in $[0, 1]$, bF and wF stand for the optimal and worst finesses grasped in the current repetitions, respectively, *SmellIndex* provides the ascending sequence of sorted $S(i)$s.

(b)   Wrapping the food: This stage models how the venous tissue structure of the SM is contracted during the search. In this regard, three parameters including the power of the waves released by the bio-oscillator, the thickness of the vein, and the speed of the cytoplasm flows are directly proportional to the concentration level of the food contacted by the vein. As explained, the SMA prioritizes different food blocks based on their concentrations. The regions with larger concentrations receive larger weights and vice versa. Thus, the position of the SM is updated toward better regions. This process is formulated in Equation (5),

$$\vec{X}^* = \begin{cases} rand \cdot (UB - LB) + LB, rand < z \\ \vec{X_b}(t) + \vec{vb} \cdot \left(W \cdot \vec{X_A}(t) - \vec{X_B}(t)\right), r < p \\ \vec{vc} \cdot \vec{X}(t), \; r \geq p \end{cases}, \tag{5}$$

in which LB and UB are the lower and upper bounds, and *rand* and r stand for the random value between 0 and 1.

(c)   Grabbling the food: The cytoplasmic flow in the veins is affected by the waves released by the biological oscillator. To simulate the variations of the SM's venous width, three vectors of $\vec{vc}$, $\vec{vb}$, and $\vec{W}$ are considered. $\vec{W}$ provides a mathematical presentation of the SM's oscillation frequency at different food concentrations. This parameter helps the SM to achieve a better food source by accelerating its movement toward high-quality ones and vice versa. The $\vec{vb}$ value randomly ranges in $[-a, a]$ and it heads to 0 as the number of iterations increases. The $\vec{vc}$ value ranges in $[-1, 1]$ and finally reaches 0. During this stage, some organic members are assigned to explore the remaining areas even if the SM reaches a more potential source compared to earlier attempts. It enables the algorithm to seek a better food block all over the area. Notably, the SM decides whether to select the proposed food source or look for another one with respect to the oscillation of $\vec{vb}$.

The Pseudo-code of the SMA is presented as Algorithm 1.

(d)   Computational complexity analysis: Considering different steps of the SMA (i.e., initialization, assessing the fitness, sorting, updating the weights, and updating the locations), the complexity of the algorithm is explained in this section. Let N and T be the number of the SM's cells and the maximum number of iterations, respectively, in a D-dimensional space. Then, $O(N)$, $O(N + N \log N)$, $O(N \times D)$, and $O(N \times D)$ are the computational complexity of initialization, fitness evaluation and sorting, weight update, and location update, respectively. Hence, the overall complexity of the SMA can be expressed as $O(N * (1 + T * N * (1 + \log N + 2 * D)))$ [55].

Algorithm 1. Pseudo-code of SMA [55].

Initialize the parameters *popsize Max_iteraition*;
Initialize the positions of slime mould $X_i (i = 1, 2, \ldots, n)$;
While $(t \leq Max\_iteraition)$
   Calculate the fitness of all slime mould;
Update bestFitness, X_b
   Calculate the W by **Equation** (3);
   For *each search portion*
   Update p, vb, vc;
   Update positions by **Equation** (5);
   End For
   $t = t + 1$;
End While
Return *bestFitness*, X_b;

2.2.2. Benchmark Trainers

Inspired by Henry's law, the HGSO is one of the most recent metaheuristic algorithms. It was developed by Hashim, Houssein, Mabrouk, Al-Atabany and Mirjalili [57]. A modified version was also designed by Hashim et al. [61]. Also, the HGSO was used by Cao et al. [62] for optimizing the parameters of a regression SVM model. In this algorithm, the position and a so-called characteristic "partial pressure" are initially assigned to each gas. The gases are then clusterized into a number of groups. The identification of the best gases is then carried out. Based on the specific rules of the algorithm, the position and solubility of each gas are updated toward raising the quality of the solution. It is worth noting that updating the worst particles is also considered as a measure for escaping from the local optimum. The algorithm is mathematically detailed in relevant studies like [63,64].

The second benchmark algorithm is the HHO that was designed by Heidari, Mirjalili, Faris, Aljarah, Mafarja and Chen [56]. The HHO has shown high applicability for various complex problems like analyzing landslide susceptibility [65] and slope stability [66]. The basis of this algorithm is the cooperative interaction between Harris' hawks for a shocking hunt that comprises tracing, encircling, approaching, and attacking. These steps are devised in three major stages. The first one is named exploration dedicated to seeking and discovering the prey. The two next stages are based on the energy of the prey. After transforming from exploration to exploitation, the attacking measures are taken in the exploitation stage. For more explanations about the HHO, please refer to [67,68].

2.3. Accuracy Indicators

To assess the performance of the developed models, two error indicators of RMSE and mean absolute error (MAE) are used. In addition, the Pearson correlation coefficient (R) is defined to reflect the agreement between the products of the networks with target values. This index can be in a range of $[-1, +1]$. Taking K as the number of data and $CSC_{i_{observed}}$ and $CSC_{i_{predicted}}$ as the expected and modeled CSC, respectively, Equations (6)–(8) formulate the RMSE, MAE, and R.

$$RMSE = \sqrt{\frac{1}{K}\sum_{i=1}^{K}\left[(CSC_{i_{observed}} - CSC_{i_{predicted}})\right]^2}, \tag{6}$$

$$MAE = \frac{1}{K}\sum_{I=1}^{K}\left|CSC_{i_{observed}} - CSC_{i_{predicted}}\right|, \tag{7}$$

$$R = \frac{\sum_{i=1}^{K}(CSC_{i_{predicted}} - \overline{CSC}_{predicted})(CSC_{i_{observed}} - \overline{CSC}_{observed})}{\sqrt{\sum_{i=1}^{K}(CSC_{i_{predicted}} - \overline{CSC}_{predicted})^2}\sqrt{\sum_{i=1}^{K}(CSC_{i_{observed}} - \overline{CSC}_{observed})^2}}. \tag{8}$$

3. Results and Discussion

3.1. Model Configuration and Training

In this work, the efficiency of the SMA scheme for the CSC modeling is examined. As explained, this algorithm plays the role of a trainer for an ANN processor. Composed of eight, seven, and one neuron(s) in the input, hidden, and output layers, respectively, a three-layer multi-layer perceptron (MLP) [69] serves as the ANN used for being hybridized by the SMA. The topology of the used ANN is obtained after trying various cases and it is presented in Figure 4. In order to form the problem function, the proposed ANN is represented by its mathematical form where it is fed by training samples. In the neurons of the hidden layer, the inputs are received from the former layer, and weight is determined for each one. The neuron then adds a bias to the value resulting from this multiplication [70]. The activation function is a significant element of the ANNs that is eventually applied to the calculations of each neuron to release the main response. In this work, Tansig and Purelin are considered the activation functions of the hidden and output layers, respectively.

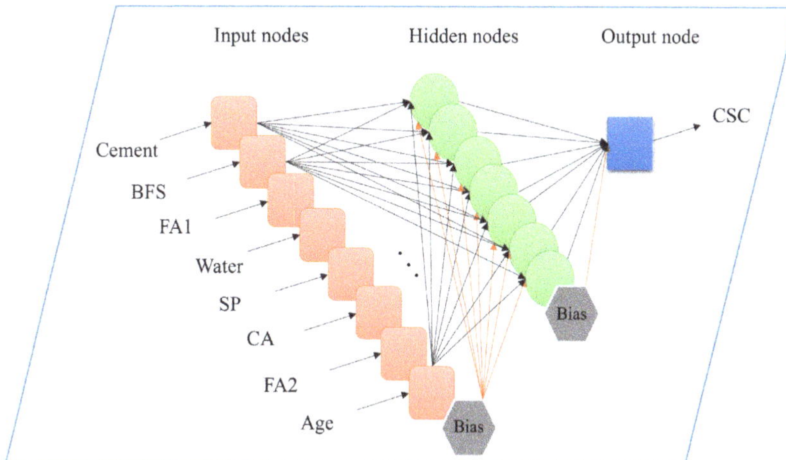

Figure 4. The used ANN topology.

Like other metaheuristic algorithms, the SMA tries to augment a random candidate solution during the implementation. It is fulfilled by an iterative procedure until either the desired goodness or the maximum number of iterations is satisfied. Figure 5 shows that each population size of the SMA has minimized the objective function (training RMSE in this case) in predicting the CSC. For this research, the values of the objective function (on the y-axis) reflect the RMSE of training in each iteration. Noticing the initial and final objective function values, this figure shows that the SMA algorithm has great potential in reducing the error of ANN training.

Table 2 provides the latest RMSEs of the tested configurations of the SMA. The same trial and error were performed for the HGSO and HHO as well. According to this table, although the final RMSEs of all population sizes are close, the differences can still reflect the eminent effect of population size. The best response of SMA, HGSO, and HHO algorithms is obtained for the population size of 400, 200, and 400, respectively. These configurations are selected to represent the SMA-ANN, HGSO-ANN, and HHO-ANN in the subsequent sections.

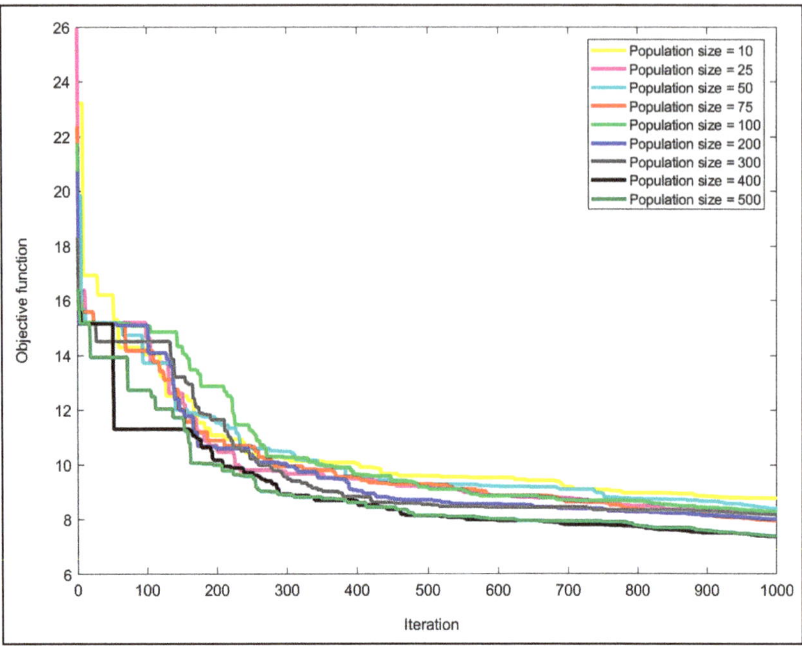

Figure 5. The results of testing different population sizes for the SMA algorithm.

Table 2. Optimal RMSEs obtained for the tested configurations.

Algorithm	Population Size								
	10	25	50	75	100	200	300	400	500
SMA	8.7830	8.1813	8.4056	7.9688	8.2738	8.0212	8.1812	7.3831	7.4055
HGSO	12.8753	11.0752	11.7703	11.4388	11.4658	9.0477	10.4068	11.0786	10.1627
HHO	11.6988	11.1059	11.4161	11.1815	10.7291	10.4342	10.7521	10.2017	10.4970

3.2. Accuracy Assessment

It was mentioned that the learning process was based on the training data. It means that the RMSEs reported in the previous section correspond to the training results. Thus, the proposed SMA-ANN achieved an RMSE of 7.3831 in grasping the CSC behavior. Also, the MAE was 5.7885. Figure 6 shows the histogram chart of the errors in this phase. The word error refers to the simple difference between the $CSC_{i_{observed}}$ and $CSC_{i_{predicted}}$ for each of the 824 data. In Figure 6, it can be seen that the error values follow a normal distribution, meaning that the higher the error magnitude, the lower the frequency. In general, this indicates desirable prediction results for all used models.

The RMSE and MAE for the benchmark models indicate a lower quality of training for the ANNs trained by the HGSO (9.0477 and 7.1566, respectively) and HHO (10.2017 and 8.2355). Since the training error represents tuning the weights and biases of the ANN, it can be deduced that the SMA algorithm found a more suitable matrix of these parameters.

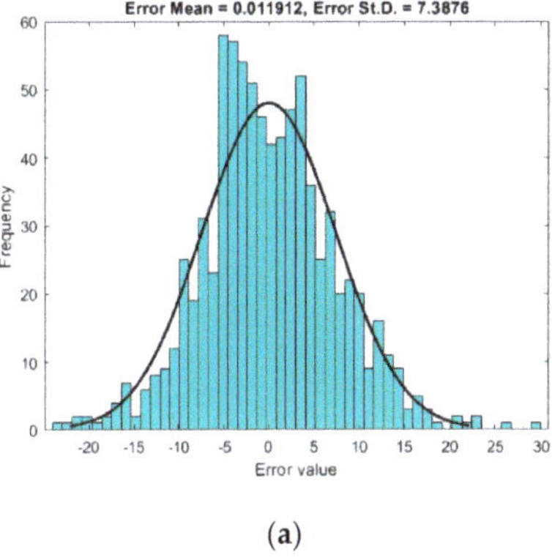

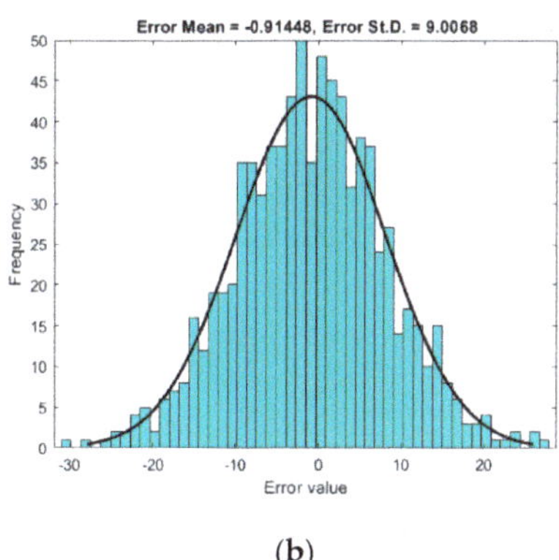

(a)

(b)

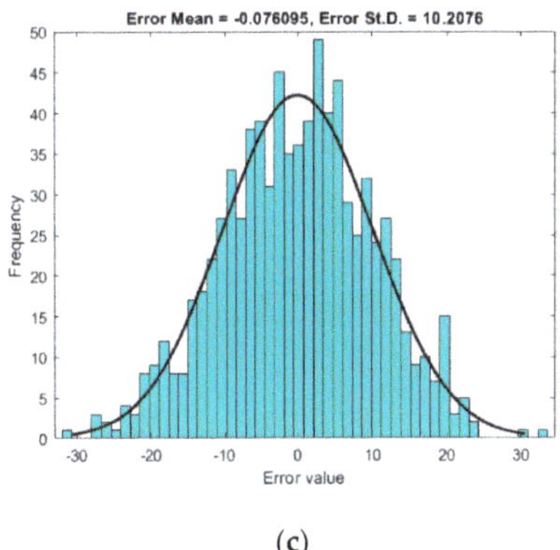

(c)

Figure 6. The histogram of training errors for (**a**) SMA-ANN, (**b**) HGSO-ANN, and (**c**) HHO-ANN.

Figure 7 shows the correlation charts of the training results. Visually, it is seen that all three charts show an acceptable correlation for the used models. However, the products of the SMA-ANN are in better agreement with the ideal situation (line Y = T). Also, the calculated R indices indicate an 89.37%, an 84.42%, and a 78.46% agreement between the expected CSCs and those estimated by the SMA-ANN, HGSO-ANN, and HHO-ANN.

The CSC pattern derived in the training phase was used to predict the CSC for 206 samples considered for evaluating the generalizability of the mapped relationship. A comparison of the modeled CSC with the estimated values is shown in Figure 8. This figure illustrates that all three models have shown good sensitivity to the changes and fluctuations in the CSC pattern.

With an RMSE of 8.1321, as well as an MAE of 6.1361, the SMA-ANN could reproduce the CSC with a good level of accuracy. Similar to the training phase, the proposed model achieved a larger accuracy in comparison with the benchmarks of the HGSO-ANN (RMSE = 9.9893 and MAE = 7.6427) and the HHO-ANN (RMSE = 11.5099 and MAE = 9.1671). Moreover, an agreement was observed between the training and testing results, meaning the better the model is trained, the more accurate outputs it produces.

Testing outputs are also depicted versus the expected CSCs in the form of correlation charts in Figure 9. Referring to the chart of the SMA-ANN, it can be said that this model can present a reliable prediction of the CSC for stranger conditions. In other words, the captured pattern is properly generalized to testing data. The calculated Rs for the models were 0.8902, 0.8278, and 0.7583, which demonstrates the superiority of the SMA over both HGSO and HHO.

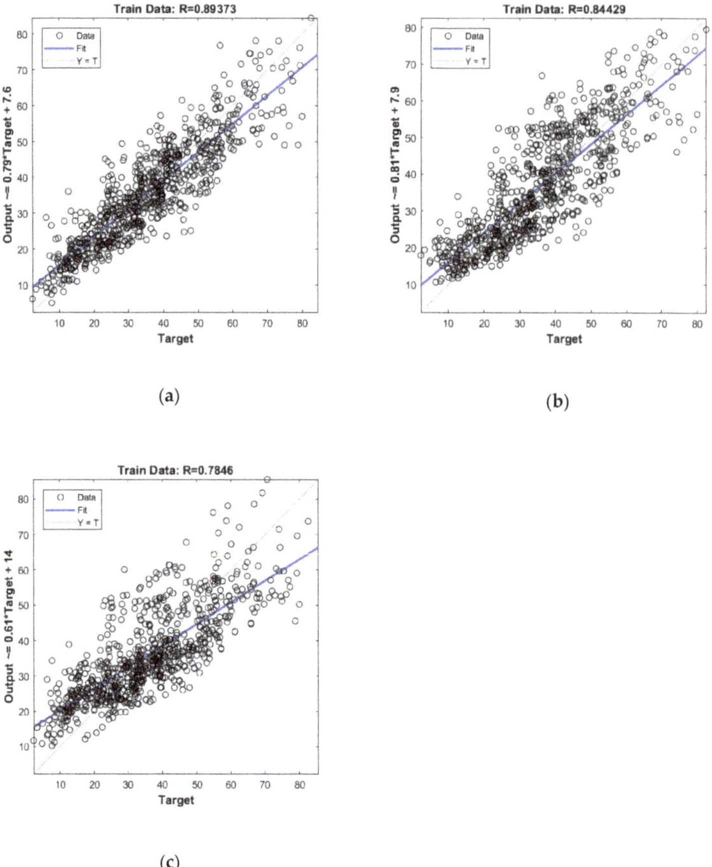

(**a**)   (**b**)

(**c**)

Figure 7. The correlation of training results for (**a**) SMA-ANN, (**b**) HGSO-ANN, and (**c**) HHO-ANN.

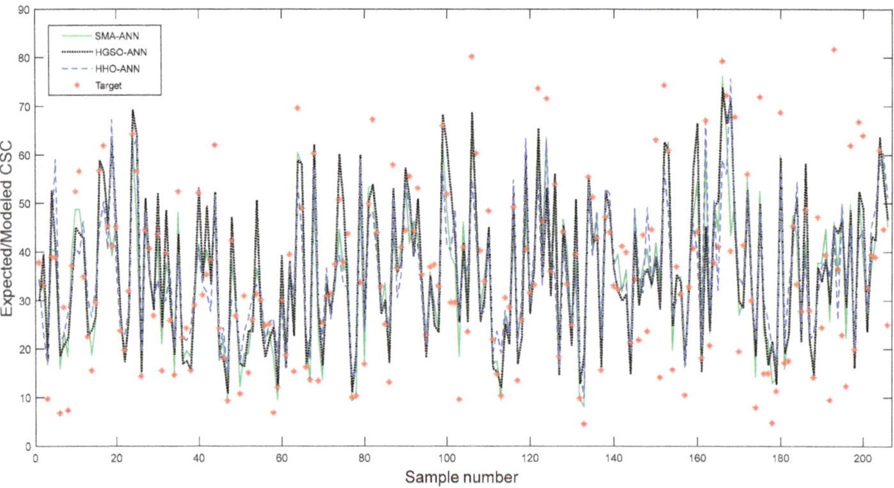

Figure 8. The testing products vs. the expected CSCs.

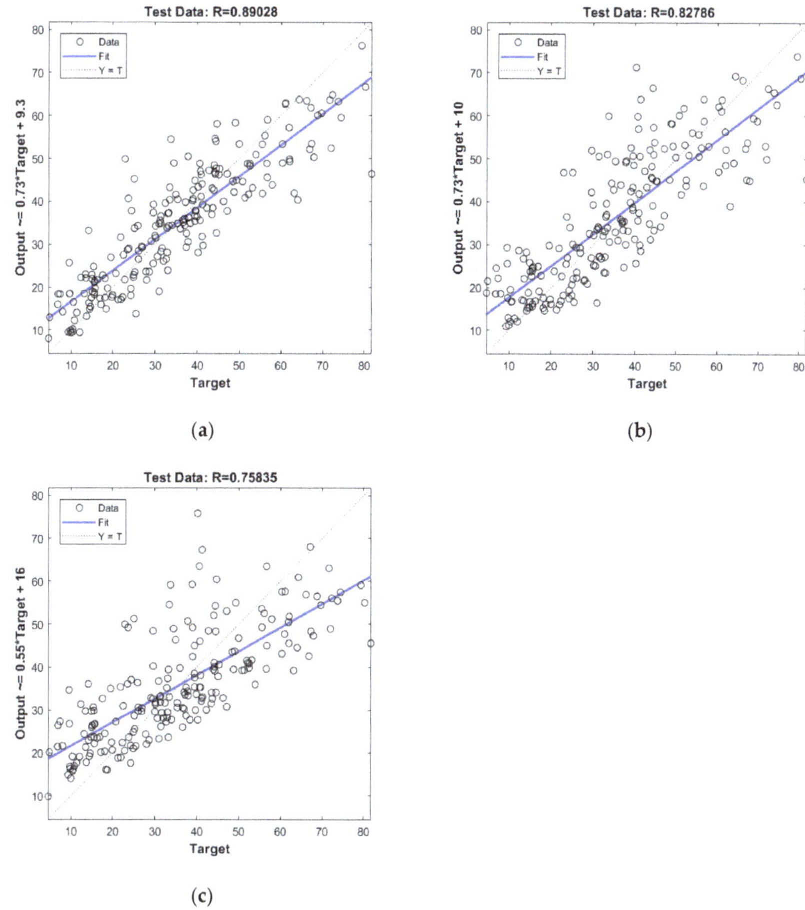

Figure 9. The correlation of testing results for (**a**) SMA-ANN, (**b**) HGSO-ANN, and (**c**) HHO-ANN.

Assessing the training and testing results together, it can be concluded that the used metaheuristic algorithms (i.e., SMA, HGSO, and HHO) performed suitable optimizations for adjusting the weights and biases of the ANN during the training process. These optimized parameters were used to create potential ANNs that could reliably predict the CSC.

3.3. A CSC Formula

As discussed, the better performance of the SMA points out the higher capability of this search scheme in tuning the ANN parameters including connecting weights and the bias of each neuron (Figure 4). The process by which the SMA-ANN estimates the CSC is summarized in Equation (9). It indicates applying the weights (i.e., *[LW]*) and bias (i.e., *[b2]*) of the output neuron to the response coming from hidden neurons. Likewise, this response is produced via multiplying the inputs (i.e., *[I]*) by the corresponding weights (i.e., *[IW]*) added to the bias vector (i.e., *[b2]*) and eventually applying $Tansig\,(x) = \frac{2}{1+e^{-2x}} - 1$ as the activation function.

$$CSC = [LW] * (Tansig\,(([IW] * [I]) + [b1])) + [b2]. \tag{9}$$

The terms used in the above equation are acquired based on Table 3:

Table 3. The parameters of the predictive formula in Equation (9).

Parameter	Value							
IW	$\begin{bmatrix} 0.1381 & -0.5092 & 0.7425 & 0.9454 & 0.9959 & 0.0459 & -0.2287 & 0.6467 \\ -0.1934 & -1.1992 & -0.8799 & -0.1856 & -0.0123 & 0.3267 & 0.1319 & -0.8830 \\ -0.1269 & 0.1029 & -1.0653 & 0.0773 & -0.5025 & -0.8111 & -1.0077 & 0.3078 \\ -0.7477 & -0.5067 & 0.4525 & -0.3054 & -0.8152 & 0.8423 & -0.6242 & 0.5573 \\ 0.1034 & 0.8107 & 0.4600 & -0.0100 & -0.8089 & 1.1472 & 0.4757 & -0.3344 \\ 0.7039 & 0.9927 & 0.1850 & 0.0316 & -0.0522 & 0.8930 & -0.8899 & -0.2825 \\ -0.1854 & 0.9990 & 0.6670 & -0.2774 & -0.2457 & -0.5999 & 0.6380 & -0.8981 \end{bmatrix}$							
I	$\begin{bmatrix} Cement \\ BFS \\ FA1 \\ water \\ SP \\ CA \\ FA2 \\ Age \end{bmatrix}$							
$b1$	$\begin{bmatrix} -1.7855 \\ 1.1903 \\ 0.5952 \\ 0.0000 \\ 0.5952 \\ 1.1903 \\ -1.7855 \end{bmatrix}$							
LW	$\begin{bmatrix} 0.4433 & 0.7756 & 0.4244 & 0.8918 & -0.9618 & 0.0239 & -0.4702 \end{bmatrix}$							
$b2$	$\begin{bmatrix} 0.7595 \end{bmatrix}$							

3.4. Importance Analysis

Owing to the fact that there are various influential parameters for determining the CSC [71–73], analyzing the importance of the input parameters is a significant step, especially when it comes to machine learning applications. To attain it, the principal component analysis (PCA) technique [74] is used within the IBM SPSS Statistics 22 environment. This technique creates some components, each representing an independent combination of the existing inputs. An eigenvalue is calculated for each component, and if the eigenvalue exceeds the threshold = 1, it means that the corresponding component is significant [75]. Figure 10 shows the calculated eigenvalues.

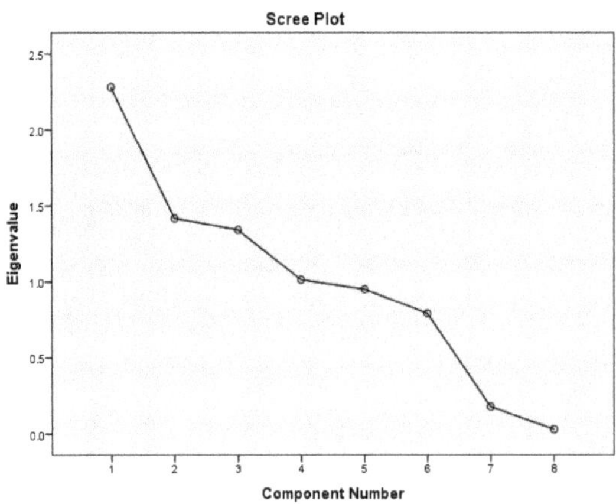

Figure 10. The scree plot of the PCA analysis.

According to this figure, four components acquired eigenvalues > 1. Based on further statistics, these four components account for an around 75.6% variation in the dataset. Table 4 presents the rotated component matrix, giving details of these components obtained from the varimax rotation method. Based on the earlier literature, two thresholds of +0.75 and −0.75 can be considered for identifying the most significant inputs [76,77]. Therefore, the inputs with loading factors lower than −0.75 and greater than +0.75 are shown with red and blue colors, respectively. Eventually, it is found that the PCA analysis suggests Cement, BFS, Water, SP, and CA as the most important inputs.

Table 4. Rotated component matrix obtained from varimax rotation method (loading factors lower than −0.75 and greater than +0.75 are shown with red and blue colors, respectively).

Inputs	Component			
	1	**2**	**3**	**4**
Cement	0.084	0.940	0.196	0.047
BFS	0.033	−0.076	−0.922	0.234
FA1	0.269	−0.662	0.384	0.032
Water	−0.878	−0.016	−0.247	0.056
SP	0.781	−0.003	0.094	0.391
CA	0.028	−0.041	0.152	−0.939
FA2	0.334	−0.265	0.470	0.342
Age	−0.613	0.168	0.257	0.205

3.5. Further Discussion and Future Studies

The models that were suggested in this work could achieve a reliable early analysis of the CSC based on the effects of mixture characteristics including cement, BFS, FA1, water, SP, CA, FA2, and age. However, it is worth discussing that the suggested model, i.e., the SMA-ANN, has achieved significant improvements with respect to some of the previous studies [43,52–54]. Table 5 compares the accuracy indices of this study to those that were commonly used in the cited studies. The respective RMSE and MAE of the SAM-ANN were 8.1321 and 6.1361, which are comparably smaller than those in the mentioned studies. Likewise, the R index was 0.89028, which is higher than the corresponding correlation indices in previous studies. This comparison indicates that the SMA-ANN outperformed several single and optimized versions of the ANN model. In other words, the results of this study add a capable methodology to the field of CSC prediction.

Table 5. Comparing accuracy indices with some previous studies.

Study	CSC Range (Mpa)	Model	RMSE	MAE	Correlation (R & R^2)
This study	[2.33, 82.60]	SMA-ANN	8.1321	6.1361	0.89028
[52]	[4.23, 96.30]	ANN-HGSO	9.5248	7.8632	0.87394
		ANN-SFO	8.5728	7.0550	0.87936
[43]	[2.33, 82.60]	ANN-SCA	10.0340	7.8248	0.80249
		ANN-CFOA	9.8392	7.6538	0.79832
[53]	[2.33, 82.60]	BP–MLP (Single ANN)	8.2675	6.2103	0.788
[54]	[2.33, 82.60]	BP–MLP (Single ANN)	8.9753	6.8112	0.7442
		SCE–MLP	8.3540	6.4657	0.7876

Nowadays, concrete is being used in many parts of the construction industry [78–80]. Hereupon, it is important to exploit efficient ways towards practical usage of the suggested models. In this sense, it can be explained that these models provide reliable, non-destructive, and cost-efficient approaches for predicting the CSC which is a non-linear and complex mechanical parameter of concrete. Engineers may use intelligent models in order to achieve a reliable design of the concrete mixture with respect to the desired CSC. Another application can be evaluating the sensitivity of the CSC to a specific mixture ingredient. For instance, engineers may be interested in assessing the variation of CSC with the changes in cement characteristics. Also, the results of the PCA analysis can be considered for this purpose, because this analysis revealed that Cement, BFS, Water, SP, and CA have the greatest influence on the CSC. To sum up, since the behavior of the CSC has been nicely understood by the models, they are applicable to predict this parameter for various mixtures without the need for conducting destructive and time-consuming laboratory tests.

Notwithstanding the advancement achieved in this study, there are some limitations that can be considered for conducting future studies. For instance, the proposed methodologies (i.e., SMA, HHO, and HGSO) can be compared to newer metaheuristic algorithms for further improving the solution. Another idea can be applying the results of the PCA analysis to reduce the dimension of the problem. More clearly, since the used dataset consists of a large number of inputs, reducing them can result in a less complicated ANN network, and enhance the prediction efficiency. Based on the PCA results, future studies are recommended to predict the CSC by considering the effect of Cement, BFS, Water, SP, and CA and disregarding FA1, FA2, and Age of the mixture. The results then can be compared to those of the present study to determine whether this idea can improve the accuracy of prediction.

4. Conclusions

In this work, a novel search method based on the foraging actions of slime mold was tested for training a popular neural system applied to the prediction of concrete compressive strength. The SMA trained an $8 \times 7 \times 1$ MLP neural network by relating the CSC to eight influential factors. The results were evaluated for the training and testing phases using three accuracy indicators. The obtained RMSEs (7.3831 and 8.1321, 9.0477 and 9.9893, and 10.2017 and 11.5099 for the training and testing phases of the SMA-ANN, HGSO-ANN, and HHO-ANN, respectively) demonstrated the superiority of the proposed technique over two recently developed colleagues in both grasping and reproducing the CSC behavior. The above 89% correlation between the expected and estimated CSCs showed that the SMA-ANN can be a promising predictive model for practical applications. Comparison with the literature disclosed the higher accuracy of the suggested SMA-ANN against some previously used models. The final solution was translated into a mathematical format to provide a reliable and convenient formula for predicting the CSC. Moreover, the PCA importance assessment revealed that Cement, BFS, Water, SP, and CA are the most important ingredients of the concrete for predicting the CSC. In the end, some ideas were suggested to cope with the limitations of the study in future projects.

Funding: This research received no funding.

Data Availability Statement: Data availability is explained in Section 2.1.

Conflicts of Interest: The author declares no conflict of interest.

References

1. Wang, J.; Tian, J.; Zhang, X.; Yang, B.; Liu, S.; Yin, L.; Zheng, W. Control of time delay force feedback teleoperation system with finite time convergence. *Front. Neurorobotics* **2022**, *16*, 877069. [CrossRef]
2. Ban, Y.; Liu, M.; Wu, P.; Yang, B.; Liu, S.; Yin, L.; Zheng, W. Depth estimation method for monocular camera defocus images in microscopic scenes. *Electronics* **2022**, *11*, 2012. [CrossRef]
3. Gu, Q.; Tian, J.; Yang, B.; Liu, M.; Gu, B.; Yin, Z.; Yin, L.; Zheng, W. A novel architecture of a six degrees of freedom parallel platform. *Electronics* **2023**, *12*, 1774. [CrossRef]

4. Zhang, C.; Yin, Y.; Yan, H.; Zhu, S.; Li, B.; Hou, X.; Yang, Y. Centrifuge modeling of multi-row stabilizing piles reinforced reservoir landslide with different row spacings. *Landslides* **2023**, *20*, 559–577. [CrossRef]
5. Zhou, S.; Lu, C.; Zhu, X.; Li, F. Preparation and characterization of high-strength geopolymer based on BH-1 lunar soil simulant with low alkali content. *Engineering* **2021**, *7*, 1631–1645. [CrossRef]
6. Jia, S.; Dai, Z.; Zhou, Z.; Ling, H.; Yang, Z.; Qi, L.; Wang, Z.; Zhang, X.; Thanh, H.V.; Soltanian, M.R. Upscaling dispersivity for conservative solute transport in naturally fractured media. *Water Res.* **2023**, *235*, 119844. [CrossRef]
7. Zhou, J.; Wang, L.; Zhong, X.; Yao, T.; Qi, J.; Wang, Y.; Xue, Y. Quantifying the major drivers for the expanding lakes in the interior Tibetan Plateau. *Sci. Bull* **2022**, *67*, 474–478. [CrossRef]
8. Pishro, A.A.; Zhang, Z.; Pishro, M.A.; Xiong, F.; Zhang, L.; Yang, Q.; Matlan, S.J. UHPC-PINN-parallel micro element system for the local bond stress–slip model subjected to monotonic loading. In *Structures*; Elsevier: Amsterdam, The Netherlands, 2022; pp. 570–597.
9. Hong, Y.; Yao, M.; Wang, L. A multi-axial bounding surface py model with application in analyzing pile responses under multi-directional lateral cycling. *Comput. Geotech.* **2023**, *157*, 105301. [CrossRef]
10. Huang, H.; Li, M.; Yuan, Y.; Bai, H. Theoretical analysis on the lateral drift of precast concrete frame with replaceable artificial controllable plastic hinges. *J. Build. Eng.* **2022**, *62*, 105386. [CrossRef]
11. Huang, H.; Yuan, Y.; Zhang, W.; Li, M. Seismic behavior of a replaceable artificial controllable plastic hinge for precast concrete beam-column joint. *Eng. Struct.* **2021**, *245*, 112848. [CrossRef]
12. Ghasemi, M.; Zhang, C.; Khorshidi, H.; Zhu, L.; Hsiao, P.-C. Seismic upgrading of existing RC frames with displacement-restraint cable bracing. *Eng. Struct.* **2023**, *282*, 115764. [CrossRef]
13. Xia, Y.; Shi, M.; Zhang, C.; Wang, C.; Sang, X.; Liu, R.; Zhao, P.; An, G.; Fang, H. Analysis of flexural failure mechanism of ultraviolet cured-in-place-pipe materials for buried pipelines rehabilitation based on curing temperature monitoring. *Eng. Fail. Anal.* **2022**, *142*, 106763. [CrossRef]
14. Li, J.; Chen, M.; Li, Z. Improved soil–structure interaction model considering time-lag effect. *Comput. Geotech.* **2022**, *148*, 104835. [CrossRef]
15. Peng, J.; Xu, C.; Dai, B.; Sun, L.; Feng, J.; Huang, Q. Numerical Investigation of Brittleness Effect on Strength and Microcracking Behavior of Crystalline Rock. *Int. J. Geomech.* **2022**, *22*, 04022178. [CrossRef]
16. Liu, C.; Cui, J.; Zhang, Z.; Liu, H.; Huang, X.; Zhang, C. The role of TBM asymmetric tail-grouting on surface settlement in coarse-grained soils of urban area: Field tests and FEA modelling. *Tunn. Undergr. Space Technol.* **2021**, *111*, 103857. [CrossRef]
17. Shafabakhsh, G.; Ahmadi, S. Evaluation of coal waste ash and rice husk ash on properties of pervious concrete pavement. *Int. J. Eng.-Trans. B Appl.* **2016**, *29*, 192–201.
18. Behnam, B.; Ronagh, H.R.; Baji, H. Methodology for investigating the behavior of reinforced concrete structures subjected to post earthquake fire. *Adv. Concr. Constr.* **2013**, *1*, 29. [CrossRef]
19. Yeh, I.-C. Modeling of strength of high-performance concrete using artificial neural networks. *Cem. Concr. Res.* **1998**, *28*, 1797–1808. [CrossRef]
20. Yeh, I.-C. Design of high-performance concrete mixture using neural networks and nonlinear programming. *J. Comput. Civ. Eng.* **1999**, *13*, 36–42. [CrossRef]
21. Prasad, B.R.; Eskandari, H.; Reddy, B.V. Prediction of compressive strength of SCC and HPC with high volume fly ash using ANN. *Constr. Build. Mater.* **2009**, *23*, 117–128. [CrossRef]
22. Duan, Z.-H.; Kou, S.-C.; Poon, C.-S. Prediction of compressive strength of recycled aggregate concrete using artificial neural networks. *Constr. Build. Mater.* **2013**, *40*, 1200–1206. [CrossRef]
23. Naderpour, H.; Rafiean, A.H.; Fakharian, P. Compressive strength prediction of environmentally friendly concrete using artificial neural networks. *J. Build. Eng.* **2018**, *16*, 213–219. [CrossRef]
24. Qiu, C.; Gong, S.; Gao, W. Three artificial intelligence-based solutions predicting concrete slump. *UPB Sci. Bull. Ser. C* **2019**, *81*, 2019.
25. Karthikeyan, J.; Upadhyay, A.; Bhandari, N.M. Artificial neural network for predicting creep and shrinkage of high performance concrete. *J. Adv. Concr. Technol.* **2008**, *6*, 135–142. [CrossRef]
26. Mohammadhassani, M.; Nezamabadi-Pour, H.; Suhatril, M.; Shariati, M. Identification of a suitable ANN architecture in predicting strain in tie section of concrete deep beams. *Struct. Eng. Mech.* **2013**, *46*, 853–868. [CrossRef]
27. Xu, L.; Cai, M.; Dong, S.; Yin, S.; Xiao, T.; Dai, Z.; Wang, Y.; Soltanian, M.R. An upscaling approach to predict mine water inflow from roof sandstone aquifers. *J. Hydrol.* **2022**, *612*, 128314. [CrossRef]
28. Zhang, Z.; Li, W.; Yang, J. Analysis of stochastic process to model safety risk in construction industry. *J. Civ. Eng. Manag.* **2021**, *27*, 87–99. [CrossRef]
29. Moayedi, H.; Mehrabi, M.; Mosallanezhad, M.; Rashid, A.S.A.; Pradhan, B. Modification of landslide susceptibility mapping using optimized PSO-ANN technique. *Eng. Comput.* **2019**, *35*, 967–984. [CrossRef]
30. Chou, J.-S.; Tsai, C.-F. Concrete compressive strength analysis using a combined classification and regression technique. *Autom. Constr.* **2012**, *24*, 52–60. [CrossRef]
31. Yaseen, Z.M.; Deo, R.C.; Hilal, A.; Abd, A.M.; Bueno, L.C.; Salcedo-Sanz, S.; Nehdi, M.L. Predicting compressive strength of lightweight foamed concrete using extreme learning machine model. *Adv. Eng. Softw.* **2018**, *115*, 112–125. [CrossRef]

32. Mousavi, S.M.; Aminian, P.; Gandomi, A.H.; Alavi, A.H.; Bolandi, H. A new predictive model for compressive strength of HPC using gene expression programming. *Adv. Eng. Softw.* **2012**, *45*, 105–114. [CrossRef]

33. Akande, K.O.; Owolabi, T.O.; Twaha, S.; Olatunji, S.O. Performance comparison of SVM and ANN in predicting compressive strength of concrete. *IOSR J. Comput. Eng.* **2014**, *16*, 88–94. [CrossRef]

34. Feng, D.-C.; Liu, Z.-T.; Wang, X.-D.; Chen, Y.; Chang, J.-Q.; Wei, D.-F.; Jiang, Z.-M. Machine learning-based compressive strength prediction for concrete: An adaptive boosting approach. *Constr. Build. Mater.* **2020**, *230*, 117000. [CrossRef]

35. Başyigit, C.; Akkurt, I.; Kilincarslan, S.; Beycioglu, A. Prediction of compressive strength of heavyweight concrete by ANN and FL models. *Neural Comput. Appl.* **2010**, *19*, 507–513. [CrossRef]

36. Vakhshouri, B.; Nejadi, S. Prediction of compressive strength of self-compacting concrete by ANFIS models. *Neurocomputing* **2018**, *280*, 13–22. [CrossRef]

37. Kandiri, A.; Golafshani, E.M.; Behnood, A. Estimation of the compressive strength of concretes containing ground granulated blast furnace slag using hybridized multi-objective ANN and salp swarm algorithm. *Constr. Build. Mater.* **2020**, *248*, 118676. [CrossRef]

38. Naseri, H.; Jahanbakhsh, H.; Hosseini, P.; Nejad, F.M. Designing sustainable concrete mixture by developing a new machine learning technique. *J. Clean. Prod.* **2020**, *258*, 120578. [CrossRef]

39. Boindala, S.P.; Arunachalam, V. Concrete Mix Design Optimization Using a Multi-objective Cuckoo Search Algorithm. In *Soft Computing: Theories and Applications*; Springer: Berlin/Heidelberg, Germany, 2020; pp. 119–126.

40. Ashrafian, A.; Shokri, F.; Amiri, M.J.T.; Yaseen, Z.M.; Rezaie-Balf, M. Compressive strength of Foamed Cellular Lightweight Concrete simulation: New development of hybrid artificial intelligence model. *Constr. Build. Mater.* **2020**, *230*, 117048. [CrossRef]

41. Golafshani, E.M.; Behnood, A.; Arashpour, M. Predicting the compressive strength of normal and High-Performance Concretes using ANN and ANFIS hybridized with Grey Wolf Optimizer. *Constr. Build. Mater.* **2020**, *232*, 117266. [CrossRef]

42. Zhang, J.; Li, D.; Wang, Y. Predicting uniaxial compressive strength of oil palm shell concrete using a hybrid artificial intelligence model. *J. Build. Eng.* **2020**, *30*, 101282. [CrossRef]

43. Akbarzadeh, M.R.; Ghafourian, H.; Anvari, A.; Pourhanasa, R.; Nehdi, M.L. Estimating Compressive Strength of Concrete Using Neural Electromagnetic Field Optimization. *Materials* **2023**, *16*, 4200. [CrossRef] [PubMed]

44. Silva, D.L.; de Jesus, K.L.M.; Villaverde, B.S.; Adina, E.M. Hybrid Artificial Neural Network and Genetic Algorithm Model for Multi-Objective Strength Optimization of Concrete with Surkhi and Buntal Fiber. In Proceedings of the 2020 12th International Conference on Computer and Automation Engineering, Sydney, NSW, Australia, 14–16 February 2020; pp. 47–51.

45. Ghazavi, M.; Bonab, S.B. Optimization of reinforced concrete retaining walls using ant colony method. *Geotech. Saf. Risk* **2011**, *2011*, 297–306.

46. García-Segura, T.; Yepes, V.; Martí, J.V.; Alcalá, J. Optimization of concrete I-beams using a new hybrid glowworm swarm algorithm. *Lat. Am. J. Solids Struct.* **2014**, *11*, 1190–1205. [CrossRef]

47. Sadowski, Ł.; Nikoo, M.; Shariq, M.; Joker, E.; Czarnecki, S. The nature-inspired metaheuristic method for predicting the creep strain of green concrete containing ground granulated blast furnace slag. *Materials* **2019**, *12*, 293. [CrossRef]

48. Duan, J.; Asteris, P.G.; Nguyen, H.; Bui, X.-N.; Moayedi, H. A novel artificial intelligence technique to predict compressive strength of recycled aggregate concrete using ICA-XGBoost model. *Eng. Comput.* **2020**, *37*, 3329–3346. [CrossRef]

49. Xue, X. Evaluation of concrete compressive strength based on an improved PSO-LSSVM model. *Comput. Concr.* **2018**, *21*, 505–511.

50. Bui, D.T.; Ghareh, S.; Moayedi, H.; Nguyen, H. Fine-tuning of neural computing using whale optimization algorithm for predicting compressive strength of concrete. *Eng. Comput.* **2019**, *37*, 701–712.

51. Zhao, Y.; Hu, H.; Song, C.; Wang, Z. Predicting compressive strength of manufactured-sand concrete using conventional and metaheuristic-tuned artificial neural network. *Measurement* **2022**, *194*, 110993. [CrossRef]

52. Wu, D.; LI, S.; Moayedi, H.; Cifci, M.A.; Li, B.N. ANN-Incorporated satin bowerbird optimizer for predicting uniaxial compressive strength of concrete. *Steel Compos. Struct.* **2022**, *45*, 281–291.

53. Moayedi, H.; Eghtesad, A.; Khajehzadeh, M.; Keawsawasvong, S.; Al-Amidi6d, M.M.; Le Van, B. Optimized ANNs for predicting compressive strength of high-performance concrete. *Steel Compos. Struct.* **2022**, *44*, 853–868.

54. Hu, P.; Moradi, Z.; Ali, H.E.; Foong, L.K. Metaheuristic-reinforced neural network for predicting the compressive strength of concrete. *Smart Struct. Syst.* **2022**, *30*, 195–207.

55. Li, S.; Chen, H.; Wang, M.; Heidari, A.A.; Mirjalili, S. Slime mould algorithm: A new method for stochastic optimization. *Future Gener. Comput. Syst.* **2020**, *111*, 300–323. [CrossRef]

56. Heidari, A.A.; Mirjalili, S.; Faris, H.; Aljarah, I.; Mafarja, M.; Chen, H. Harris hawks optimization: Algorithm and applications. *Future Gener. Comput. Syst.* **2019**, *97*, 849–872. [CrossRef]

57. Hashim, F.A.; Houssein, E.H.; Mabrouk, M.S.; Al-Atabany, W.; Mirjalili, S. Henry gas solubility optimization: A novel physics-based algorithm. *Future Gener. Comput. Syst.* **2019**, *101*, 646–667. [CrossRef]

58. Howard, F.L. The life history of Physarum polycephalum. *Am. J. Bot.* **1931**, *18*, 116–133. [CrossRef]

59. Latty, T.; Beekman, M. Food quality and the risk of light exposure affect patch-choice decisions in the slime mold Physarum polycephalum. *Ecology* **2010**, *91*, 22–27. [CrossRef]

60. Beekman, M.; Latty, T. Brainless but multi-headed: Decision making by the acellular slime mould Physarum polycephalum. *J. Mol. Biol.* **2015**, *427*, 3734–3743. [CrossRef]

61. Hashim, F.A.; Houssein, E.H.; Hussain, K.; Mabrouk, M.S.; Al-Atabany, W. A modified Henry gas solubility optimization for solving motif discovery problem. *Neural Comput. Appl.* **2019**, *32*, 10759–10771. [CrossRef]
62. Cao, W.; Liu, X.; Ni, J. Parameter Optimization of Support Vector Regression Using Henry Gas Solubility Optimization Algorithm. *IEEE Access* **2020**, *8*, 88633–88642. [CrossRef]
63. Shehabeldeen, T.A.; Abd Elaziz, M.; Elsheikh, A.H.; Hassan, O.F.; Yin, Y.; Ji, X.; Shen, X.; Zhou, J. A Novel Method for Predicting Tensile Strength of Friction Stir Welded AA6061 Aluminium Alloy Joints Based on Hybrid Random Vector Functional Link and Henry Gas Solubility Optimization. *IEEE Access* **2020**, *8*, 79896–79907. [CrossRef]
64. Yıldız, B.S.; Yıldız, A.R.; Pholdee, N.; Bureerat, S.; Sait, S.M.; Patel, V. The Henry gas solubility optimization algorithm for optimum structural design of automobile brake components. *Mater. Test.* **2020**, *62*, 261–264. [CrossRef]
65. Bui, D.T.; Moayedi, H.; Kalantar, B.; Osouli, A.; Pradhan, B.; Nguyen, H.; Rashid, A.S.A. A novel swarm intelligence—Harris hawks optimization for spatial assessment of landslide susceptibility. *Sensors* **2019**, *19*, 3590. [CrossRef]
66. Moayedi, H.; Osouli, A.; Nguyen, H.; Rashid, A.S.A. A novel Harris hawks' optimization and k-fold cross-validation predicting slope stability. *Eng. Comput.* **2019**, *37*, 369–379. [CrossRef]
67. Chen, H.; Jiao, S.; Wang, M.; Heidari, A.A.; Zhao, X. Parameters identification of photovoltaic cells and modules using diversification-enriched Harris hawks optimization with chaotic drifts. *J. Clean. Prod.* **2020**, *244*, 118778. [CrossRef]
68. Chen, H.; Heidari, A.A.; Chen, H.; Wang, M.; Pan, Z.; Gandomi, A.H. Multi-population differential evolution-assisted Harris hawks optimization: Framework and case studies. *Future Gener. Comput. Syst.* **2020**, *111*, 175–198. [CrossRef]
69. Pinkus, A. Approximation theory of the MLP model in neural networks. *Acta Numer.* **1999**, *8*, 143–195. [CrossRef]
70. Nguyen, H.; Mehrabi, M.; Kalantar, B.; Moayedi, H.; Abdullahi, M.a.M. Potential of hybrid evolutionary approaches for assessment of geo-hazard landslide susceptibility mapping. *Geomat. Nat. Hazards Risk* **2019**, *10*, 1667–1693. [CrossRef]
71. Fang, B.; Hu, Z.; Shi, T.; Liu, Y.; Wang, X.; Yang, D.; Zhu, K.; Zhao, X.; Zhao, Z. Research progress on the properties and applications of magnesium phosphate cement. *Ceram. Int.* **2022**, *49*, 4001–4016. [CrossRef]
72. Han, Y.; Shao, S.; Fang, B.; Shi, T.; Zhang, B.; Wang, X.; Zhao, X. Chloride ion penetration resistance of matrix and interfacial transition zone of multi-walled carbon nanotube-reinforced concrete. *J. Build. Eng.* **2023**, *72*, 106587. [CrossRef]
73. Shi, T.; Liu, Y.; Hu, Z.; Cen, M.; Zeng, C.; Xu, J.; Zhao, Z. Deformation Performance and Fracture Toughness of Carbon Nanofiber-Modified Cement-Based Materials. *ACI Mater. J.* **2022**, *119*, 119–128.
74. Abdi, H.; Williams, L.J. Principal component analysis. *Wiley Interdiscip. Rev. Comput. Stat.* **2010**, *2*, 433–459. [CrossRef]
75. Kim, J.-O.; Ahtola, O.; Spector, P.E.; Kim, J.-O.; Mueller, C.W. *Introduction to Factor Analysis: What It Is and How to Do It*; Sage: Thousand Oaks, CA, USA, 1978.
76. Liu, C.-W.; Lin, K.-H.; Kuo, Y.-M. Application of factor analysis in the assessment of groundwater quality in a blackfoot disease area in Taiwan. *Sci. Total Environ.* **2003**, *313*, 77–89. [CrossRef] [PubMed]
77. Azid, A.; Juahir, H.; Toriman, M.E.; Kamarudin, M.K.A.; Saudi, A.S.M.; Hasnam, C.N.C.; Aziz, N.A.A.; Azaman, F.; Latif, M.T.; Zainuddin, S.F.M. Prediction of the level of air pollution using principal component analysis and artificial neural network techniques: A case study in Malaysia. *Water Air Soil Pollut.* **2014**, *225*, 2063. [CrossRef]
78. Wang, M.; Yang, X.; Wang, W. Establishing a 3D aggregates database from X-ray CT scans of bulk concrete. *Constr. Build. Mater.* **2022**, *315*, 125740. [CrossRef]
79. Huang, Y.; Zhang, W.; Liu, X. Assessment of diagonal macrocrack-induced debonding mechanisms in FRP-strengthened RC beams. *J. Compos. Constr.* **2022**, *26*, 04022056. [CrossRef]
80. Huang, H.; Li, M.; Zhang, W.; Yuan, Y. Seismic behavior of a friction-type artificial plastic hinge for the precast beam–column connection. *Arch. Civ. Mech. Eng.* **2022**, *22*, 201. [CrossRef]

Article

GIS-Based Risk Assessment of Structure Attributes in Flood Zones of Odiongan, Romblon, Philippines

Jerome G. Gacu [1,2,3], Cris Edward F. Monjardin [1,2,4,*], Kevin Lawrence M. de Jesus [1,4] and Delia B. Senoro [1,2,4]

1 School of Graduate Studies, Mapua University, Manila 1002, Philippines
2 School of Civil, Environmental, and Geological Engineering, Mapua University, Manila 1002, Philippines
3 Civil Engineering Department, College of Engineering and Technology, Romblon State University, Liwanag, Odiongan, Romblon 5505, Philippines
4 Resiliency and Sustainable Development Center, Yuchengco Innovation Center, Mapua University, Manila 1002, Philippines
* Correspondence: cefmonjardin@mapua.edu.ph

Abstract: Flood triggered by heavy rains and typhoons leads to extensive damage to land and structures putting rural communities in crucial condition. Most of the studies on risk assessment focus on environmental factors, and building attributes have not been given attention. The five most expensive typhoon events in the Philippines were recorded in 2008–2013, causing USD 138 million in damage costs. This indicates the lack of tool/s that would aid in the creation of appropriate mitigation measure/s and/or program/s in the country to reduce damage caused by typhoons and flooding. Hence, this study highlights a structure vulnerability assessment approach employing the combination of analytical hierarchy process, physical structure attributes, and existing flood hazard maps by the local government unit. The available flood hazard maps were layered into base maps, and building attributes were digitized using a geographic information system. The result is an essential local scale risk map indicating the building risk index correlated to the structural information of each exposed structure. It was recorded that of 3094 structures in the community, 370 or 10.25% were found to be at moderate risk, 3094 (76.79%) were found to be high risk, and 503 (12.94%) were very high risk. The local government unit can utilize the resulting maps and information to determine flood risk priority areas to plan flood mitigation management strategies and educate people to improve the structural integrity of their houses. A risk map gives people an idea of what to improve in their houses to reduce their vulnerability to natural disasters. Moreover, the result of the study provides direction for future studies in the country to reduce loss and enhance structure resiliency against flooding.

Keywords: AHP; building attributes; flood; risk assessment; GIS

Citation: Gacu, J.G.; Monjardin, C.E.F.; de Jesus, K.L.M.; Senoro, D.B. GIS-Based Risk Assessment of Structure Attributes in Flood Zones of Odiongan, Romblon, Philippines. *Buildings* **2023**, *13*, 506. https://doi.org/10.3390/buildings13020506

Academic Editors: Tom Lahmer, Ehsan Harirchian and Viviana Novelli

Received: 14 January 2023
Revised: 3 February 2023
Accepted: 9 February 2023
Published: 13 February 2023

1. Introduction

Only a few studies analyzed the risks of flood disasters based on buildings attributes, and risk assessments mostly focused on earthquakes [1], seismic areas [2], and debris flow [3]. Some studies have focused on tackling aftermath scenarios such as building damage loss [4], building damage assessment [5], building asset value [6], level of damage [7], and building vulnerability [8] that lack primitive information about the risk level these structures were at prior to the occurrence of the disaster. Some approaches have also been introduced in describing the risk of building systems that require sufficient and comprehensive knowledge, for example, using LiDAR data [9], vulnerability curve [8], indicator-based methodology [10], rapid visual screening [11], and a probabilistic approach [1]. Another study [12] of mapping the risk of a building to flood was conducted by combining flood frequency analysis, estimation of inundation depth, and damage to loss estimation, which involve substantial information that is seldom available.

Extreme weather conditions stem from the tropical climate change effect, and many uncontrolled anthropogenic activities have converted major causes of flooding in the human ecosystem [13,14]. Floods are caused by inadequate natural paths and drainages to control the overflowing of waters during excessive rainfall or super typhoons [15]; the capacity of these natural drainages was also affected by the rapid change of land use in the area [16] that reduces land infiltration. Flooding is a devastating natural disaster that frequently happens in the Philippines [17], causing severe damage to land, numerous buildings, and rural communities [18–20]. The flood events that happen in several Asian countries, including China [18], Taiwan [21], Vietnam [22], Indonesia [23,24], Malaysia [25], and Sri Lanka [26], cause losses to both life and property annually. Significant losses, due to flooding, of billions of dollars were also reported in European countries, including Germany [27], Serbia [28], and Greece [29]. Situated in a region with a typical climate and geophysical tempest, the Philippines inevitably suffers from different calamities and is a hotbed of disaster [30]. The country is exposed to typhoons yearly, and flooding is one of the most frequently occurring natural hazards that put lives and properties at risk for affected communities [16,31]. Table 1 presents the list of the costliest typhoon events that happened in the country.

Table 1. Summary of the costliest typhoon events in the Philippines.

No.	Typhoon Name	Year	Damage Cost (in USD)	Reference
1	Bopha	2012	753.6 million	Yu et al., 2013 [32]
2	Haiyan	2013	714.3 million	Seriño et al., 2021 [33]
3	Parma	2009	487.5 million	Nolasco-Javier et al., 2015 [34]
4	Nesat	2011	267.9 million	Porio et al., 2121 [35]
5	Fengshen	2008	241 million	Bagsit et al., 2014 [36]

The global community demands action to focus on and achieve disaster risk reduction for people exposed to such occurrences. To accomplish this, society needs to understand the nature of flooding and typhoons' risks to these communities and their homes [16].

Natural hazards risk assessment involves different data on the built environment such as cities, buildings, urban spaces, walkways, roadways, etc., in terms of land use and land cover [37]. The flood risk assessment (FRA) identifies at-risk communities and supports mitigation decisions to augment investment benefits. High-resolution flood models and precise lot information are essential for flood risk analysis to estimate dependable outcomes for planning, preparedness, and decision-making applications [38]. However, the quantitative building risk assessment is still less studied, as the focus of most of the risk studies is mainly on people. Previous research studies focused on flood disasters' impacts on people, land, and agriculture. There are few studies about the vulnerability of a structure itself against disaster and the possible effects of flooding in the design of structures were not considered. Structures serve as the first line of defense against disasters so there is really a need to focus a risk assessment on structures too and this could help local government provide some regulations in issuing building permits in high-risk zones to ensure the resiliency of structures that are to be constructed. With the use of advanced technologies nowadays, such as remote sensing and geographical information system [39], it has become much easier to collect and analyze location-based data and combine them with other spatial data to provide significant results. Due to these technologies, it is feasible to obtain data on building parameters in a large community, making a risk assessment of buildings much easier and faster [40].

Exposure identification of lives and properties is vital in any risk assessment connected to a natural disaster. Many of the houses constructed in rural areas in the Philippines are made of light materials that are always vulnerable to the effects of flooding disasters [41]. The exposure of residential buildings as the smallest unit in rural and urban spaces to pluvial flooding might result in damages if they were not designed properly [40,42]. The building design is one of the most critical parameters in flood risk assessment [43]. Basement windows, doors, and underground garage entrances are familiar places where floodwater can accumulate in a building. Many buildings have already considered preparing and designing their structures to cope with the threat of flooding, but others are still not giving any importance to it. Therefore, evaluating the risk level a building is exposed to is very important to educate people on how to properly design their houses to be more resilient against flooding. Their houses are basically their first line of protection against this disaster. Identification of at what risk level a particular building would require information related to its structural integrity such as heights of the doors, windows, installation, materials used, and age of the structure, to name a few [44].

Comprehensive FRA would require reliable flood hazard data to identify locations very susceptible to flooding, which significantly affects the risk computation [45]. The Sendai Framework was used in the study to determine the risk level [46] structures are at by combining buildings' hazard, vulnerability, and exposure attributes. This framework is usually used in disaster risk reduction management (DRRM) assessment and provides quantifiable parameters for estimating risk levels in a community. However, this study used the Sendai Framework to focus on building exposure and is not specific to people. Identifying the risk level of a certain structure could also raise awareness among people on how to improve the resiliency of their houses against disasters. The compilation and evaluation of disaster damages under the Sendai Framework develop an understanding of the efficiency of implemented disaster risk reduction policies by the local government [47]. In addition, studies [48–50] have been conducted utilizing the framework considering different parameters from its components (hazard, vulnerability, and exposure).

The geographic information system is vital in flood risk assessment because the evaluation process requires spatial information [51]. This tool was proven to be effective in combining spatial information and gathered field data to provide better results. It saves human, physical, and financial resources in mapping flood disaster information [6]. In the Philippines, GIS is one of the leading technologies used nationwide for modeling and mapping flood hazards together with remote sensing (RS) [52]. Several studies successfully utilized GIS in assessing flood risk in building and rural housing in Canada [53] and China [54].

The multicriteria decision-making (MCDM) or multicriteria analysis (MCA) proves that it can hold distinctive assessments on identifying factors or parameters of a composite decision, organize the aspects into a hierarchical tree, and analyze the relationship of elements for the identified risk [55]. Many methodologies have been recommended for MCA, but the analytical hierarchy process (AHP) is the most popular and reliable tool for resolving flood risk assessment studies [56]. Several studies commenced in different Asian regions that utilized AHP in FRA studies, including Central Asia [57], South Asia [58–60], West Asia [61,62], East Asia [63–66], and Southeast Asia [67–70]. Local studies in some areas of the Philippines, such as Romblon [49], Quezon Province [71], and Davao Oriental [72], made use of AHP as part of their risk assessment methodology. The popularity of using AHP-based research studies caused it to be more straightforward in creating a model of indecision without compromising the subjective and objective features of the valuation procedure [73].

The preceding local study [49], illustrated and focused only on the spatial distribution of flooding in the municipality. With the result of this study, the local government unit can use the maps to determine flood risk priority areas and inform people about the current state of their buildings/houses against the effect of flooding. Furthermore, the result of the study delivers direction for upcoming research in the country to condense loss and improve building resiliency. The GIS-based flood risk assessment approach focuses on buildings and not on people by evaluating building attributes. A building flood risk map was developed to identify the level of risk the structures in the community are at. It could be used to educate people on how to make their houses more resilient to floods and could help LGU to revise its land use plans. This study answers the gap in doing risk assessment focusing entirely on the vulnerability of structures which could provide references to reduce loss of life and property, and enhance community resiliency during flooding.

2. Materials and Methods

2.1. Study Area

The municipality of Odiongan, with coordinates of 12°24′4.88″ N, 121°59′2.17″ E is one of the progressive municipalities in Romblon. The said municipality, representing 12.11% of the entire Islands Province of Romblon, has a land area of 185.67 square kilometers. The study focused on the town proper and nearby barangay in the low-lying plains [51]. Figure 1 shows the imagery map of Odiongan with its barangay administrative boundaries.

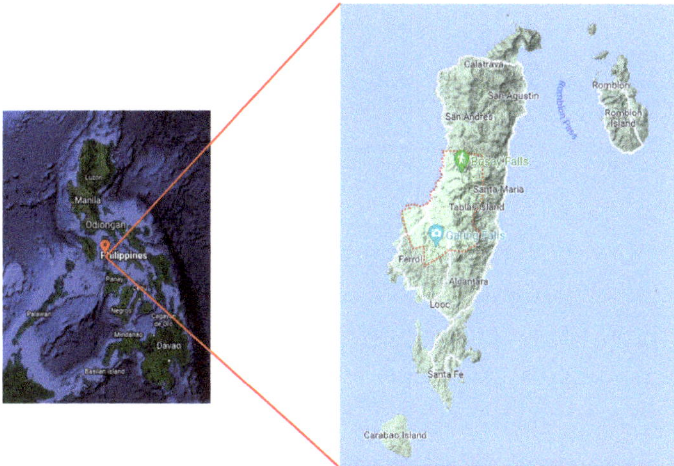

Figure 1. Location of the study area, Odiongan, Romblon, Philippines (12°24′4.88″ N, 121°59′2.17″ E).

According to the Corona climatic categorization system, Romblon has a climate type III, typically dry from January to May but without a clearly defined wet and dry season. The annual precipitation for the municipality of Odiongan is 1811 mm, wherein July has the highest average precipitation while February has the lowest [74]. Based on the 2022 cities and municipalities competitive index of the Department of Trade and Industry (DTI), the municipality of Odiongan ranked 81st out of 512 1st to 2nd class municipalities in terms of the infrastructure criterion, which includes components such as road network, distance to ports, availability of essential utilities, vehicles, education, health, local government unit investment, accommodation capacity, information technology capacity, and financial technology capacity. Considering the resiliency criterion, the municipality ranked 170th out of 512 1st and 2nd class municipalities in the country, which includes components such as a land use plan, disaster risk reduction plan, annual disaster drill, early warning system, budget for disaster risk reduction management, local risk assessments, emergency infrastructure, utilities, employed population, and sanitary system [75].

2.2. Methods of Assessing Flood Risk on Buildings

In this study, the Sendai Framework was employed to determine parameters for assessing flood risk levels focusing mainly on building exposure using the spatial and gathered field data. The conceptual framework used in this study is presented in Figure 2, showing the processes used and the relationships of identified parameters.

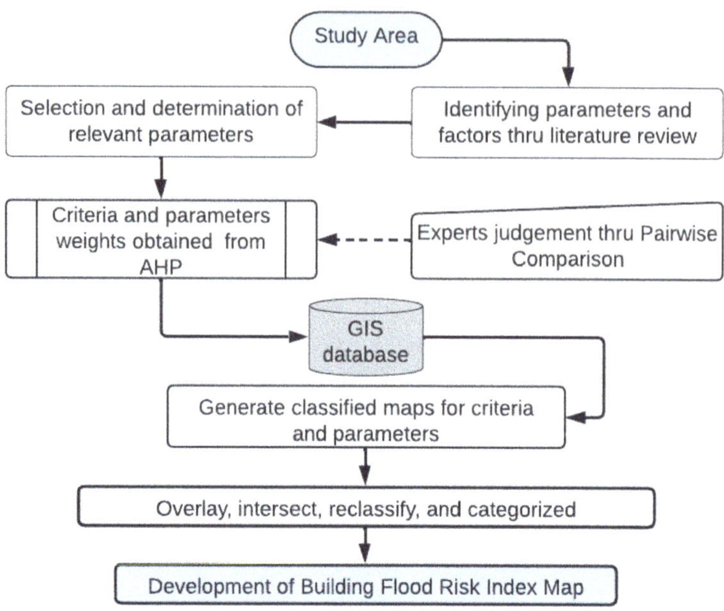

Figure 2. The study's methodological framework in assessing building flood risk.

Figure 2 presents the framework used in the study. Parameters in the assessment were identified through the literature review as shown in Table 2, the weights in determining hazard, vulnerability, and exposure indices were identified using an analytical hierarchy process with the help of experts consulted in this field. Data were collected through field data collection and remote sensing techniques. All these data were overlayed to produce a flood risk map focusing on the buildings that considered the three components of risk assessment, such as (a) hazard, (b) vulnerability, and (c) exposure [76].

Table 2. Different parameters, data types, duration/year, and sources were used in the study.

Parameters	Data Type	Duration/ Year	Source	References
Hazard Parameters				
Annual Average Rainfall	Interpolated Climatological Normal using Isohyetal Method	2020	Philippine Atmospheric, Geophysical and Astronomical Services Administration (PAGASA)	[77–79]
Slope	Generated from IfSAR Digital Terrian Model (DTM) using slope tool in GIS	2013	National Mapping and Resource Information Authority (NAMRIA)	[51]
Elevation	Generated from IfSAT DTM using field contour tool from GIS	2013	NAMRIA	[78]
Flood Depth	Raster file derived from existing FRA study in Odiongan, Romblon	2022	Previous study	[78]

Table 2. *Cont.*

Parameters	Data Type	Duration/Year	Source	References
Vulnerability Parameters				
Average Income	Average income per barangay	2020	Previous study	[51]
Gender Ratio	Barangay men-to-women gender ratio	2020	Previous study	[51]
Land Cover	Land cover map	2020	NAMRIA	[80,81]
Roofing Material	Roofing material per building	2022	Field data	[82,83]
Flooring Material	Flooring material per building	2022	Field data	[82]
Interior Walling Material	Wall material per building	2022	Field data	[82]
Number of Floors	Number of floors per building	2022	Field data	[84–86]
Types of Fencing Material	Fencing material per building	2022	Field data	[86]
Age of Building	Age of building structure	2022	Field data	[84–86]
Total Height of Building	Earth to roof height per building	2022	Field data	[82]
Types of Windows	Window types per building	2022	Field data	[82]
Distance to River	Shapefile clipped from water courses (river) map	2015	United Nations Office for the Coordination of Humanitarian Affairs (OCHA) Philippines	[78,86]
Exposure Parameters				
Building Density	Area of buildings in 50 m by 50 m land area	2022	Field data	[78,87]
Number of Buildings	Number of buildings in 50 m by 50 m land area	2022	Field data	[51]
Use of Building	List of building types according to use	2022	Field data	[80,88]

The study used annual average rainfall, slope, elevation, and flood height for hazard parameters that pertain to the parameters that promote menace to buildings. Vulnerability parameters, on the other hand, consider the ability of the building and the people living in it to cope with the identified hazard. The following were used as parameters in determining the vulnerability of buildings: average income, gender ratio, land cover/use, roofing material, flooring material, walling material, number of floors, types of fencing material, age of the building, the total height of the building, types of windows, and distance to river bodies. Last is the exposure parameters, which consider the elements that might be hazardous, including population density, number of households, and buildings. All parameters with the data type and sources are shown in Table 2.

2.2.1. Building Flood Hazard Parameters

The risk assessment for buildings against flooding should be constructed based on relevant flood hazard indicators defining the impact on the surrounding areas. The parameters are as follows:

1. The study utilized the annual average rainfall map from the previous study [58], where rainfall was plotted on a base map with different stations using the isohyetal

method. The study used the climatological normal records from long-term averages over 30 years of PAGASA weather stations with corresponding coordinates.

2. The slope map was prepared using the IfSAR DTM from NAMRIA and the spatial tool in the GIS application platform [89]. IfSAR has a pixel size of 5 m × 5 m which is enough to provide accurate measurement of the slope in the area.

3. The elevation is one of the most significant factors contributing to flood hazards. Due to gravity, water flows from higher to lower elevations; therefore, low-lying areas are more prone to experience higher and longer flood duration [90]. The GIS Field Contour tool was used to process the IfSAR DTM from NAMRIA.

4. The flood susceptibility map is the most crucial parameter in hazard assessment [91]. The MGB map and modeled flood map in the study of Gacu et al. [51] were used as data for flood height. These flood susceptibility maps were based on a 100 years return period with four susceptibility levels as follows: low susceptibility to areas that could experience flood height of less than 0.5 m, moderate susceptibility for flood heights between 0.5 and 1 m, high susceptibility where flood heights are expected to reach 1 to 2 m, and last is very high susceptibility for those areas that could experience flood height of greater than 2 m [92].

2.2.2. Building Flood Vulnerability Parameters

Vulnerability parameters considered in this study include the evaluation of the resiliency of the building based on its design and materials used. Human factors were also considered as part of the vulnerability assessment to identify the capability to repair or improve the integrity of the building against flood disasters. Existing demographics, spatial maps, and field data were gathered through various government agencies. Field data gathering was conducted to investigate the actual state of the structures in the area.

1. Demographic data from barangay profiles, such as average income and gender ratio, were considered in the study to determine the ability of the people to provide maintenance and repair of their houses. Average income provides the capability of the people to spend on repair and maintenance while the gender ratio gives an idea of how many members living in that building could provide the manpower needed for the repairs.

2. Building attributes were surveyed manually to collect data such as building materials used, structural orientation, age, and physical dimension. The assessment focused on the building's most used material/attribute. House-to-house surveys were conducted strategically to capture data that would best represent the totality of the community. These data were digitized using GIS software to combine with other spatial data and produce a deeper assessment of building vulnerability.

2.2.3. Building Flood Exposure Parameters

Exposure parameters pertain to life and property features that could be exposed to flooding events. However, the focus of this study is to have a deeper exposure assessment of buildings to flood hazards. The determined exposure elements were building density, number of buildings, and type of building. Data for these exposure elements were taken from the field data.

2.3. Assessment of Parameters Using AHP

Identified parameters for hazard, vulnerability, and exposure components were assessed using pairwise comparison and AHP to determine the weights of each parameter based on experts' judgment [93]. Ten (10) experts in the field of DRRM participated in assessing the relevance of one parameter over the other presented in a matrix. Each parameter was graded by experts using the pairwise comparison to identify how significant it is over the other. A nine-point intensity matrix was used in the questionnaire to identify the degree of significance of one parameter over the other.

For the derivation of overall relative weights, the relative significance was calculated with the normalized values for each criterion and parameter. Normalized values for each criterion and indicator in their respective matrices originated by dividing each cell into its column, resulting in a total column of one (1) for each criterion and indicator. Weights were computed by getting the mean of the rows of the matrix. The final relative weights of the indicators were described by computing the product's linear combination (L.C.) between the relative weight of each criterion and the indicator for the specific criterion. The decision makers choose the best according to the indicators' overall weights if the experts' knowledge is recognized as consistent. Equation (1) shows the mathematical expression for the number of combinations.

$$C = \left\{ C_j | j = 1, 2, \dots, n \right\} \tag{1}$$

The pairwise comparison on n criteria can be simplified using the matrix (A) in which every element is the quotient of weights of the criteria given in Equation (2).

$$A = \begin{bmatrix} a_{11} & a_{12} & \cdots & a_{1n} \\ a_{21} & a_{22} & \cdots & a_{2n} \\ \vdots & \vdots & \cdots & \vdots \\ a_{n1} & a_{n2} & \cdots & a_{nn} \end{bmatrix}, \ a_{ii} = 1, \ a_{ji} = \frac{1}{a_{ji}}, \ a_{ij} \neq 0 \tag{2}$$

For the last mathematical process, relative weights per matrix were identified and normalized. The right eigenvector imparts the relative weights (w) following the highest eigenvalue ($\lambda_{\max}$) as in Equation (3).

$$A_w = \lambda_{\max} \tag{3}$$

If the pairwise comparisons were consistent, the matrix A has rank one and $\lambda_{\max} = n$; the weights can be generalized by normalizing any of the rows or columns of A. The relativeness between the items determines the consistency, and the consistency ratio (CI) is assumed by Equation (4).

$$CI = \frac{(\lambda_{\max} - n)}{(n - 1)} \tag{4}$$

The final consistency ratio (CR), which permits the decision maker to accomplish whether the assessments are passably accurate, is computed as the CI divided by the random index (RI) quotient, as shown in Equation (5).

$$CR = \frac{CI}{RI} \tag{5}$$

In the last calculation, this step tells if the proportion exceeds 0.1; the judgment is considered inconsistent. So, a consistency ratio must be below 0.1 or 10%. The process is repeated if the judgment is unpredictable until the CR is within the wanted percentage, then the decision maker develops a conclusion concerning the assessment results.

2.4. Development of Building Flood Risk Map

AHP and building flood risk assessment results were put into maps for better presentation and appreciation. According to Rincón et al., maps' flood risk helps people appreciate its value [82,94]. In determining the priorities among the decision elements, feature weights were assigned to each parameter. Levels were reclassified and normalized into one (1) for the least important and five (5) for the most important. After the identification of weights, data of each parameter were combined and overlayed in a map using GIS. A risk level for each element was then developed after combining the weights, data, and attributes. The process proceeded to overlay the three (3) criteria maps (hazard, vulnerability, and exposure) with equal weights producing the building flood risk map. The resulting maps were validated based on existing flood assessments and historical flooding event records.

3. Results

3.1. Spatial Mapping of Risk Parameters

Parameters identified from the literature reviews and collected data were processed and presented into maps. Data for each parameter were gathered from different government agencies, previous studies, and field surveys. Spatial maps have a 1:30,000 scale for better visualization. There were 3976 buildings/structures assessed in this study that were analyzed spatially to determine the risk level of each against flooding.

3.1.1. Building Flood Hazard Parameters

Hazard parameters in this study were identified from the literature review and listed as the following: average annual rainfall [77–79], slope [51], elevation, and flood height [78]. Figure 3 shows the map of hazard parameters for the town proper of Odiongan, Romblon. Figure 3a presents the annual average rainfall using the isohyetal method in three (3) levels in which the rain intensifies from east to west of the study area. The majority of Odiongan municipality experiences an annual rainfall of 2230 mm where 3250 structures are affected, this most likely causing flooding problems in the area. There were also 560 structures that were inside the area experiencing annual average rainfall of 2240 mm and then 157 with 2220 mm.

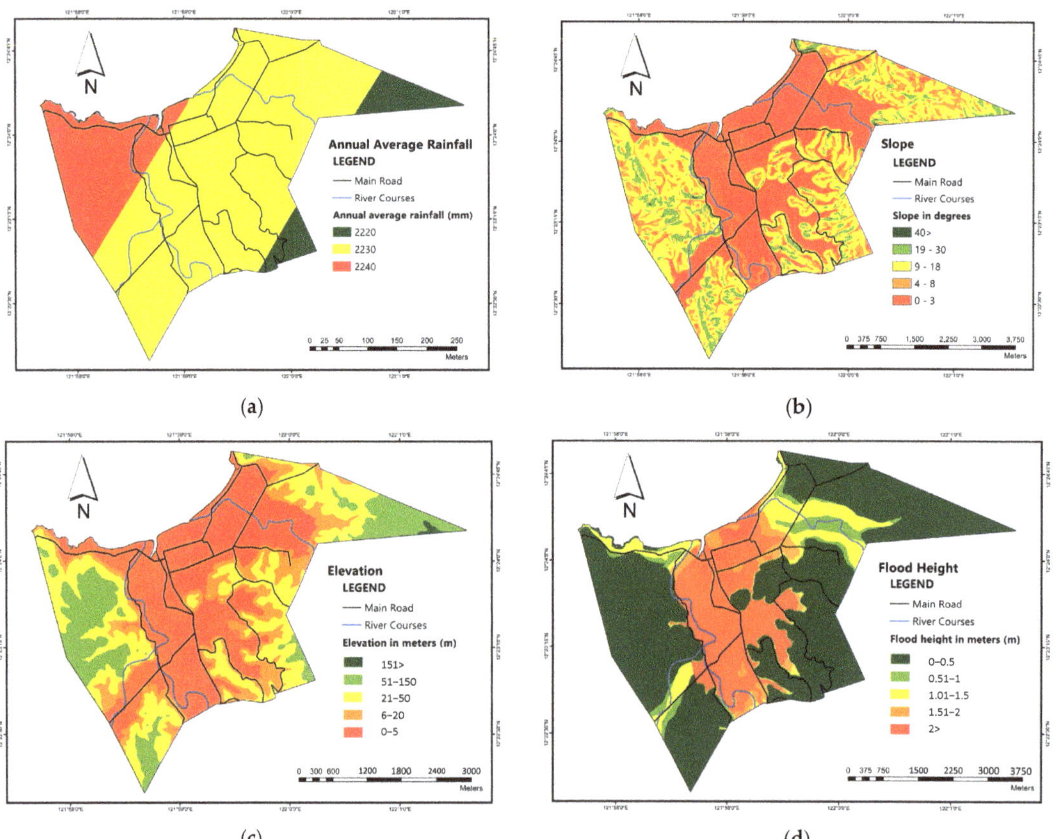

(a)

(b)

(c)

(d)

Figure 3. Generated maps from ArcMap in building flood hazard parameters: (**a**) average annual rainfall map, (**b**) slope map, (**c**) elevation, and (**d**) flood height map of the municipality of Odiongan.

Slope maps of the study area were categorized into five levels: <40, 19–30, 9–18, 4–8, and 0–3 with legends of green as high elevation to red as low elevation as shown in Figure 3b. The town proper itself where the majority of the residential and commercial buildings are located has a slope of 0–3 degrees covering 2762 buildings, a lower slope provides slower water velocity that could lead to water stagnation and a higher chance of flooding.

Figure 3c on the other hand shows the elevation map of the project site. Similar to slope, elevation basically presents how high the ground is with respect to mean sea level and is directly correlated to the occurrence of flooding and mean sea level is considered to be the end point of all the water flowing downstream. Area with lower elevations experiences a higher probability of flooding compared to those located in higher elevations. The elevation map was classified into five (5) categories: 0–5 m, 6–20 m, 21–50 m, 51–150 m, and 151 m and above. A total of 2855 (71.97%) buildings in the area are located in regions with a ground elevation of 0–5 m above mean sea level which puts them at risk to experience flooding.

A flood hazard map developed using hydraulic modeling combined with the susceptibility map of the Mines and Geosciences Bureau is presented in Figure 3d, five hazard levels were considered in the study, such as (1) 0–0.5 m low, (2) 0.5–1 m moderate, (3) 1.01–1.5 m high, (4) 1.51–2 m very high, and (5) 2 m and above is considered extremely high. The map shows essential information on flood height in the town proper of Odiongan. Areas near the main river experience flood heights of above 2 m which puts 1720 (43.38%) buildings near it at the most risk. There were also 1121 structures that experience flood levels from 0.5 m to 2 m. Complete records of these parameters are presented in Table A1.

3.1.2. Building Flood Vulnerability Parameters

Vulnerability analysis of buildings was based on demographics and building attributes related to being resilient of structure to flood. Figure 4 shows the spatial maps developed for the identified vulnerability parameters such as average income, gender ratio, land cover, roofing material, flooring material, walling material, number of floors, types of fencing material, age of the building, the total height of the building, types of window material, and distance to river network. Data used in these maps were gathered through actual surveys of building attributes on site and available records from local government units.

Figure 4a,b show the study area's average annual income and gender ratio per barangay zone of the municipality of Odiongan. The majority of the people in the area have an average annual income between PHP 250,000 and 499,999 that are living in 2921 (73.63%) buildings and could be seen located in low-lying areas and near the river. This was followed by residents living in 681 buildings that earn between PHP 60,000 and 99,999 located on the upstream part but still near the river while communities with a total of 365 buildings and earning less than PHP 40,000 annually are located on the left side of the town and are a little far from the river. Data showed that many economic activities happen in flat and low-lying areas since transportation is much easier which shows why many people tend to live there even if exposed to higher risk due to flooding. Average annual income was considered in the study as this gives people the capability to provide repairs and improvements to their houses for protection against flooding. The men-to-women-gender ratio was also considered as part of the vulnerability parameters since having more men is favorable as they could do heavy work and could perform repairs of their houses if damaged by disasters. Values do not differ across the municipality ranging from 0.8393 to 1.0377. There is a good ratio of men and women in the municipality; however, it can be seen at the municipal center and areas near the sea that the ratio is lower meaning more women are in those areas compared to men.

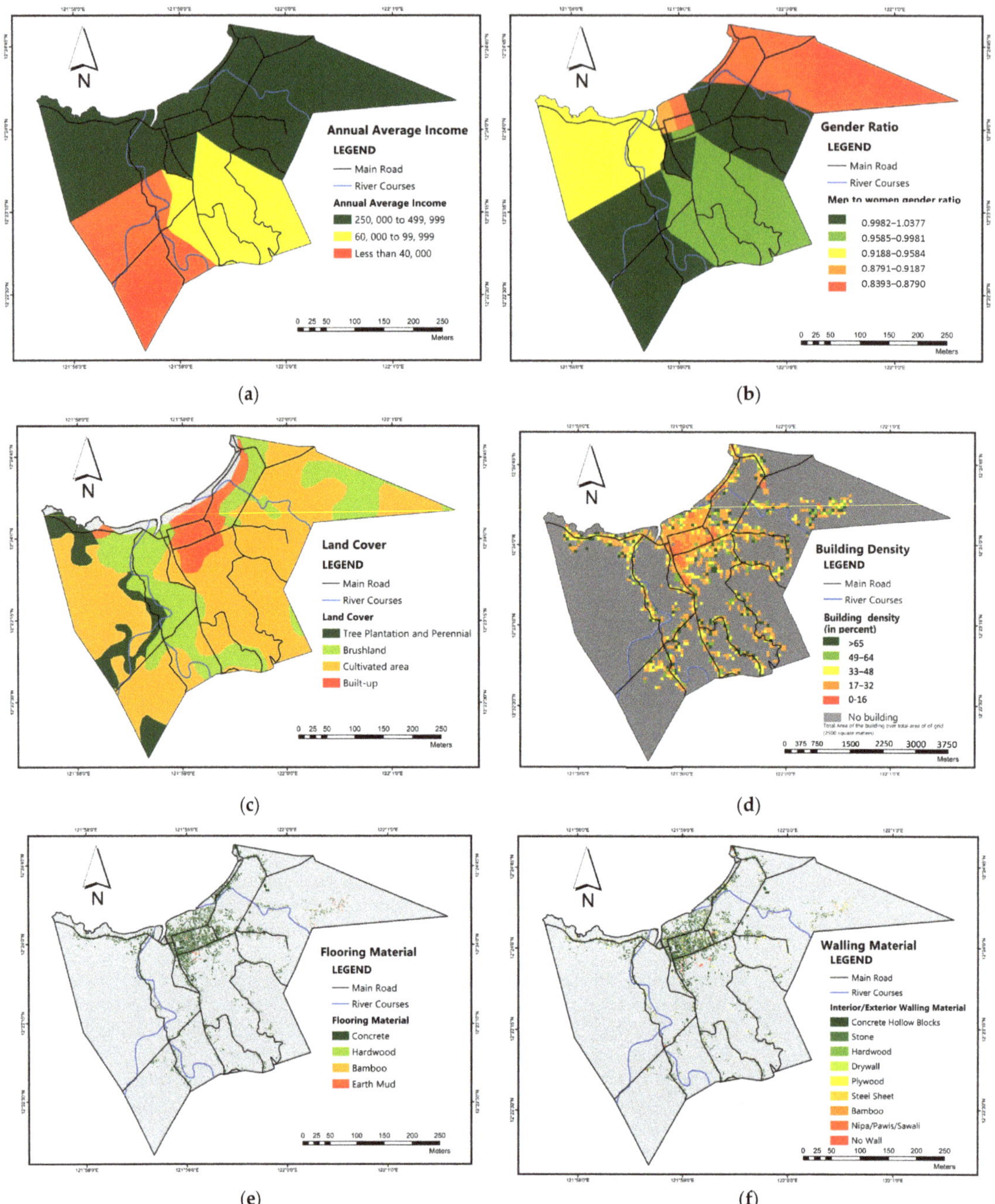

Figure 4. *Cont.*

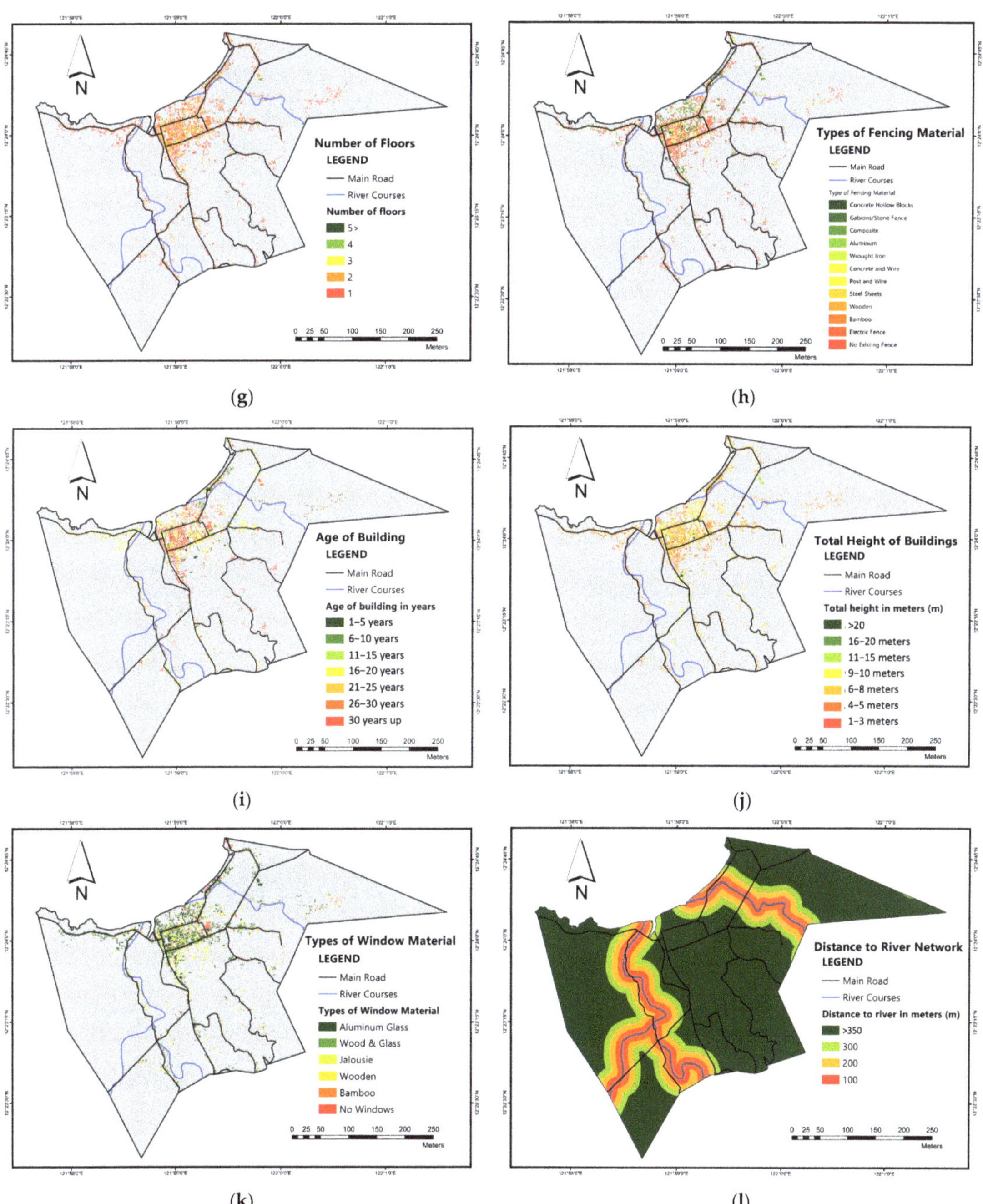

Figure 4. Building flood vulnerability parameters in the conduct of study: (**a**) annual average income, (**b**) gender ratio, (**c**) land cover, (**d**) types of roofing material, (**e**) types of flooring material, (**f**) interior/exterior walling material, (**g**) number of floors, (**h**) types of fencing material, (**i**) age of the building, (**j**) total height of the building, (**k**) types of window material, and (**l**) distance to river network.

Land cover was considered part of the analysis of the structure's vulnerability in this study. The location where a building was constructed is very important to identify how vulnerable it is to flooding. Developed areas are more exposed to frequent flooding due to pavements of road that reduces infiltration capability of the ground and economic activities. The study area's land cover classifications were extracted from the Comprehensive Land Use Plan (CLUP) of the municipality. The following four classifications were considered: built-up, cultivated area, brushland and tree plantation, and perennial. The map shown in Figure 4c indicates that a total of 1425 (35.92%) structures are located in built-up areas which is basically near the river and in low-lying areas, this shows poor planning that puts people and structures at risk of flooding.

Materials used in the construction of each building are an integral part of its resiliency against flooding disasters. Structures built using light materials are at higher risk to be damaged by disasters compared to those built in concrete. An actual field survey was conducted in this study to determine materials used in the construction of buildings and evaluate their resiliency to flooding. Roofing material (Figure 4d), flooring material (Figure 4e), wall material (Figure 4f), number of floors (Figure 4g), types of fencing material (Figure 4h), age of the building (Figure 4i), the total height of the building (Figure 4j), and types of window material (Figure 4k) were the parameters drawn into maps with corresponding categories and classifications. There were three thousand nine hundred sixty-seven (3967) facilities, establishments, and residential houses considered in the field data around the Poblacion (Ligaya, Liwanag, Liwayway, Tabin-Dagat, and Dapawan) and others close to town barangays such us Tulay, Bangon, and Poctoy.

Figure 4d provides information about the roofing materials of buildings in the area, 95.92% or 3805 buildings have metal sheets as their roofing material, which is a good sign that people are really investing in their houses. The flooring of these buildings was also assessed as presented in Figure 4e, a total of 3746 buildings have already used concrete as their flooring which is much stronger to resist the effect of flooding. Materials used in the wall of buildings were also evaluated, 77.84% of all the buildings have used concrete hollow blocks and the remaining 22.16% are still using light materials such as bamboo, nipa, and plywood, this is presented in Figure 4f. Figure 4g, on the other hand, presents the number of floors each building have, a building that has two or more floors is more resilient to flooding compared to those with one floor. Having multiple-level structures provides people with a place to go if the ground floor was flooded. The majority or 2763 houses in the area have only one floor which puts them at risk if a high level of flood happens. Types of fencing materials used by each house were also considered in the study as this provides protection against debris brought by flood water. A total of 2807 (70.76%) buildings do not even have a fence installed as shown in Figure 4h. The ages of buildings were also asked during the field survey and the majority or 34.51% of all the buildings were already above 30 years old. Figure 4i provides the spatial location of buildings' age which could also be a basis for local government to conduct structural integrity assessment for the safety of their community. The total height of the building as presented in Figure 4j was also measured in the study which is basically correlated to the number of floors; similarly, 49.03% (1945) of all the buildings have a height of less than 5 m. Figure 4k presents the spatial data for the type of window material used in buildings, 43.48% or 1725 buildings had aluminum and glass as their window which is resilient to the effect of flood but there was still 26.64% that used light materials such as bamboo and plywoods. The number of buildings near the river network was also determined in the study, those buildings near the river are at the most risk of flooding compared to those who are located further. There are even some who literally built their house on the riverbanks themselves. Figure 4l shows the distances of the building to the river network in four levels 100 m, 200 m, 300 m, and less than 350 m, 19.26% of all the buildings have a distance of at least 200 m from the river. These structures are the first ones to be affected by flooding disasters in the area by location. Complete records of vulnerability assessment of structures in the area could be seen in Table A2.

3.1.3. Building Flood Exposure Parameters

Figure 5a–c show the spatial data of the three (3) identified parameters for building exposure against flood. Building density, number of structures, and type of building use were considered as the exposure parameters. Building density was computed by getting the percent total area occupied by buildings inside a 50×50 m grid land area. The majority of the grids have a building density of 0–16% which is 65.32% of all the grids created in the map as shown in Figure 5a. This shows that buildings are quite far from each other and not crowding in a certain area. There were still some crowded places located inside Poblacion with a building density of greater than 65% which is 1.32% of the total grid. The Poblacion area basically is where economic activity in the municipality is centered. The number of buildings were computed by counting the number of structures in a 50 m by 50 m grid as shown in Figure 5b, 73.02% of the total grids have 1–5 structures located within. Houses and structures in provinces are naturally built far from each other unlike in the city where it is crowded everywhere. The highest number of buildings was observed in the center part of the town showing a count of 19 to 23 buildings which is in Poblacion. The use of the building was also determined during actual field data collection and used the following categories: residential, residential/commercial, institutional, agricultural, industrial, and infrastructure. Abandoned buildings were also considered for their unknown use. The majority of the buildings were residential with 2911 or 73.38% of the total. Complete records of the data gathered are presented in Table A3. Residential buildings are where people usually live and the ones that should be evaluated to ensure the safety of the people.

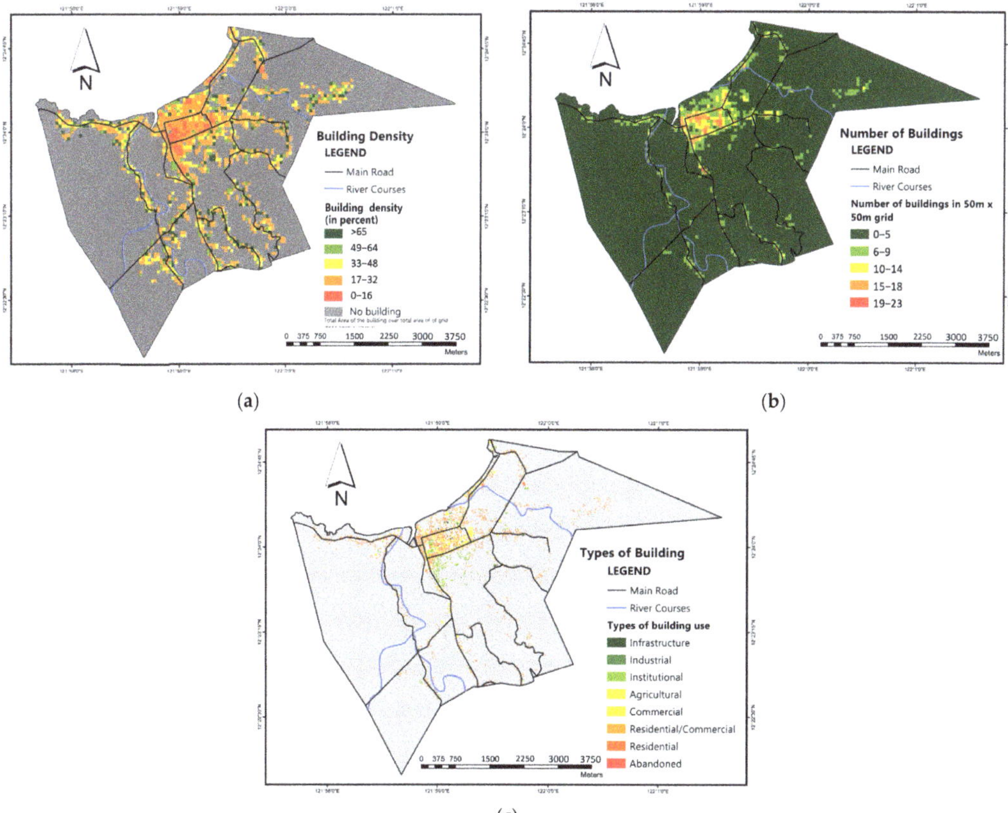

Figure 5. Result maps for building flood exposure parameters using ArcMap: (**a**) building density, (**b**) number of buildings per grid, and (**c**) type of building use.

3.2. Identification of Parameter's Weight Using AHP

Influencing parameters were evaluated and assessed using AHP and pairwise comparison techniques. The decision was separated into components and was shown in a hierarchy diagram (Figure 6) of at least three (3) levels: goal, criteria, and parameters.

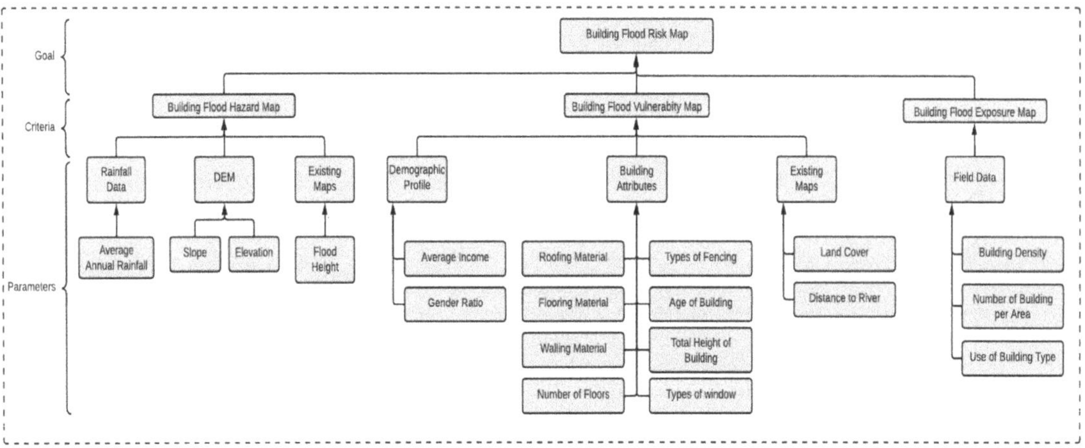

Figure 6. Hierarchy tree for AHP framework for building flood risk analysis.

The hierarchy tree consists of the uppermost place, the primary goal of developing a building flood risk map. The next level was the criteria for generating the goal compose of the definition of risk according to the Sendai Framework (hazard, vulnerability, and exposure), this framework was applied here specifically to building exposure to hazards. The lowest level included the parameters per criteria, substantially contributing to defining the criteria maps.

The featured class and weight were assigned for each parameter, which was reclassified and normalized using a geographic information system as shown in Table 3. Designated values depend on the type of level or category. As per results based on ten (10) experts' inputs and calculations in AHP, the final weights (percentage weights in Table 3) were identified for each and were ensured to give the CI requirement for the multicriteria analysis to be reflected valid. Flood depth had the highest weight in the building flood hazard parameters with 41.98%, almost half of the total, followed by the elevation at 30.07% then annual average rainfall at 14.6%, and lastly, slope at 13.35%. Flood depth as expected had the highest weight among flood hazard parameters since it directly provides information about the intensity of flooding in a certain area. On the other hand, twelve parameters were considered to represent the vulnerability of buildings against flooding and distance to river had the highest weight with 27.37%, followed by the total height of the building (11.91%), number of floors (11.55%), land cover (11.18%), age of the building (7.33%), roofing material (5.86%), flooring material (5.54%), wall material (4.42%), average income (4.18%), type of windows (4.02%), fencing material (3.97%), and the lowest weight of 2.67% for the gender ratio. Distance to river had the highest weight since structures near a river are expected to be exposed frequently to flooding while the height of the building and number of floors came in second and third since structures with higher height and with second floors are safer to live in. Building materials also had higher weights which showed their importance to building resiliency against flooding. Three exposure parameters were considered in the study namely building density, use of the building, and the total number of buildings with weights of 43.74%, 31.29%, and 24.97%, respectively. Building density in an area had the highest weight as it provides information on how crowded structures are in an area that could be at risk. This was followed by the type of building, residential houses are greatly affected by flooding which should be considered to be at a higher risk level compared to other commercial and industrial establishments. The last parameter was the total number of buildings which provides information on the number of buildings that might be exposed to flood which is also an important parameter in exposure assessment.

Table 3. Parameters with feature class, feature weight, and percentage weight.

Parameters	Percentage Weights (%)
Building Flood Hazard Parameters	
Annual Average Rainfall	14.6
Slope	13.35
Elevation	30.07
Flood Depth	41.98
Building Flood Vulnerability Parameters	
Average Income	4.18
Gender Ratio	2.67
Land Cover	11.18
Roofing Material	5.86
Flooring Material	5.54
Interior/Exterior Walling Material	4.42
Number of Floors	11.55
Types of Fencing Material	3.97
Age of Building	7.33
Total Height of Building	11.91
Types of Windows	4.02
Distance to River	27.37
Building Flood Exposure Parameters	
Building Density	43.74
Number of Buildings per Area	24.97
Use of Building	31.29

3.3. Development of Building Flood Risk Map

The Sendai Framework indicates that risk is a combination of hazard, vulnerability, and exposure indices. All indices should be present to determine risk level. Figure 7a–c shows the resulting spatial map for building hazard, vulnerability, and exposure indices. All spatial maps generated from hazard, vulnerability, and exposure parameters were processed and combined together to produce the final indices of each. The weights of each parameter identified from AHP were considered in the computation of these indices. Data from all spatial maps in Figures 3–5 were combined using a 50 m grid size. Each grid extracted the data from all the spatial maps and these data with weights were used to compute the indices. There were five index levels for each: very low (green), low (yellow-green), moderate (yellow), high (orange), and very high (red) as shown in Figure 7a–c.

The resulting building flood hazard index map is presented in Figure 7a, covering 2, 424, 813, 1076, and 1652 buildings identified as very low, low, moderate, high, and very high, respectively. It was spotted that areas in the hazard index with high values were mainly affected by the flood depth map, where river networks were located. It is alarming that 68.77% of all the buildings in the municipality of Odiongan have a hazard index of high to very high.

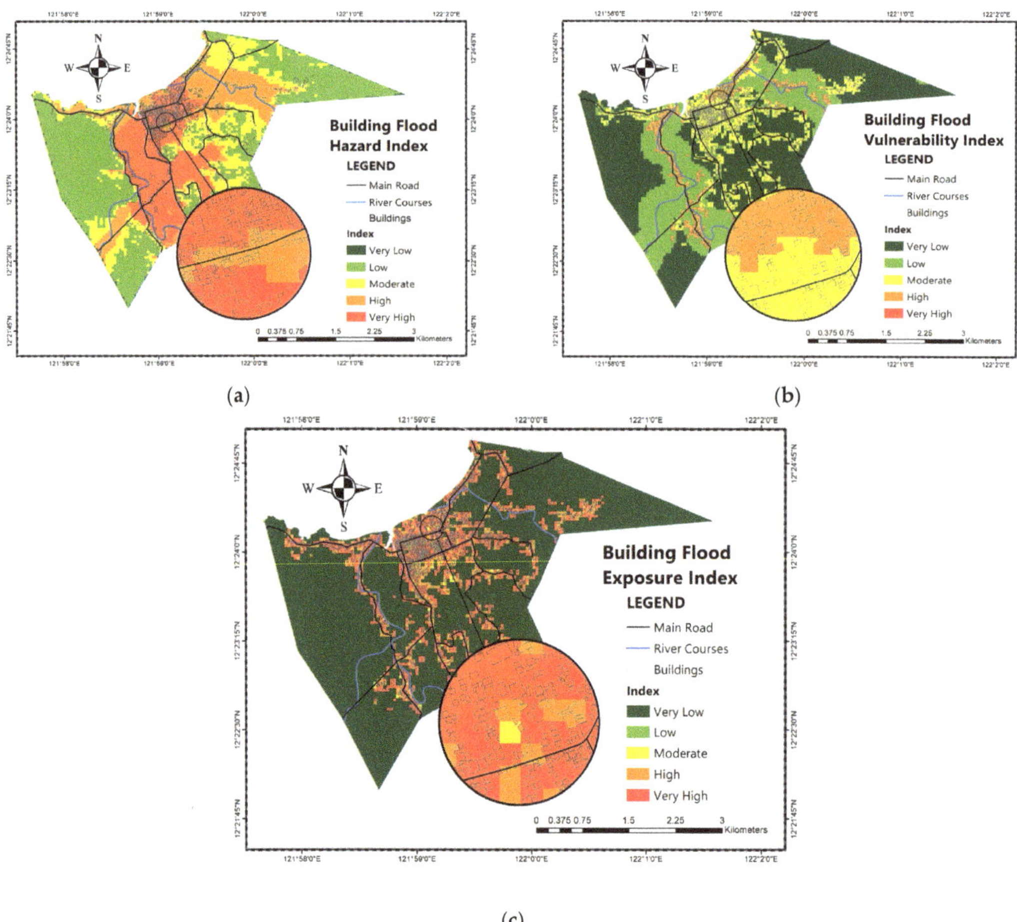

Figure 7. Generated maps from the different parameters in (**a**) building flood hazard, (**b**) flood vulnerability, and (**c**) flood exposure of the town proper of Odiongan, Romblon.

Figure 7b on the other hand shows the building vulnerability map. This spatial map was obtained by combining all twelve (12) parameters with five categories that connote very low vulnerability (0), low (221), moderate (2825), high (920), and very high (1). It could be seen that 71.21% of all the buildings have a vulnerability index level of moderate. This means that the majority of the buildings built in the area have moderate capability to adapt to the effect of flooding disasters. This is something that could be improved by educating the people.

Figure 7c displays the flood exposure map generated considering three (3) parameters: building density, the total number of buildings, and building use. Exposure level was presented using the following categories very low (0), low (11), moderate (1216), high (2590), and very high (15). It can be seen that 69.07% of all the buildings have a high to very high exposure index. A detailed count of buildings categorized at each level of hazard, vulnerability, and exposure index is presented in Table A4.

The generated results of building flood risk assessment were put into a map, as shown in Figure 8.

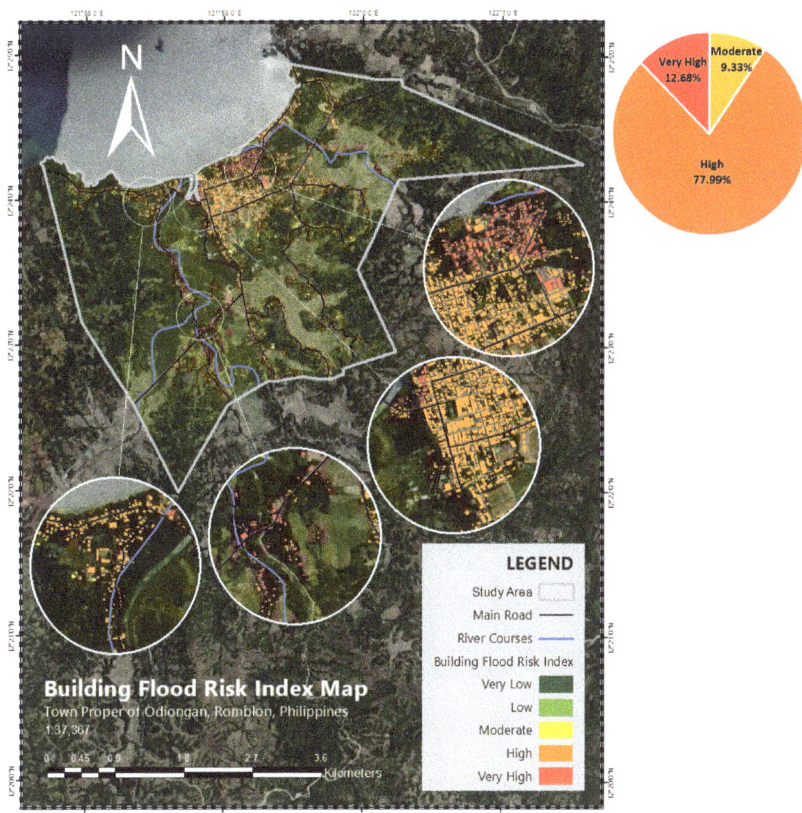

Figure 8. Building flood risk map with imagery and 1:15,000 scale zoomed-in map of concentrated areas with moderate to high risk of the town proper of Odiongan, Romblon.

The building flood risk map was categorized into five (5) levels (very low, low, moderate, high, and very high). The risk levels for each building/structure in the area were identified and presented in Figure 8. Out of 3967 buildings, there were 0 buildings at very low and low risk, 370 (9.33%) at moderate risk, 3094 (77.99%) at high risk, and 503 (12.68%) at very high risk. Buildings categorized to be high risk to very high risk are concentrated in the Poblacion area, specifically Ligaya, Liwanag, Liwayway, Tabin-Dagat, Budiong, and Dapawan barangay zone, these areas were located at the center of the municipality and basically located at high flood hazard zones, too. Structures in the area had high-risk levels not only because it is located in flood-prone areas but also because of the structural integrity of their buildings. The result of this study could help the community understand the risk level their structure/building/houses are at so they could do needed renovations and improvements to adapt to flooding disasters. This map could also aid the local government in updating their land use planning to avoid building structures in high-risk areas. They could also use this map to identify what type of building should be allowed to be constructed in the area based on the flood risk level.

4. Discussion

The study highlighted the risk assessment of floods in the building, incorporating its attributes and materials as primary building features in risk assessment. In recent years, it has become more common to integrate the AHP with GIS and remote sensing, allowing a variety of parameters such as hydrological, geographic, and socioeconomic data to be taken into account with no limits on the number of input criteria [95]. Using the AHP to determine the parameter's weights, it was observed that the trend of building flood hazard attributes is FD > E > AAR > S. This is the trend in terms of percentage weights calculated with flood depth (41.98%) which is the most influential parameter. While slope (14.6%) is the least significant feature among the parameters of buildings. Flood depth basically indicated the possible flood height that could occur in the area considering different rainfall scenarios. In the study of Gacu et al. [51], flood depth had a high percentage in risk assessment focusing on people at 21.88%. This study showed that risk assessment focusing on people has comparable results to risk assessment focusing on building features. Building flood vulnerability parameters on the other hand have trend DR > THB > NF > LC > AB > RM > FM > IEWM > AI > TW > TFM > GR with distance to the river (27.37%) as the most significant parameter in terms of the percentage weights. While the gender ratio (2.67%) is the least important parameter in the building flood vulnerability assessment. Distance to the river evidently is the most influential hazard parameter. The recorded relationship in this study is similar to the work of Jamrussri et al. [96] in Thailand which showed that the highest evacuation was observed in zones nearest to the river. Lastly, for the building flood exposure parameters, the trend of the relative importance in terms of the percentage weights was observed to be BD > UB > NBA wherein the building density (43.74%) is the most influential parameter and the number of buildings per area (24.97%) is the least significant parameter. Building density is the ratio of the building floor area to the land area of concern. This provides information on how crowded the specific area. This result is comparable to the study of Duan et al., [97] that showed global flood risks were high in crowded areas around the world. This information will aid in policy-making related to the planning and development of a small or big government unit.

Building attributes and materials used as parameters combined with topographical and disaster-related data support the development of a building flood risk map. This map is essential in decision-making for city planning and climate change adaptation measures by classifying risk on buildings and determining the priority of risky buildings. The results of the study indicate that the use of AHP is suitable for local risk studies, and this is similar to the findings of Swain et al. in 2020 [98].

5. Conclusions

In this research, a building flood risk zone-based technique was developed using AHP-GIS to produce a trustworthy risk assessment of building characteristics. This study used local field data to propose a hybrid multicriteria approach. The study assessed the risk level of structures against flooding in the town proper of Odiongan, Romblon, considering building attributes and materials following the Sendai Framework in risk assessment. The study utilized GIS to map field data and perform spatial analysis. The following hazard parameters were considered such as annual average rainfall, slope, elevation, and flood depth. Building vulnerability index on the other hand includes average income, gender ratio, land cover, roofing material, flooring material, interior/exterior walling material, number of floors, types of fencing material, age of the building, the total height of the building, types of windows, and distance to the river. The exposure index covered three parameters: building density, use of a building, and the number of buildings. Each parameter was compared to one another by pairwise comparison to compute the final weights using AHP based on experts' decisions. In the hazard index, flood depth had the highest weight with 41.98% which pertains mainly to possible flood height that could occur in the area, while the building's distance to the river had the highest weight with 27.37%, which corresponds to the records of LGU that most of the residents evacuating during a disaster were those living near the rivers. Lastly, for the exposure index, building density had the highest weight of 43.74% which is a measure of how much land area are being occupied by buildings in a certain zone. Combining all these indices provides the risk level of buildings against flooding in the municipality of Odiongan. No buildings were categorized to be at very low and low risk, while 370 (10.25%) buildings were found to be at moderate risk, 3094 (76.79%) at high risk, and 503 (12.94%) at very high risk. The majority of the buildings that have risk levels from high to very high are located in the Poblacion area where high flood level is also expected to occur and the resiliency of structures are also low. This building flood risk map is essential for the LGU's planning and development strategy, to improve their zoning plan and requirements for issuing building permits and ensure that structures are resilient enough against disaster. Educating people about their structures' risk level would be an eye-opener for them to make necessary improvements to their buildings. Commonly, risk assessment focused on people but buildings/structures are the first line of defense of the community against disaster which should also be taken into consideration. The methodology developed for evaluating building risk level is a first step in making or helping communities be more resilient against natural disasters. This study could serve as a reference to conduct building risk assessments applicable to countries with similar conditions.

Author Contributions: Conceptualization, C.E.F.M., J.G.G., and K.L.M.d.J.; methodology, C.E.F.M. and J.G.G.; software, C.E.F.M., J.G.G., and K.L.M.d.J.; validation, C.E.F.M. and J.G.G.; formal analysis, C.E.F.M. and K.L.M.d.J.; investigation, C.E.F.M. and J.G.G.; resources, C.E.F.M. and D.B.S.; data curation, C.E.F.M. and J.G.G.; writing—original draft preparation, J.G.G.; writing—review and editing, C.E.F.M., K.L.M.d.J., and D.B.S.; visualization, C.E.F.M. and K.L.M.d.J.; supervision, C.E.F.M.; project administration, C.E.F.M.; funding acquisition, C.E.F.M. All authors have read and agreed to the published version of the manuscript.

Funding: This research received no external funding.

Data Availability Statement: All data are contained in the manuscript.

Acknowledgments: This is to acknowledge the support in-kind of Mapua University and Romblon State University.

Conflicts of Interest: The authors declare no conflict of interest.

Appendix A

Table A1. Feature count of the building flood hazard parameters.

Parameters	Feature Class	Number of Buildings	Percentage %
Building Flood Hazard Parameters			
Annual Average Rainfall	2220 mm	157	3.96
	2230 mm	3250	81.93
	2240 mm	560	14.12
Slope	40 degrees>	10	0.25
	19–30 degrees	186	4.69
	9–18 degrees	666	16.79
	4–8 degrees	343	8.65
	1–3 degrees	2762	69.62
Elevation	151 m>	7	0.18
	51–150 m	14	0.35
	21–50 m	757	19.08
	6–20 m	334	8.42
	0–5 m	2855	71.97
Flood Depth	0–0.5 m	1125	28.36
	0.51–1 m	265	6.68
	1.01–1.5 m	324	8.17
	1.51–2 m	532	13.41
	2 m>	1721	43.38

Table A2. Feature count of the building flood vulnerability parameters.

Parameters	Feature Class	Number of Buildings	Percentage %
Building Flood Vulnerability Parameters			
Average Income	250,000 to 499,999	2921	73.63
	60,000 to 99,999	681	17.17
	Less than 40,000	365	9.20
Gender Ratio	0.9982–1.0377	1699	42.83
	0.3585–0.9981	823	20.75
	0.9188–0.9584	371	9.35
	0.8791–0.9187	331	8.34
	0.8393–0.8790	743	18.73
Land Cover	Tree Plantation and Perennial	197	4.97
	Brushland	1057	26.64
	Cultivated area	1288	32.47
	Built-up	1425	35.92
Roofing Material	Concrete	132	3.33
	Metal Sheet	3805	95.92
	Nipa/Pawid	21	0.53
	No Roof	9	0.23
Flooring Material	Concrete	3746	94.43
	Hardwood	9	0.23
	Bamboo	49	1.24
	Earth Mud	163	4.11
Interior/Exterior Walling Material	Concrete Hollow Blocks	3088	77.84
	Stone	3	0.08
	Hardwood	28	0.71
	Drywall	128	3.23
	Plywood	447	11.27
	Steel Sheet	93	2.34
	Bamboo	27	0.68
	Nipa/Pawid/Sawali	107	2.70
	No Wall	46	1.16
Number of Floors	>5	6	0.15
	4	36	0.91
	3	142	3.58
	2	1020	25.71
	1	2763	69.65

Table A2. *Cont.*

Parameters	Feature Class	Number of Buildings	Percentage %
Building Flood Vulnerability Parameters			
Types of Fencing Material	Concrete Hollow Blocks	156	3.93
	Gabions/Stone Fence	1	0.03
	Composite	441	11.12
	Aluminum	4	0.10
	Wrought Iron	225	5.67
	Concrete and Wire	17	0.43
	Post and Wire	197	4.97
	Steel Sheets	6	0.15
	Wooden	60	1.51
	Bamboo	52	1.31
	Electric Fence	1	0.03
	No Existing Fence	2807	70.76
Age of Building	1–5 years	222	5.60
	6–10 years	303	7.64
	11–15 years	503	12.68
	16–20 years	815	20.54
	21–25 years	712	17.95
	26–30 years	43	1.08
	30 years up	1369	34.51
Total Height of Building	>20 m	7	0.18
	16–20 m	32	0.81
	11–15 m	185	4.66
	9–10 m	301	7.59
	6–8 m	1497	37.74
	4–5 m	1782	44.92
	1–3 m	163	4.11
Types of Windows	Aluminum Glass	1725	43.48
	Wood and Glass	11	0.28
	Jalousie	1174	29.59
	Wooden	217	5.47
	Bamboo	488	12.30
	No Windows	352	8.87
Distance to River	>350 m	10,581,002.65	69.59
	300 m	1,696,785.76	11.16
	200 m	1,502,012.00	9.88
	100 m	1,425,639.62	9.38

Table A3. Feature count of building flood exposure parameters.

Parameters	Feature Class	Number of Buildings	Percentage %
Building Flood Exposure Parameters			
Use of Building	Infrastructure	52	1.31
	Industrial	17	0.43
	Institutional	244	6.15
	Agricultural	4	0.10
	Commercial	722	18.20
	Residential/Commercial	6	0.15
	Residential	2911	73.38
	Abandoned	11	0.28
Building Density per Grid	0–16%	65.32%	
	17–32%	20.52%	
	33–64%	12.84%	
	>65%	1.32%	
Number of Buildings per Grid	1–5	73.02%	
	6–10	19.23%	
	11–14	4.62%	
	15–18	2.54%	
	19–23	0.60%	

Table A4. Feature count of computed building exposure, vulnerability, and hazard index level.

Index Level	Exposure	Vulnerability	Hazard
Very Low	2	0	0
Low	424	221	11
Moderate	813	2825	1216
High	1076	920	2590
Very High	1652	1	150
Total features		3967	

References

1. Monjardin-Quevedo, J.G.; Valenzuela-Beltran, F.; Reyes-Salazar, A.; Leal-Graciano, J.M.; Torres-Carrillo, X.G.; Gaxiola-Camacho, J.R. Probabilistic Assessment of Buildings Subjected to Multi-Level Earthquake Loading Based on the PBSD Concept. *Buildings* **2022**, *12*, 1942. [CrossRef]
2. Boukri, M.; Farsi, M.N.; Mebarki, A.; Belazougui, M.; Ait-Belkacem, M.; Yousfi, N.; Guessoum, N.; Benamar, D.A.; Naili, M.; Mezouar, N.; et al. Seismic vulnerability assessment at urban scale: Case of Algerian buildings. *Int. J. Disaster Risk Reduct.* **2018**, *31*, 555–575. [CrossRef]
3. Luo, H.Y.; Fan, R.L.; Wang, H.J.; Zhang, L.M. Physics of building vulnerability to debris flows, floods and earth flows. *Eng. Geol.* **2020**, *271*, 105611. [CrossRef]
4. Afifi, Z.; Chu, H.J.; Kuo, Y.L.; Hsu, Y.C.; Wong, H.K.; Zeeshan Ali, M. Residential flood loss assessment and risk mapping from high-resolution simulation. *Water* **2019**, *11*, 751. [CrossRef]
5. Shrestha, B.B.; Kawasaki, A.; Zin, W.W. Development of flood damage assessment method for residential areas considering various house types for Bago Region of Myanmar. *Int. J. Disaster Risk Reduct.* **2021**, *66*, 102602. [CrossRef]
6. De Risi, R.; De Paola, F.; Turpie, J.; Kroeger, T. Life Cycle Cost and Return on Investment as complementary decision variables for urban flood risk management in developing countries. *Int. J. Disaster Risk Reduct.* **2018**, *28*, 88–106. [CrossRef]
7. Gautam, D.; Dong, Y. Multi-hazard vulnerability of structures and lifelines due to the 2015 Gorkha earthquake and 2017 central Nepal flash flood. *J. Build. Eng.* **2018**, *17*, 196–201. [CrossRef]

8. Mück, M.; Taubenböck, H.; Post, J.; Wegscheider, S.; Strunz, G.; Sumaryono, S.; Ismail, F.A. Assessing building vulnerability to earthquake and tsunami hazard using remotely sensed data. *Nat. Hazards* **2013**, *68*, 97–114. [CrossRef]
9. Huţanu, E.; Mihu-Pintilie, A.; Urzica, A.; Paveluc, L.E.; Stoleriu, C.C.; Grozavu, A. Using 1D HEC-RAS modeling and LiDAR data to improve flood hazard maps accuracy: A case study from Jijia Floodplain (NE Romania). *Water* **2020**, *12*, 1624. [CrossRef]
10. Kaoje, I.U.; Rahman, M.Z.A.; Idris, N.H.; Tam, T.H.; Sallah, M.R.M. Physical flood vulnerability assessment of buildings in Kota Bharu, Malaysia: An indicator-based approach. *Int. J. Disaster Resil. Built Environ.* **2020**, *12*, 413–424. [CrossRef]
11. Mohamad, I.I.; Yunus, M.M.; Harith, N.S.H. Assessment of building vulnerability by integrating rapid visual screening and geographic information system: A case study of Ranau township. In *IOP Conference Series: Materials Science and Engineering*; IOP Publishing: Bristol, UK, 2019; Volume 527, p. 012042.
12. Oubennaceur, K.; Chokmani, K.; Nastev, M.; Lhissou, R.; El Alem, A. Flood risk mapping for direct damage to residential buildings in Quebec, Canada. *Int. J. Disaster Risk Reduct.* **2019**, *33*, 44–54. [CrossRef]
13. Nkeki, F.N.; Bello, E.I.; Agbaje, I.G. Flood Risk Mapping and Urban Infrastructural Susceptibility Assessment Using a G.I.S. and Analytic Hierarchical Raster Fusion Approach in the Ona River Basin, Nigeria. *Int. J. Disaster Risk Reduct.* **2022**, *77*, 103097. [CrossRef]
14. Monjardin, C.E.F.; Bacuel, A.C.; Rubin, N.K.; Tiongson, M.A.J.; Valdecanas, G.; Yamat, R.U. Effect of climate change to Ambuklao reservoir, simulation of El Niño and La Niña. In *AIP Conference Proceedings*; AIP Publishing LLC.: Melville, NY, USA, 2018; Volume 2045, p. 020064.
15. Osei, B.K.; Ahenkorah, I.; Ewusi, A.; Fiadonu, E.B. Assessment of Flood Prone Zones in the Tarkwa Mining Area of Ghana Using a GIS-Based Approach. *Environ. Chall.* **2021**, *3*, 100028. [CrossRef]
16. Caja, C.C.; Ibunes, N.L.; Paril, J.A.; Reyes, A.R.; Nazareno, J.P.; Monjardin, C.E.; Uy, F.A. Effects of land cover changes to the quantity of water supply and hydrologic cycle using water balance models. In *MATEC Web of Conferences*; EDP Sciences: Les Ulis, France, 2018; Volume 150, p. 06004.
17. Monjardin, C.E.F.; Transfiguracion, K.M.; Mangunay, J.P.J.; Paguia, K.M.; Uy, F.A.A.; Tan, F.J. Determination of river water level triggering flood in Manghinao River in Bauan, Batangas, Philippines. *J. Mech. Eng.* **2021**, *18*, 181–192. [CrossRef]
18. Zhang, Q.; Zhang, J.; Jiang, L.; Liu, X.; Tong, Z. Flood disaster risk assessment of rural housings—A case study of Kouqian Town in China. *Int. J. Environ. Res. Public Health* **2014**, *11*, 3787–3802. [CrossRef] [PubMed]
19. Stephenson, V.; Finlayson, A.; Miranda Morel, L. A risk-based approach to shelter resilience following flood and typhoon damage in rural Philippines. *Geosciences* **2018**, *8*, 76. [CrossRef]
20. Marín-García, D.; Rubio-Gómez-Torga, J.; Duarte-Pinheiro, M.; Moyano, J. Simplified automatic prediction of the level of damage to similar buildings affected by river flood in a specific area. *Sustain. Cities Soc.* **2023**, *88*, 104251. [CrossRef]
21. Chen, Y.R.; Yeh, C.H.; Yu, B. Integrated application of the analytic hierarchy process and the geographic information system for flood risk assessment and flood plain management in Taiwan. *Nat. Hazards* **2011**, *59*, 1261–1276. [CrossRef]
22. Van, C.T.; Tuan, N.C.; Son, N.T.; Tri, D.Q.; Tran, D.D. Flood vulnerability assessment and mapping: A case of Ben Hai-Thach Han River basin in Vietnam. *Int. J. Disaster Risk Reduct.* **2022**, *75*, 102969. [CrossRef]
23. Marfai, M.A.; King, L. Monitoring land subsidence in Semarang, Indonesia. *Environ. Geol.* **2007**, *53*, 651–659. [CrossRef]
24. Febrianto, H.; Fariza, A.; Hasim, J.A.N. Urban flood risk mapping using analytic hierarchy process and natural break classification (Case study: Surabaya, East Java, Indonesia). In Proceedings of the 2016 International Conference on Knowledge Creation and Intelligent Computing (KCIC), Manado, Indonesia, 15–17 November 2016; IEEE: New York, NY, USA, 2016; pp. 148–154.
25. Tehrany, M.S.; Pradhan, B.; Jebur, M.N. Spatial prediction of flood susceptible areas using rule based decision tree (DT) and a novel ensemble bivariate and multivariate statistical models in GIS. *J. Hydrol.* **2013**, *504*, 69–79. [CrossRef]
26. Weerasinghe, K.M.; Gehrels, H.; Arambepola, N.M.S.I.; Vajja, H.P.; Herath, J.M.K.; Atapattu, K.B. Qualitative flood risk assessment for the Western Province of Sri Lanka. *Procedia Eng.* **2018**, *212*, 503–510. [CrossRef]
27. Maiwald, H.; Schwarz, J.; Kaufmann, C.; Langhammer, T.; Golz, S.; Wehner, T. Innovative vulnerability and risk assessment of urban areas against flood events: Prognosis of structural damage with a new approach considering flow velocity. *Water* **2022**, *14*, 2793. [CrossRef]
28. Gigović, L.; Pamučar, D.; Bajić, Z.; Drobnjak, S. Application of GIS-interval rough AHP methodology for flood hazard mapping in urban areas. *Water* **2017**, *9*, 360. [CrossRef]
29. Stefanidis, S.; Stathis, D. Assessment of flood hazard based on natural and anthropogenic factors using analytic hierarchy process (AHP). *Nat. Hazards* **2013**, *68*, 569–585. [CrossRef]
30. Lagmay, A.M.F.A.; Racoma, B.A.; Aracan, K.A.; Alconis-Ayco, J.; Saddi, I.L. Disseminating near-real-time hazards information and flood maps in the Philippines through Web-GIS. *J. Environ. Sci.* **2017**, *59*, 13–23. [CrossRef]
31. Talisay, B.A.M.; Puno, G.R.; Amper, R.A.L. Flood hazard mapping in an urban area using combined hydrologic-hydraulic models and geospatial technologies. *Glob. J. Environ. Sci. Manag.* **2019**, *5*, 139–154.
32. Yu, K.D.S.; Tan, R.R.; Santos, J.R. Impact estimation of flooding in Manila: An inoperability input-output approach. In Proceedings of the 2013 IEEE Systems and Information Engineering Design Symposium, Charlottesville, VA, USA, 26 April 2013; IEEE: New York, NY, USA, 2013; pp. 47–51.
33. Seriño, M.N.V.; Cavero, J.A.; Cuizon, J.; Ratilla, T.C.; Ramoneda, B.M.; Bellezas, M.H.I.; Ceniza, M.J.C. Impact of the 2013 super typhoon haiyan on the livelihood of small-scale coconut farmers in Leyte island, Philippines. *Int. J. Disaster Risk Reduct.* **2021**, *52*, 101939. [CrossRef]

34. Nolasco-Javier, D.; Kumar, L.; Tengonciang, A.M.P. Rapid appraisal of rainfall threshold and selected landslides in Baguio, Philippines. *Nat. Hazards* **2015**, *78*, 1587–1607. [CrossRef]
35. Porio, E. Climate Resilience Initiative in Metro Manila: Participatory Community Risk Assessment and Power in Community Interventions. In *International Clinical Sociology*; Springer: Cham, Switzerland, 2021; pp. 257–276.
36. Bagsit, F.U.; Suyo, J.G.B.; Subade, R.F.; Basco, J. Do adaptation and coping mechanisms to extreme climate events differ by gender? The case of flood-affected households in Dumangas, Iloilo, Philippines. *Asian Fish. Sci.* **2014**, *25*, 111–118.
37. Chen, K.; Blong, R. Extracting building features from high resolution aerial imagery for natural hazards risk assessment. In Proceedings of the IEEE International Geoscience and Remote Sensing Symposium, Toronto, ON, Canada, 24–28 June 2002; IEEE: New York, NY, USA, 2022; Volume 4, pp. 2039–2041.
38. Yildirim, E.; Just, C.; Demir, I. Flood risk assessment and quantification at the community and property level in the State of Iowa. *Int. J. Disaster Risk Reduct.* **2022**, *77*, 103106. [CrossRef]
39. Kakooei, M.; Ghorbanian, A.; Baleghi, Y.; Amani, M.; Nascetti, A. Remote sensing technology for postdisaster building damage assessment. In *Computers in Earth and Environmental Sciences*; Elsevier: Amsterdam, The Netherlands, 2022; pp. 509–521.
40. Li, Z.; Wang, L.; Shen, J.; Ma, Q.; Du, S. A Method for Assessing Flood Vulnerability Based on Vulnerability Curves and Online Data of Residential Buildings—A Case Study of Shanghai. *Water* **2022**, *14*, 2840. [CrossRef]
41. Wu, J.; Ye, M.; Wang, X.; Koks, E. Building asset value mapping in support of flood risk assessments: A case study of Shanghai, China. *Sustainability* **2019**, *11*, 971. [CrossRef]
42. Mobini, S.; Nilsson, E.; Persson, A.; Becker, P.; Larsson, R. Analysis of pluvial flood damage costs in residential buildings–a case study in Malmö. *Int. J. Disaster Risk Reduct.* **2021**, *62*, 102407. [CrossRef]
43. Kwak, Y.J.; Natsuaki, R.; Yun, S.H. Effect of Building Orientation on Urban Flood Mapping Using ALOS-2 Amplitude Images. In Proceedings of the IGARSS 2018-2018 IEEE International Geoscience and Remote Sensing Symposium, Valencia, Spain, 22–27 July 2018; IEEE: New York, NY, USA, 2018; pp. 2350–2353.
44. Feng, Y.; Xiao, Q.; Brenner, C.; Peche, A.; Yang, J.; Feuerhake, U.; Sester, M. Determination of building flood risk maps from LiDAR mobile mapping data. *Comput. Environ. Urban Syst.* **2022**, *93*, 101759. [CrossRef]
45. Pham, B.T.; Luu, C.; Van Phong, T.; Nguyen, H.D.; Van Le, H.; Tran, T.Q.; Ta, H.T.; Prakash, I. Flood risk assessment using hybrid artificial intelligence models integrated with multi-criteria decision analysis in Quang Nam Province, Vietnam. *J. Hydrol.* **2021**, *592*, 125815. [CrossRef]
46. Abe, Y.; Zodrow, I.; Johnson, D.A.; Silerio, L. Risk informed and resilient development: Engaging the private sector in the era of the Sendai Framework. *Prog. Disaster Sci.* **2019**, *2*, 100020. [CrossRef]
47. Wilkins, A.; Pennaz, A.; Dix, M.; Smith, A.; Vawter, J.; Karlson, D.; Tokar, S.; Brooks, E. Challenges and opportunities for Sendai framework disaster loss reporting in the United States. *Prog. Disaster Sci.* **2021**, *10*, 100167. [CrossRef]
48. Criado, M.; Martínez-Graña, A.; San Román, J.S.; Santos-Francés, F. Flood risk evaluation in urban spaces: The study case of Tormes River (Salamanca, Spain). *Int. J. Environ. Res. Public Health* **2019**, *16*, 5. [CrossRef]
49. Lyu, H.M.; Sun, W.J.; Shen, S.L.; Zhou, A.N. Risk assessment using a new consulting process in fuzzy AHP. *J. Constr. Eng. Manag.* **2020**, *146*, 04019112. [CrossRef]
50. Puno, G.R.; Amper, R.A.L.; Talisay, B.A.M. Flood simulation using geospatial and hydrologic models in Manupali Watershed, Bukidnon, Philippines. *J. Bio. Environ. Sci.* **2018**, *12*, 294–303.
51. Gacu, J.G.; Monjardin, C.E.F.; Senoro, D.B.; Tan, F.J. Flood Risk Assessment Using GIS-Based Analytical Hierarchy Process in the Municipality of Odiongan, Romblon, Philippines. *Appl. Sci.* **2022**, *12*, 9456. [CrossRef]
52. Santillan, J.R.; Marqueso, J.T.; Makinano-Santillan, M.; Serviano, J.L. Beyond flood hazard maps: Detailed flood characterization with remote sensing, gis and 2D modelling. *Int. Arch. Photogramm. Remote Sens. Spat. Inf. Sci.* **2016**, *42*, 315. [CrossRef]
53. Chakraborty, L.; Thistlethwaite, J.; Minano, A.; Henstra, D.; Scott, D. Leveraging hazard, exposure, and social vulnerability data to assess flood risk to indigenous communities in Canada. *Int. J. Disaster Risk Sci.* **2021**, *12*, 821–838. [CrossRef]
54. Liu, S.; Zheng, W.; Zhou, Z.; Zhong, G.; Zhen, Y.; Shi, Z. Flood Risk Assessment of Buildings Based on Vulnerability Curve: A Case Study in Anji County. *Water* **2022**, *14*, 3572. [CrossRef]
55. Boroushaki, S.; Malczewski, J. Using the fuzzy majority approach for GIS-based multicriteria group decision-making. *Comput. Geosci.* **2010**, *36*, 302–312. [CrossRef]
56. Ouma, Y.O.; Tateishi, R. Urban flood vulnerability and risk mapping using integrated multi-parametric AHP and GIS: Methodological overview and case study assessment. *Water* **2014**, *6*, 1515–1545. [CrossRef]
57. Karatayev, M.; Kapsalyamova, Z.; Spankulova, L.; Skakova, A.; Movkebayeva, G.; Kongyrbay, A. Priorities and challenges for a sustainable management of water resources in Kazakhstan. *Sustain. Water Qual. Ecol.* **2017**, *9*, 115–135. [CrossRef]
58. Waqas, H.; Lu, L.; Tariq, A.; Li, Q.; Baqa, M.F.; Xing, J.; Sajjad, A. Flash flood susceptibility assessment and zonation using an integrating analytic hierarchy process and frequency ratio model for the Chitral District, Khyber Pakhtunkhwa, Pakistan. *Water* **2021**, *13*, 1650. [CrossRef]
59. Ghosh, G.C.; Khan, M.J.H.; Chakraborty, T.K.; Zaman, S.; Kabir, A.E.; Tanaka, H. Human health risk assessment of elevated and variable iron and manganese intake with arsenic-safe groundwater in Jashore, Bangladesh. *Sci. Rep.* **2020**, *10*, 5206. [CrossRef]
60. Hoque, M.A.A.; Tasfia, S.; Ahmed, N.; Pradhan, B. Assessing spatial flood vulnerability at Kalapara Upazila in Bangladesh using an analytic hierarchy process. *Sensors* **2019**, *19*, 1302. [CrossRef]

61. Mahmoody Vanolya, N.; Jelokhani-Niaraki, M. The use of subjective–objective weights in GIS-based multi-criteria decision analysis for flood hazard assessment: A case study in Mazandaran, Iran. *GeoJournal* **2021**, *86*, 379–398. [CrossRef]
62. Al-Abadi, A.M.; Shahid, S.; Al-Ali, A.K. A GIS-based integration of catastrophe theory and analytical hierarchy process for mapping flood susceptibility: A case study of Teeb area, Southern Iraq. *Environ. Earth Sci.* **2016**, *75*, 687. [CrossRef]
63. Cai, S.; Fan, J.; Yang, W. Flooding risk assessment and analysis based on GIS and the TFN-AHP method: A case study of Chongqing, China. *Atmosphere* **2021**, *12*, 623. [CrossRef]
64. Chen, H.; Ito, Y.; Sawamukai, M.; Tokunaga, T. Flood hazard assessment in the Kujukuri Plain of Chiba Prefecture, Japan, based on GIS and multicriteria decision analysis. *Nat. Hazards* **2015**, *78*, 105–120. [CrossRef]
65. Liu, W.C.; Hsieh, T.H.; Liu, H.M. Flood risk assessment in urban areas of southern Taiwan. *Sustainability* **2021**, *13*, 3180. [CrossRef]
66. Narimani, R.; Jun, C.; Shahzad, S.; Oh, J.; Park, K. Application of a novel hybrid method for flood susceptibility mapping with satellite images: A case study of Seoul, Korea. *Remote Sens.* **2021**, *13*, 2786. [CrossRef]
67. Kittipongvises, S.; Phetrak, A.; Rattanapun, P.; Brundiers, K.; Buizer, J.L.; Melnick, R. AHP-GIS analysis for flood hazard assessment of the communities nearby the world heritage site on Ayutthaya Island, Thailand. *Int. J. Disaster Risk Reduct.* **2020**, *48*, 101612. [CrossRef]
68. Luu, C.; Von Meding, J. A flood risk assessment of Quang Nam, Vietnam using spatial multicriteria decision analysis. *Water* **2018**, *10*, 461. [CrossRef]
69. Fariza, A.; Rusydi, I.; Hasim, J.A.N.; Basofi, A. Spatial flood risk mapping in east Java, Indonesia, using analytic hierarchy process—Natural breaks classification. In Proceedings of the 2017 2nd International conferences on Information Technology, Information Systems and Electrical Engineering (ICITISEE), Yogyakarta, Indonesia, 1–2 November 2017; IEEE: New York, NY, USA, 2017; pp. 406–411.
70. Usman Kaoje, I.; Abdul Rahman, M.Z.; Idris, N.H.; Razak, K.A.; Wan Mohd Rani, W.N.M.; Tam, T.H.; Mohd Salleh, M.R. Physical flood vulnerability assessment using geospatial indicator-based approach and participatory analytical hierarchy process: A case study in Kota bharu, Malaysia. *Water* **2021**, *13*, 1786. [CrossRef]
71. Monjardin, C.E.F.; Tan, F.J.; Uy, F.A.A.; Bale, F.J.P.; Voluntad, E.O.; Batac, R.M.N. Assessment of the existing drainage system in Infanta, Quezon province for flood hazard management using analytical hierarchy process. In Proceedings of the 2020 IEEE Conference on Technologies for Sustainability (SusTech), Santa Ana, CA, USA, 23–25 April 2020; IEEE: New York, NY, USA, 2020; pp. 1–7.
72. Cabrera, J.S.; Lee, H.S. Flood risk assessment for Davao Oriental in the Philippines using geographic information system-based multi-criteria analysis and the maximum entropy model. *J. Flood Risk Manag.* **2020**, *13*, e12607. [CrossRef]
73. Danumah, J.H.; Odai, S.N.; Saley, B.M.; Szarzynski, J.; Thiel, M.; Kwaku, A.; Kouame, F.K.; Akpa, L.Y. Flood risk assessment and mapping in Abidjan district using multi-criteria analysis (AHP) model and geoinformation techniques, (cote d'ivoire). *Geoenviron. Disasters* **2016**, *3*, 10. [CrossRef]
74. Senoro, D.B.; Monjardin, C.E.F.; Fetalvero, E.G.; Benjamin, Z.E.C.; Gorospe, A.F.B.; de Jesus, K.L.M.; Ical, M.L.G.; Wong, J.P. Quantitative Assessment and Spatial Analysis of Metals and Metalloids in Soil Using the Geo-Accumulation Index in the Capital Town of Romblon Province, Philippines. *Toxics* **2022**, *10*, 633. [CrossRef]
75. Department of Trade and Industry Cities and Municipalities Competitive Index—Odiongan, Romblon. Available online: https://cmci.dti.gov.ph/lgu-profile.php?lgu=Odiongan (accessed on 21 December 2022).
76. Allen, S.K.; Ballesteros-Canovas, J.; Randhawa, S.S.; Singha, A.K.; Huggel, C.; Stoffel, M. Translating the concept of climate risk into an assessment framework to inform adaptation planning: Insights from a pilot study of flood risk in Himachal Pradesh, Northern India. *Environ. Sci. Policy* **2018**, *87*, 1–10. [CrossRef]
77. Bertsch, R.; Glenis, V.; Kilsby, C. Building level flood exposure analysis using a hydrodynamic model. *Environ. Model. Softw.* **2022**, *156*, 105490. [CrossRef]
78. Lin, J.; He, X.; Lu, S.; Liu, D.; He, P. Investigating the influence of three-dimensional building configuration on urban pluvial flooding using random forest algorithm. *Environ. Res.* **2021**, *196*, 110438. [CrossRef] [PubMed]
79. Rahadianto, H.; Fariza, A.; Hasim, J.A.N. Risk-level assessment system on Bengawan Solo River basin flood prone areas using analytic hierarchy process and natural breaks: Study case: East Java. In Proceedings of the 2015 International Conference on Data and Software Engineering (ICoDSE), Yogyakarta, Indonesia, 25–26 November 2015; IEEE: New York, NY, USA, 2015; pp. 195–200.
80. Waghwala, R.K.; Agnihotri, P.G. Flood risk assessment and resilience strategies for flood risk management: A case study of Surat City. *Int. J. Disaster Risk Reduct.* **2019**, *40*, 101155. [CrossRef]
81. Xiao, Y.; Yi, S.; Huang, Y. A study on flood risk assessment approach based on multi-source geographic data integration and GIS methods: A case study of lower Han River region. In Proceedings of the 2015 23rd International Conference on Geoinformatics, Wuhan, China, 19–21 June 2015; IEEE: New York, NY, USA, 2015; pp. 1–5.
82. Chen, W.; Zhang, L. Building vulnerability assessment in seismic areas using ensemble learning: A Nepal case study. *J. Clean. Prod.* **2022**, *350*, 131418. [CrossRef]
83. Shah, M.F.; Kegyes-B, O.K.; Ray, R.P.; Ahmed, A.; Al-ghamadi, A. Vulnerability assessment of residential buildings in Jeddah: A methodological proposal. *GEOMATE J.* **2018**, *14*, 134–141. [CrossRef]
84. Arrighi, C.; Mazzanti, B.; Pistone, F.; Castelli, F. Empirical flash flood vulnerability functions for residential buildings. *SN Appl. Sci.* **2020**, *2*, 904. [CrossRef]
85. Ferreira, T.M.; Santos, P.P. An integrated approach for assessing flood risk in historic city centres. *Water* **2020**, *12*, 1648. [CrossRef]

86. Zhen, Y.; Liu, S.; Zhong, G.; Zhou, Z.; Liang, J.; Zheng, W.; Fang, Q. Risk Assessment of Flash Flood to Buildings Using an Indicator-Based Methodology: A Case Study of Mountainous Rural Settlements in Southwest China. *Front. Environ. Sci.* **2022**, *10*, 931029. [CrossRef]

87. Park, K.; Lee, M.H. The development and application of the urban flood risk assessment model for reflecting upon urban planning elements. *Water* **2019**, *11*, 920. [CrossRef]

88. Arrighi, C.; Rossi, L.; Trasforini, E.; Rudari, R.; Ferraris, L.; Brugioni, M.; Franceschini, S.; Castelli, F. Quantification of flood risk mitigation benefits: A building-scale damage assessment through the RASOR platform. *J. Environ. Manag.* **2018**, *207*, 92–104. [CrossRef]

89. Sandoval, J.A.; Tiburan Jr, C.L. Identification of potential artificial groundwater recharge sites in Mount Makiling Forest Reserve, Philippines using GIS and Analytical Hierarchy Process. *Appl. Geogr.* **2019**, *105*, 73–85. [CrossRef]

90. Shaikh, A.A.; Pathan, A.I.; Waikhom, S.I.; Agnihotri, P.G.; Islam, M.N.; Singh, S.K. Application of latest HEC-RAS version 6 for 2D hydrodynamic modeling through GIS framework: A case study from coastal urban floodplain in India. *Model. Earth Syst. Environ.* **2022**, 1–17. [CrossRef]

91. Sy, B.; Frischknecht, C.; Dao, H.; Consuegra, D.; Giuliani, G. Flood hazard assessment and the role of citizen science. *J. Flood Risk Manag.* **2019**, *12*, e12519. [CrossRef]

92. M.G.B. Geohazard Maps. Available online: https://region4b.mgb.gov.ph/28-geohazard-maps/98-geohazard-maps#romblon-2 (accessed on 21 December 2022).

93. Hammami, S.; Zouhri, L.; Souissi, D.; Souei, A.; Zghibi, A.; Marzougui, A.; Dlala, M. Application of the GIS based multi-criteria decision analysis and analytical hierarchy process (AHP) in the flood susceptibility mapping (Tunisia). *Arab. J. Geosci.* **2019**, *12*, 653. [CrossRef]

94. Rincón, D.; Khan, U.T.; Armenakis, C. Flood risk mapping using GIS and multi-criteria analysis: A greater Toronto area case study. *Geosciences* **2018**, *8*, 275. [CrossRef]

95. Dung, N.B.; Long, N.Q.; Goyal, R.; An, D.T.; Minh, D.T. The role of factors affecting flood hazard zoning using analytical hierarchy process: A review. *Earth Syst. Environ.* **2022**, *6*, 697–713. [CrossRef]

96. Jamrussri, S.; Toda, Y. Available Flood Evacuation Time for High-Risk Areas in the Middle Reach of Chao Phraya River Basin. *Water* **2018**, *10*, 1871. [CrossRef]

97. Duan, Y.; Xiong, J.; Cheng, W.; Li, Y.; Wang, N.; Shen, G.; Yang, J. Increasing Global Flood Risk in 2005–2020 from a Multi-Scale Perspective. *Remote Sens.* **2022**, *14*, 5551. [CrossRef]

98. Swain, K.C.; Singha, C.; Nayak, L. Flood susceptibility mapping through the GIS-AHP technique using the cloud. *ISPRS Int. J. Geo-Inf.* **2020**, *9*, 720. [CrossRef]

 *buildings*

Article

Comparative Analysis of the 2023 Pazarcık and Elbistan Earthquakes in Diyarbakır

Ibrahim Baran Karasin

Department of Construction Technology, Diyarbakir Vocational School of Technical Sciences, Dicle University, Diyarbakır 21280, Türkiye; baran.karasin@dicle.edu.tr or barankarasin@gmail.com

Abstract: Türkiye is prone to earthquakes due to its location on various tectonic plates, which can lead to a loss of lives and property. Recently, on 6 February 2023, two major earthquakes hit Pazarcık and Elbistan in Türkiye, causing widespread destruction on the East Anatolian Fault (EAF) zone. Even Diyarbakır, a distant province from the epicentre, was severely affected, highlighting the need to evaluate Turkish earthquake codes. As part of this evaluation, a structural analysis was conducted on earthquake-damaged and collapsed buildings in Diyarbakır. The study analysed three buildings with different levels of damage and six collapsed buildings as case studies. The seismic parameters of the earthquakes were compared to the values in the two recent earthquake hazard maps used in Türkiye's codes, as well as the Eurocode 8 damage limit values obtained from pushover analysis. The results revealed significant differences between the current seismic values of earthquakes and the current peak ground acceleration (PGA) values specified in the Turkish Earthquake Design Regulations. Additionally, the selected buildings showed inadequate structural behaviours, with significant differences between the expected and actual seismic performances with respect to the PGA values as one of the most important earthquake characteristics.

Keywords: Pazarcık-Elbistan earthquakes; Diyarbakir; pushover analysis; damage limits; Eurocode

1. Introduction

Situated at the convergence of multiple active tectonic plates and traversing the Alpine-Himalayan seismic belt, Türkiye has been a witness to a series of devastating earthquakes throughout both its historical and instrumental periods. The intricate interplay of tectonic forces from the Arabian and African plates has exacerbated the region's vulnerability, resulting in substantial human and infrastructural losses. Of paramount significance are Türkiye's pivotal fault systems—the North Anatolian Fault (NAF) and the East Anatolian Fault (EAF) zones—each bearing the latent potential to unleash highly catastrophic seismic events. This seismic potential has, over time, inflicted profound human casualties and extensive material devastation. The chronicles of past local earthquakes serve as invaluable records, lending predictive insights for potential seismic occurrences and guiding the formulation of seismic design codes. Marking a pivotal juncture, 6 February 2023 etched a distressing chapter in Türkiye's seismic narrative, bearing witness to two monumental earthquakes that etched their names as indelible signatures within the recent annals of seismic chronicles. Among these impactful events, the Pazarcık Earthquake of 6 February 2023, and its synchronous counterpart, the Elbistan earthquake, take centre stage.

Türkiye has developed and utilised earthquake hazard maps to estimate the potential risk and to inform seismic design codes. The last two earthquake hazard maps, both the previous and current versions, are compared in terms of their seismic parameters. These comparisons reveal differences in the peak ground acceleration values and expected target structural displacement values [1]. The investigation of the effects of earthquakes on engineering structures in earthquake-prone parts of the Earth is a unique tool for determining the effects of the next ground motion [2]. In this case, much research in this area [3–15] has

Citation: Karasin, I.B. Comparative Analysis of the 2023 Pazarcık and Elbistan Earthquakes in Diyarbakır. *Buildings* **2023**, *13*, 2474. https://doi.org/10.3390/buildings13102474

Academic Editor: Binsheng (Ben) Zhang

Received: 19 August 2023
Revised: 24 September 2023
Accepted: 25 September 2023
Published: 29 September 2023

investigated the damage assessment and sustainability of RC buildings by using fragility curves and proposed a new model to evaluate damages for structures. On the other hand, the seismic response of buildings is also an important topic for earthquake engineering. Determination of the seismic response of structures has been an important issue not only for reinforced concrete structures but also for steel structures. There are many innovative and novel studies in this field [16,17]. Faizah and Amaliah [18] investigated the seismic situations in thirty-four cities in Indonesia by comparing the values of the spectral response parameters (S_{DS} and S_{D1}) according to the 2012 and 2019 Indonesian earthquake codes. Avcil et al. [19] have made comparisons of the target displacements in different seismic zones under the effects of different soil conditions in their work. Wei et al. [20] have proposed an evaluation method for the seismic damage of bridges. They used the maximum and residual drifts as engineering demand parameters (EDPs). In the study by Khanmohammadi et al. [21], the dynamic properties of forty-six reinforced concrete and steel buildings affected by the Sarpol-e Zahab earthquake (Mw 7.3) were determined by ambient vibration tests. Ghasemi et al. [22] investigated the seismic performance of RC systems with cable bracings. The findings of the work indicated that the PGA capacity of the RC building increases as the number of braces rises. Zhou et al. [23] investigated the seismic performance and collapse mechanism of a five-storey reinforced concrete structure along the ground fissure using the pushover analysis method. Mazza [24] has proposed a displacement-damage based design procedure with a computer-aided tool called DAMPERS (Damage Protection of Earthquake Resistant Structures). Harirchian et al. [25] aimed to bridge the gap between rapid visual screening (RVS) and multi-criteria decision-making (MCDM) methods by using the codes from India, Türkiye and the Federal Emergency Management Agency (FEMA) in their study. Eroglu et al. [26] measured the sensitivity of seismic hazard assessments using different declustering techniques. Accordingly, the recently compiled earthquake catalogue of Türkiye was declustered using three declustering algorithms in their study.

This article delves into the analysis of these earthquakes and their effects on the province of Diyarbakır located in south-eastern Türkiye. It compares seismic parameters with recent earthquake hazard maps and evaluates the effectiveness of earthquake codes. Structural analysis of earthquake-damaged buildings reveals the differences between the expected and actual seismic performances. At its core, this case study seeks to analyse the seismic parameters of the Pazarcık and Elbistan earthquakes involving a comparative assessment between these seismic parameters and the data presented in the two recent earthquake hazard maps utilized within the Turkish context. Furthermore, the comparison extends to encompass the Eurocode 8 damage limit values, derived from pushover analysis conducted on a typical reinforced concrete building.

In light of the disparities observed between the seismic characteristics of the earthquakes and the PGA values stipulated in the Turkish Earthquake Design Regulations, a vital evaluation of the efficacy of the existing Turkish earthquake codes is undertaken. Three buildings exhibiting slight, moderate and extensive damages were selected, along with six buildings that collapsed. The seismic parameters are analysed in accordance with both the 2018 and 2007 seismic codes of Türkiye, enabling a comparison to highlight potential advancements or deficiencies in the seismic design criteria.

In this study, the geographical location and seismic parameters of the structures damaged in Diyarbakır province in the Pazarcık and Elbistan earthquakes were determined for the first time, along with an attempt to reveal the effect of structural analysis by comparing the Eurocode and Turkish regulations in terms of limit values. The aim of the study is to compare the measured and predicted PGA values and the target displacements to be obtained based on them. In this way, the effects of the earthquakes on Diyarbakır and whether these effects on the structural analysis are adequately represented will be revealed.

2. Materials and Methods

The location data of the earthquakes with respect to the Bogazici University Kandilli Observatory and Earthquake Monitoring Center (KOERI) [27], Geofon Data Center (GFZ) [28], Disaster and Emergency Management Presidency, Republic of Türkiye (AFAD) [29] and United States Geological Survey (USGS) [30] are shown in Figure 1. On the other hand, the current earthquake hazard map of Türkiye and the location of Diyarbakır province's Türkiye Earthquake Hazard Maps Interactive Web Application (TEHMIWA) [29] is given in Figure 2.

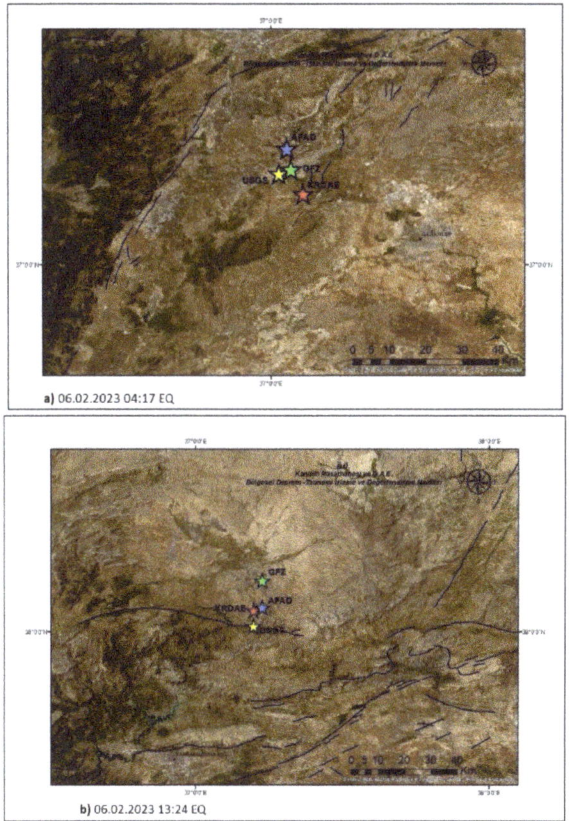

Figure 1. The location data of the earthquakes: (**a**) Pazarcık; (**b**) Elbistan.

The focal depths of the earthquakes were estimated between 5 to 10 km by the seismological centres and classified as shallow-focused earthquakes. The location and magnitude data of these two earthquakes according to the centres are provided in Table 1.

The seismic intensity map projection of the 6 February 2023 04:17 Kahramanmaraş Earthquake is shown in Figure 3. The maximum intensity of 11–12 around the striking rupture fault(s) can be observed. It is noted that for Diyarbakir province, far away from the epicentre, it was about 6–7.

At the Çermik station (coded 2107), which is 234.922 km (R_{jb} = 92.53) away from the epicentre of the Pazarcık earthquake and the closest station to Diyarbakır, the minimum acceleration of about 0.04 g was recorded in the vertical direction, while the maximum acceleration of 0.11 g was measured in the east–west direction. According to the AFAD data, the peak ground acceleration value during the Pazarcık earthquake was 2.017 g in the east–west direction at the Pazarcık station (coded 4614 (NAR) R_{jb} = 1.02). Since the maximum intensities are in the E–W direction, the PGV, Arias and Houser intensities are only given in the E–W direction.

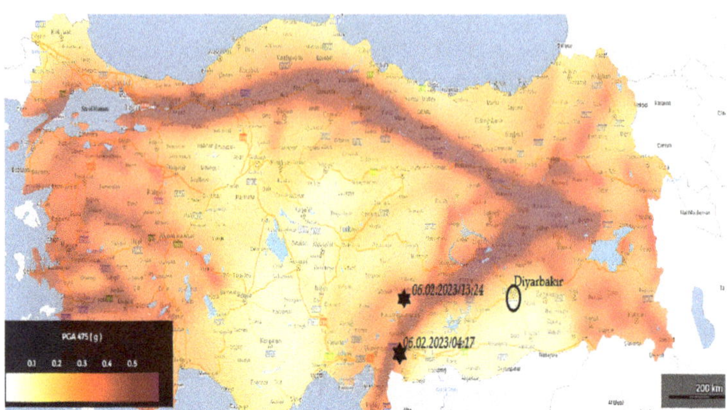

Figure 2. Location data of 6 February 2023 earthquakes on TEHMIWA [29].

Table 1. Location and magnitude data of the earthquakes that occurred on 6 February 2023 given by different seismological centres.

Institute	6 February 2023 04:17		6 February 2023 13:24	
	Magnitude (M_w)	Location	Magnitude (M_w)	Location
KOERI(KRDAE)	7.7	37.17-37.08	7.6	38.07-37.20
AFAD	7.7	37.28-37.04	7.6	38.08-37.23
GFZ	7.7	37.27-37.05	7.6	38.17-37.23
USGS	7.8	37.22-37.02	7.5	38.02-37.20

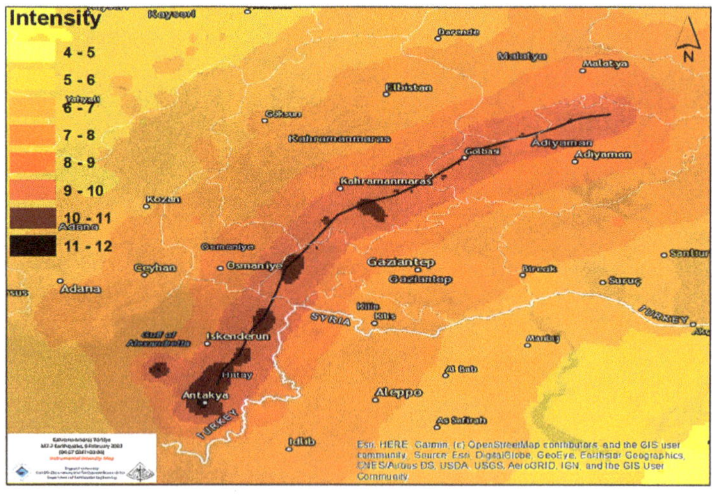

Figure 3. 6 February 2023 04:17 earthquake intensity map [31].

During the Elbistan earthquake, the Göksun station (coded 4612 (R_{epi} = 66.68)) recorded the highest acceleration value of 0.63 g in the north–south direction. Meanwhile, the Çermik station (R_{epi} = 198.48) in Diyarbakır recorded the highest value of 0.047 g in the east–west direction. These values are given in Tables 2 and 3 for both earthquakes. Figures 4 and 5 show the accelerations and spectra obtained with the data of station 2107 in Diyarbakır province for both the Pazarcık and Elbistan earthquakes. Since the maximum intensities are in the N–S direction, the PGV, Arias and Houser intensities are only given in the N–S direction.

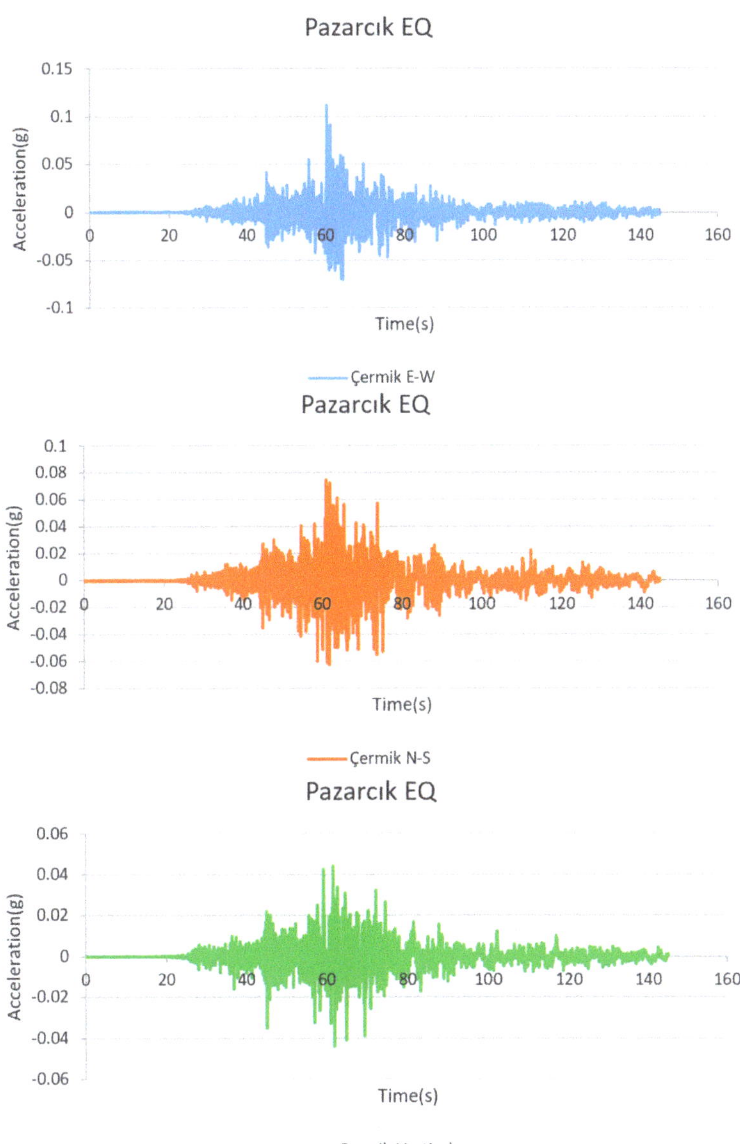

Figure 4. Acceleration records at the station 2107 in Diyarbakır province (Pazarcık EQ).

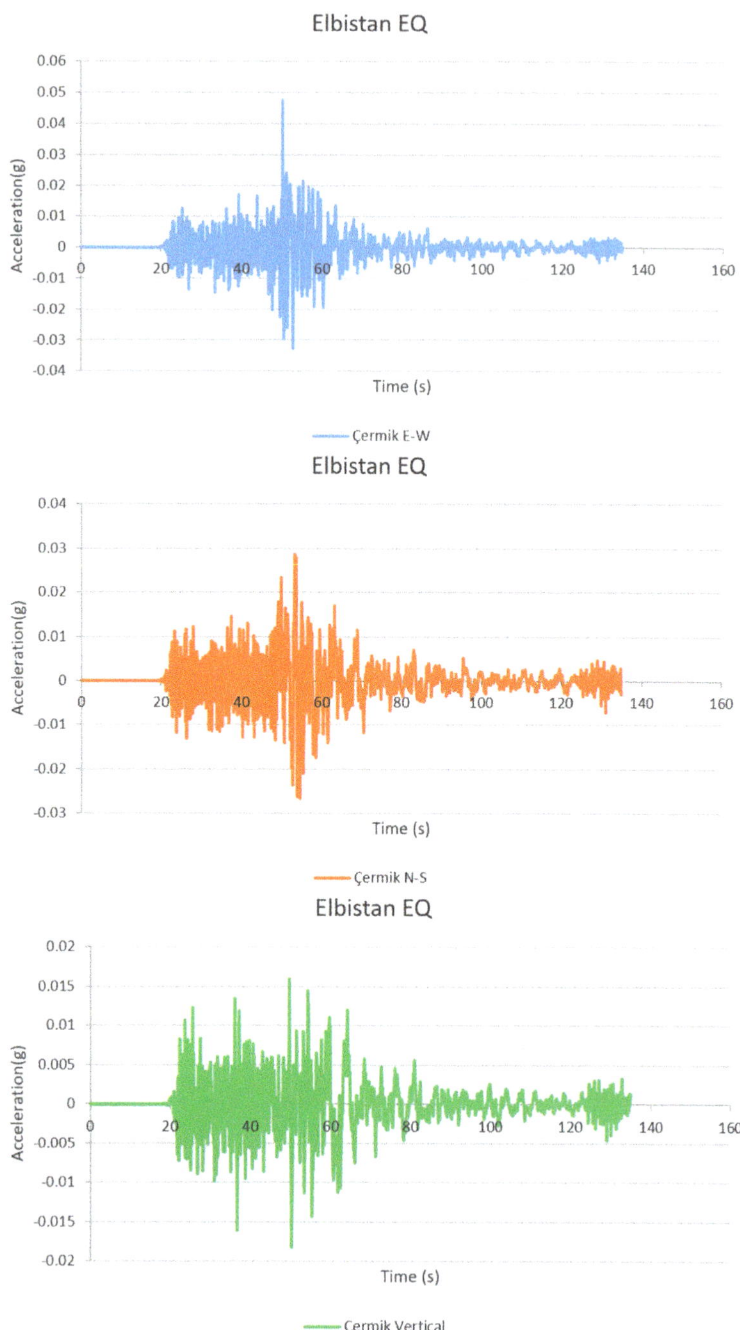

Figure 5. Acceleration records at the station 2107 in Diyarbakır province (Elbistan EQ).

Table 2. Pazarcık EQ records.

Location	PGA (cm/sn^2) (N–S)	PGA (cm/sn^2) (E–W)	PGA (cm/sn^2) (Vertical)	PGV (cm/sn)	Arias Intensity (cm/sn)	Housner Intensity (cm)
Pazarcık	2016.99	2039.20	1582.62	78.64	8181.83	261.44
Çermik	74.76	112.27	44.24	26.07	28.53	74.15

Table 3. Elbistan EQ records.

Location	PGA (cm/sn^2) (N–S)	PGA (cm/sn^2) (E–W)	PGA (cm/sn^2) (Vertical)	PGV (cm/sn)	Arias Intensity (cm/sn)	Housner Intensity (cm)
Göksun	635.45	523.21	494.91	170.78	417.03	393.13
Çermik	28.64	47.61	18.20	10.93	4.93	30.62

It is noted that, based on the earthquake intensities and acceleration records taken into consideration for Diyarbakir province, there may be some moderate damage resulting from the relatively long duration of the earthquake. Past earthquakes have demonstrated that numerous factors, including earthquake duration, near-fault and far-fault effects [32] and soil–structure interaction, can significantly impact earthquake damage. However, in the specific case of Diyarbakır, the distance from the earthquake's epicentre and the relatively lower values of the measured earthquake parameters (such as Arias and Houser intensities and PGA and PGV values) than the obtained values from the EQ epicentre indicate the severity of the structural damage observed. However, because of substandard dwelling construction aspects, the structural damage and loss of lives resulting from these earthquakes extended to unacceptable levels.

2.1. Investigation of the Collapsed Buildings in Diyarbakır

Earthquake amplitudes decrease in parallel with the distance–decay relationship as one moves away from the earthquake focal point, and accordingly, lower values were expected to be observed in Diyarbakır. However, the unexpected disproportionate destruction, damage and loss of lives in response to these low acceleration values raise some questions. In the observations made in the field after the earthquake, the presence of many structural defects, especially low-strength concrete and poor steel reinforcement, are noteworthy. In Diyarbakır province, the locations of a total of 18 buildings, including the building shown in Figures 6–8, 5 other collapsed buildings, 9 damaged (slightly, moderately and extensively damaged) buildings and 3 undamaged buildings, were marked with the help of the GPS on-site detection.

Figure 6. Collapsed buildings in Diyarbakır.

Figure 7. Shear damage of the beam and inappropriate stirrups.

Figure 8. Retrofitting of the ground floor only and the failure of the old column.

Figure 8 shows the undamaged ground floor columns of the completely collapsed building. The slender column rising after the column with an exceptionally large cross-section might indicate that this building was subjected to an insufficient retrofitting process. Some damage examples of inappropriate shear walls and columns are illustrated in Figures 9 and 10.

According to the official Provincial Damage Assessment Reports, the numbers of the extensively damaged or collapsed, moderately damaged and slightly damaged buildings were recorded as 8602, 11,209 and 113,223, respectively, with hundreds of casualties [33]. The common cause of the collapse of the mentioned collapsed or extensive damaged structures can be considered according to many constructional defects, especially low concrete strength.

Figure 9. Extensively damaged columns.

Figure 10. Damage to the short column and shear wall with poor concrete.

2.2. Structural Analyses

Local seismic parameters were obtained with the help of the Türkiye Earthquake Hazard Map created by TBEC2018 [34] with micro zonation. Previous regulation TSC-2007 [35] considered the local seismic parameters constant due to the regional approach. In order to compare these parameters, a sample reinforced concrete structure was modelled in SeismoStruct v.2023 software in accordance with the current regulations. Using pushover analysis, the damage limit values determined in Eurocode 8 [36] were obtained and compared for different PGA values of the investigated structure.

The analysis steps of pushover analysis as a performance-based evaluation method that estimates the structural responses to seismic loads by applying a series of increasing lateral loads are shown in Figure 11. The method is particularly useful in identifying the weak points and failure mechanisms of structures and helps in determining appropriate retrofitting strategies [37–39].

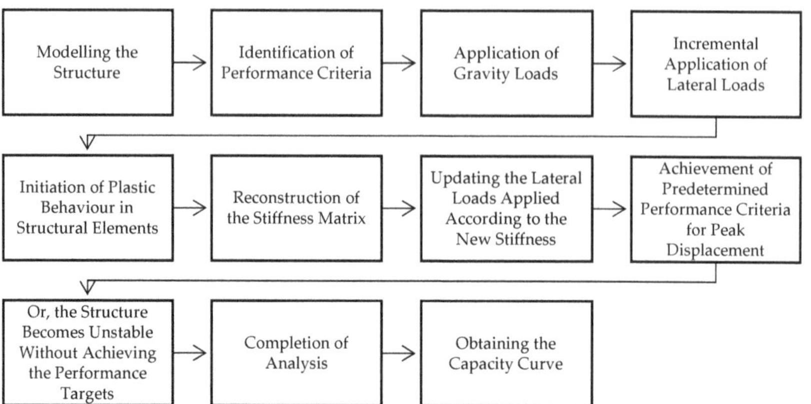

Figure 11. Analysis steps of pushover analysis.

The initial step involves assuming a particular pattern for the lateral load, followed by conducting a static analysis of the structural model under this load pattern in a pushover analysis. A distributed load pattern has been used with a 0.3 m target displacement in this study. The evaluation of the earthquake impact and the devastating effects on buildings can be performed not only by extracting its various peaks and cumulative parameters but also by calculating various types of linear and nonlinear seismic spectra. Furthermore, pushover analysis can be performed for various simplified cases of buildings in an effort to estimate the response that they would exhibit during the earthquake. The calculated parameters can provide some hints about the destructiveness of the earthquake and how the buildings could be designed to be able to resist such earthquakes in the future [40].

In the present study, the seismic parameters of the earthquakes are calculated to provide explanations about the large unexpected structural destructiveness in Diyarbakir. Structural analysis of a reinforced concrete building in Diyarbakır was performed using both the measured and current acceleration values according to the last two earthquake hazard maps. The results of this analysis reveal significant differences between the expected target structural displacement values, highlighting the importance of continuous updates to these maps for accurate seismic risk assessment.

For a comprehensive understanding of the Pazarcık Earthquake's impact on Diyarbakır, earthquake-damaged and collapsed buildings in the province were analysed. Three buildings, each with minor, moderate, and severe damage, and six buildings that collapsed during the earthquake were selected for this purpose. The sample building's 2D view, 3D view, and blueprint of the sample RC model can be seen in Figure 12. The numerical specifications of the sample building are given in Table 4. The fundamental period of the structure was 0.8271 s; the maximum base shear was 6557.27 kN; the elastic stiffness K_elas and the effective stiffness K-eff were obtained as 120,492.20 kN/m and 63,600.91 kN/m, respectively.

The latest version of the Turkish Building Earthquake Code (TBEC-2018) has introduced three additional levels of ground motion (DD-1, DD-3, DD-4) compared to the previous edition. In the previous code, only the standard earthquake ground motion level with a recurrence period of 475 years and a 10% probability of exceedance in 50 years (DD-2) was considered in TSC-2007. It is important to note that the first two levels (DD-1 and DD-2) of the four earthquake levels correspond to the design earthquakes in ASCE-07 [41]. The current code outlines four different levels of ground motion, which can be found in Table 5.

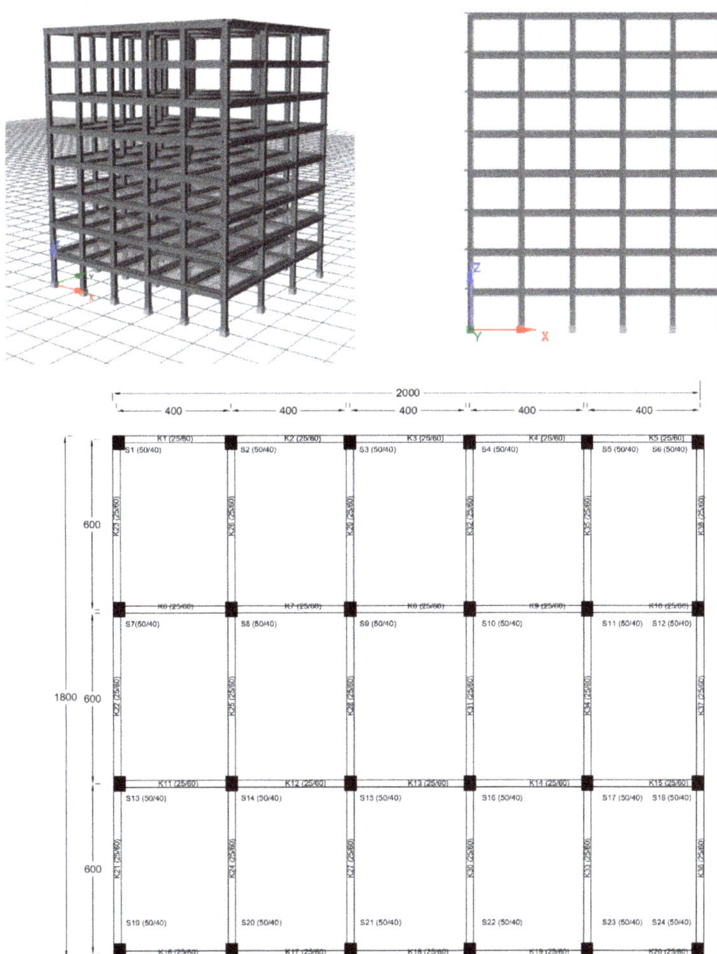

Figure 12. 2D and 3D views with blueprint of the sample RC model.

Table 4. Structural specifications of the sample building.

Parameter	Value	Parameter	Value
Concrete grade	C25	Transverse reinforcement (columns)	$\Phi 10/100$
Reinforcement grade	S420	Transverse reinforcement (beam)	$\Phi 10/150$
Beams	250 mm × 600 mm	Steel material model	Menegotto–Pinto
Height of floor	120 mm	Constraint type	Rigid diaphragm
Height of each storey	3 m	Local ground type	ZC
Cover thickness	25 mm	Incremental load	10 kN
Columns	500 mm × 400 mm	Permanent Load (Slabs and Infills)	7 kN/m
Longitudinal Reinforcement (columns)	Corners 4Φ16 Top bottom side 4Φ16 Left right side 4Φ16	Damping	5%
Target-displacement (8-storey)	0.30 m	Importance class	IV

Table 5. Earthquake ground motion levels (TBEC-2018).

Earthquake Level	Repetition Period (Year)	Probability of Exceedance in 50 Years	Description
DD-1	2475	2%	Largest earthquake ground motion
DD-2	475	10%	Standard design earthquake ground motion
DD-3	72	50%	Frequent earthquake ground motion
DD-4	43	68%	Service earthquake ground motion

TEHMIWA is a user-friendly and practical web application designed for use by the Disaster and Emergency Management Agency of Türkiye (AFAD). It enables the calculations of earthquake parameters for any location using various earthquake ground motion levels, local ground conditions, and latitude and longitude data. The application takes into account the probability of exceedance and local ground conditions to obtain the short-period map spectral acceleration coefficient (S_S) and the map spectral acceleration coefficient (S_1) for the 1 s period. The design spectral acceleration coefficients S_{DS} and S_{D1} are determined using equations that consider the spectral acceleration coefficients (S_S and S_1) and local ground coefficients (F_S, F_1).

$$S_{DS} = S_S \times F_S \tag{1}$$

$$S_{D1} = S_1 \times F_1 \tag{2}$$

Tables 6 and 7 provide the local ground effect coefficients for the short period (F_S) and 1.0 s (F_1), respectively. These coefficients are being used for the first time in the current code, with greater emphasis on the local ground effect.

Table 6. Local ground effect coefficients for the short-period zone (Fs) (TBEC-2018).

Local Soil Class	Local Ground Impact Coefficients for Short-Period Zone (F_S)					
	$S_S < 0.25$	$S_S = 0.50$	$S_S = 0.75$	$S_S = 1.00$	$S_S = 1.25$	$S_S > 1.50$
ZA	0.8	0.8	0.8	0.8	0.8	0.8
ZB	0.9	0.9	0.9	0.9	0.9	0.9
ZC	1.3	1.3	1.2	1.2	1.2	1.2
ZD	1.6	1.4	1.2	1.1	1	1
ZE	2.4	1.7	1.3	1.1	0.9	0.8
ZF	Site-specific ground behaviour analysis will be carried out.					

Table 7. Local ground effect coefficients for a period of 1.0 s (F_1) (TBEC-2018).

Local Soil Class	Local Ground Impact Coefficients for 1 s Period Zone (F_1)					
	$S_1 < 0.10$	$S_1 = 0.20$	$S_1 = 0.30$	$S_1 = 0.40$	$S_1 = 0.50$	$S_1 > 0.60$
ZA	0.8	0.8	0.8	0.8	0.8	0.8
ZB	0.8	0.8	0.8	0.8	0.8	0.8
ZC	1.5	1.5	1.5	1.5	1.5	1.4
ZD	2.4	2.2	2	1.9	1.8	1.7
ZE	4.2	3.3	2.8	2.4	2.2	2
ZF	Site-specific ground behaviour analysis will be carried out.					

To allow for accurate comparisons, ZC was selected as the standard soil class based on TBEC-2018. This soil class will remain consistent for all parameters requiring a local soil class, and its characteristics can be found in Table 8.

Table 8. Local soil class type ZC (TBEC-2018).

Local Soil Class	Soil Type	$(V_S)_{30}$ [m/s]	Upper Average at 30 m $(N_{60})_{30}$ [Pulse/30 cm]	$(c_u)_{30}$ [kPa]
ZC	Very tight sand, gravel and hard clay layers or weathered, very cracked weak rocks	360–760	>50	>250

The spectral acceleration coefficients are only compared for the DD-2 ground motion level. This is because the previous code only used ground motion levels with a recurrence period of 475 years and a 10% probability of exceedance in 50 years. Table 9 shows the comparison of the spectral acceleration coefficients based on the last two seismic design codes. There were no vertical values in the previous code, so no comparisons were made in that direction.

Table 9. The comparison of the spectral acceleration coefficients with ground-type ZC.

| DD-2 Building | Spectral Acceleration Coefficients | | | | | Horizontal | | | | 2018/2007 | | Vertical | | |
| | All Ground Types TSC-2007 | | ZC TBEC-2018 | | S_{DS} 2018/2007 | TSC-2007 | | TBEC-2018 | | ZC | | TSC-2007 | TBEC-2018 | |
	S_{DS}	$0.40S_{DS}$	S_{DS}	$0.40S_{DS}$		T_A	T_B	T_A	T_B	T_A	T_B	T_{AD} T_{BD}	T_{AD}	T_{BD}
No Damage 1	1	0.4	0.422	0.1688	0.422	0.15	0.60	0.095	0.476	0.63	0.79		0.032	0.159
No Damage 2	1	0.4	0.417	0.1668	0.417	0.15	0.60	0.096	0.478	0.64	0.80		0.032	0.159
No Damage 3	1	0.4	0.428	0.1712	0.428	0.15	0.60	0.095	0.473	0.63	0.79		0.032	0.158
Slight Damage 1	1	0.4	0.421	0.1684	0.421	0.15	0.60	0.095	0.474	0.63	0.79		0.032	0.158
Slight Damage 2	1	0.4	0.426	0.1704	0.426	0.15	0.60	0.094	0.471	0.63	0.79		0.031	0.157
Slight Damage 3	1	0.4	0.420	0.168	0.420	0.15	0.60	0.095	0.475	0.63	0.79		0.032	0.158
Moderate Damage 1	1	0.4	0.417	0.1668	0.417	0.15	0.60	0.096	0.478	0.64	0.80		0.032	0.159
Moderate Damage 2	1	0.4	0.416	0.1664	0.416	0.15	0.60	0.095	0.476	0.63	0.79	There is no vertical spectrum	0.032	0.159
Moderate Damage 3	1	0.4	0.415	0.166	0.415	0.15	0.60	0.095	0.477	0.63	0.80		0.032	0.159
Extensive Damage 1	1	0.4	0.419	0.1676	0.419	0.15	0.60	0.095	0.477	0.63	0.80		0.032	0.159
Extensive Damage 2	1	0.4	0.437	0.1748	0.437	0.15	0.60	0.094	0.47	0.63	0.78		0.031	0.157
Extensive Damage 3	1	0.4	0.398	0.1592	0.398	0.15	0.60	0.097	0.486	0.65	0.81		0.032	0.162
Collapsed 1	1	0.4	0.408	0.1632	0.408	0.15	0.60	0.096	0.481	0.64	0.80		0.032	0.16
Collapsed 2	1	0.4	0.403	0.1612	0.403	0.15	0.60	0.097	0.484	0.65	0.81		0.032	0.161
Collapsed 3	1	0.4	0.412	0.1648	0.412	0.15	0.60	0.096	0.48	0.64	0.80		0.032	0.16
Collapsed 4	1	0.4	0.413	0.1652	0.413	0.15	0.60	0.096	0.479	0.64	0.80		0.032	0.16
Collapsed 5	1	0.4	0.402	0.1608	0.402	0.15	0.60	0.097	0.485	0.65	0.81		0.032	0.162
Collapsed 6	1	0.4	0.407	0.1628	0.407	0.15	0.60	0.097	0.483	0.65	0.81		0.032	0.161

Damage limits, which represent the point at which a structure can no longer withstand applied loads, play a crucial role in assessing the performance of buildings during earthquakes. In the case of the Pazarcık Earthquake, an evaluation was conducted to determine the damage limits of the selected buildings in Diyarbakır. This assessment provided valuable information regarding the responses of local structures to the seismic activities and the extent of the damage they sustained.

The limit states given in Eurocode 8, which is used worldwide for damage estimation, were also taken into consideration in the study. Detailed descriptions of these limit states are given in Table 10.

The local seismic parameters of these selected buildings were analysed according to the 2018 Building Earthquake Code of Türkiye. This analysis revealed significant differences between the expected target structural displacement values resulting from the structural analysis.

The seismic values of the Pazarcık and Elbistan earthquakes were compared with the current peak ground acceleration values specified in the current Turkish Earthquake Design Regulation. This comparison demonstrated differences between the measured and proposed peak ground accelerations for some earthquakes.

Table 10. Limit states in Eurocode 8 (Part 3) (CEN 2004) [36].

Limit State	Description	Return Period (Year)
Limit state of damage limitation (DL)	Only lightly damaged, damage to non-structural components is economically repairable.	225
Limit state of significant damage (SD)	Significantly damaged, some residual strength and stiffness, non-structural components damaged, uneconomic to repair.	475
Limit state of near collapse (NC)	Extensively damaged, very low residual strength and stiffness, large permanent drift but still standing.	2475

These data are from TBEC (2018) (Turkish Earthquake Building Code). DD1 is 2% in 50-year ground motion level. A comparison with DD2, i.e., 10% in 50-year ground motion level, is given in Figure 13.

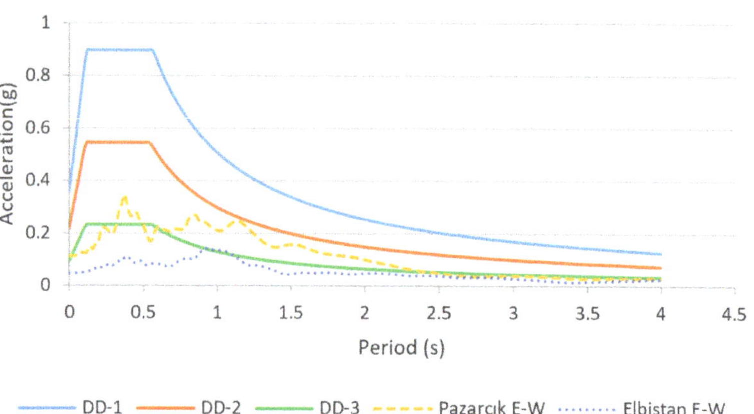

Figure 13. Comparison of the spectral acceleration and specification elastic spectrum capacity values of the station 2107 in Diyarbakır province.

3. Results

The local seismic parameters of the selected buildings in Diyarbakır, as analysed according to the Türkiye Building Earthquake Code (TBEC2018) and Turkish Seismic Code 2007 (TSC2007), were compared with the Eurocode 8 specifications. This comparison provided insights into the effectiveness of the current design regulations in adequately addressing the seismic risks in the region.

The seismic parameters of the chosen buildings in Diyarbakır were assessed based on the local seismic data and analysed in accordance with TBEC2018 and TSC2007. In order to evaluate the adequacy of the current design regulations in addressing seismic risks, a comprehensive comparison was made with the Eurocode 8 specifications in terms of limit states. This comparative analysis has clarified the compatibility between the local seismic conditions and the prescribed design criteria, highlighting the areas where improvements or adjustments may be required. By analysing the similarities and differences between the two codes, valuable insights have been gained into the effectiveness of the current design codes in ensuring the structural strength of buildings in the face of seismic events in the region.

The seismic performances of various buildings in Diyarbakır were evaluated based on the recorded ground motion parameters. In the analysis, four different damage levels were considered with three buildings each: no damage, slight damage, moderate damage and severe damage. Ground motion parameters including DD1, DD2 and DD3 were compared with the TSC 2007 design code and the PGA (peak ground acceleration) values obtained from the first and second earthquakes. The analysed buildings are marked on the city map as shown in Figure 14.

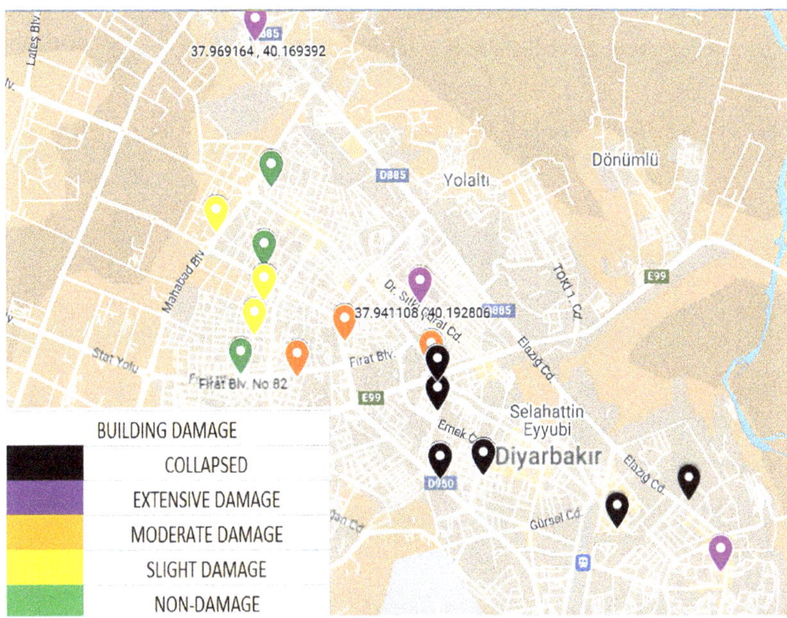

Figure 14. Map of the damaged buildings and their locations.

Among the analysed buildings, those categorised as undamaged did not show relatively lower PGA values in all three components as expected. It would be expected as the damage level increased from slight damage to extensive damage that the PGA values would generally increase, indicating a higher level of ground shaking and potential structural damage. Instead, an undamaged building may have higher PGA values than a collapsed one, as shown in Tables 11 and 12.

Table 11. PGA values of the analysed damaged buildings (g).

Buildings	DD1	DD2	DD3	TSC 2007	1st EQ PGA	2nd EQ PGA	Soil Type	Importance Class	Damping
No Damage 1	0.259 g	0.145 g	0.062 g						
No Damage 2	0.256 g	0.143 g	0.061 g						
No Damage 3	0.262 g	0.146 g	0.063 g						
Slight Damage 1	0.258 g	0.144 g	0.062 g						
Slight Damage 2	0.260 g	0.146 g	0.063 g						
Slight Damage 3	0.257 g	0.143 g	0.062 g	0.3 g	0.119 g	0.049 g	C	2	0.05
Moderate Damage 1	0.256 g	0.143 g	0.061 g						
Moderate Damage 2	0.255 g	0.142 g	0.061 g						
Moderate Damage 3	0.254 g	0.142 g	0.061 g						
Extensive Damage 1	0.256 g	0.143 g	0.062 g						
Extensive Damage 2	0.267 g	0.149 g	0.065 g						
Extensive Damage 3	0.245 g	0.136 g	0.059 g						

Table 12. PGA values of the analysed collapsed buildings (g).

Buildings	DD1	DD2	DD3	TSC 2007	1st EQ PGA	2nd EQ PGA
Collapsed 1	0.250 g	0.140 g	0.060 g			
Collapsed 2	0.248 g	0.138 g	0.060 g			
Collapsed 3	0.252 g	0.141 g	0.061 g	0.3 g	0.119 g	0.049 g
Collapsed 4	0.253 g	0.141 g	0.061 g			
Collapsed 5	0.247 g	0.138 g	0.059 g			
Collapsed 6	0.250 g	0.139 g	0.060 g			

The seismic performances of each building were assessed against the prescribed design criteria and provided findings on the effectiveness of the existing seismic design codes in addressing the seismic risks in the region. The obtained damage limit values are shown in Tables 13 and 14.

Table 13. Eurocode 8 damage limits for the PGA values of TBEC2018.

PGA Values	TBEC 2018-DD1			TBEC 2018-DD2			TBEC 2018-DD3		
Buildings/Damage Limits (m)	DL	SD	NC	DL	SD	NC	DL	SD	NC
No Damage 1	0.100	0.129	0.223	0.056	0.072	0.125	0.024	0.031	0.053
No Damage 2	0.099	0.127	0.221	0.055	0.071	0.123	0.024	0.030	0.053
No Damage 3	0.101	0.130	0.226	0.057	0.073	0.126	0.024	0.031	0.054
Slight Damage 1	0.100	0.128	0.222	0.056	0.072	0.124	0.024	0.031	0.053
Slight Damage 2	0.101	0.129	0.224	0.057	0.073	0.126	0.024	0.031	0.054
Slight Damage 3	0.100	0.128	0.221	0.055	0.071	0.123	0.024	0.031	0.053
Moderate Damage 1	0.099	0.127	0.221	0.055	0.071	0.123	0.024	0.030	0.053
Moderate Damage 2	0.099	0.127	0.220	0.055	0.071	0.122	0.024	0.030	0.053
Moderate Damage 3	0.098	0.126	0.219	0.055	0.071	0.122	0.024	0.030	0.053
Extensive Damage 1	0.099	0.127	0.221	0.055	0.071	0.123	0.024	0.030	0.053
Extensive Damage 2	0.103	0.133	0.230	0.058	0.074	0.128	0.025	0.032	0.056
Extensive Damage 3	0.095	0.122	0.211	0.053	0.068	0.117	0.023	0.029	0.051
Collapsed 1	0.097	0.124	0.215	0.054	0.070	0.121	0.023	0.030	0.052
Collapsed 2	0.096	0.123	0.214	0.053	0.069	0.119	0.023	0.030	0.052
Collapsed 3	0.098	0.125	0.217	0.055	0.070	0.121	0.024	0.031	0.054
Collapsed 4	0.098	0.126	0.218	0.055	0.070	0.121	0.024	0.031	0.054
Collapsed 5	0.096	0.123	0.213	0.053	0.069	0.119	0.023	0.029	0.051
Collapsed 6	0.097	0.124	0.215	0.054	0.069	0.120	0.023	0.030	0.052

Table 14. Eurocode 8 damage limits for the PGA values of TSC2007 and earthquakes.

Codes	TSC 2007			1st EQ			2nd EQ		
Buildings/Damage Limits (m)	DL	SD	NC	DL	SD	NC	DL	SD	NC
All Studied Buildings	0.116	0.149	0.258	0.046	0.059	0.103	0.019	0.024	0.042

The data in Tables 13 and 14 include the results obtained based on pushover analysis of the sample building. The analysed sample building is an eight-storey reinforced concrete building, and the results of the analysis show the maximum allowable (calculated) values for different earthquake levels. The peak ground acceleration (PGA) values calculated for different earthquake levels, and the damage limit values calculated accordingly, can be used for comparisons. Higher damage limit values are expected for DD1, DD2 and DD3, which have a higher earthquake level, respectively. On the other hand, damage limit values calculated according to the TSC 2007 earthquake regulation are also included in the data. According to the TSC 2007 regulation, Diyarbakır province is considered a second-degree earthquake zone, so the PGA values are constant, and therefore the damage limit values are also constant.

Comparing the damage limit values calculated for different earthquake levels given in the TBEC2018 regulation with the damage limit values calculated according to the TSC

2007 regulation can provide an estimate of the damage potential of buildings. It can be expected that locations with higher PGA values may cause more damage, and, accordingly, damage limit values may increase. In the current code, none of the locations analysed in this study have PGA values higher than the old code TSC2007. Therefore, even if the buildings were constructed in accordance with the old code, they could be expected to respond to earthquake loads with minimal damage. Table 15 shows the damage limits according to the pushover analysis for the studied buildings.

Table 15. TBEC 2018 damage limits for different earthquake levels.

Buildings/Damage Limits (m)	DD-1	DD-2	DD-3
No Damage 1	0.170	0.100	0.043
No Damage 2	0.169	0.099	0.043
No Damage 3	0.171	0.101	0.043
Slight Damage 1	0.169	0.099	0.043
Slight Damage 2	0.171	0.100	0.043
Slight Damage 3	0.169	0.099	0.043
Moderate Damage 1	0.169	0.099	0.043
Moderate Damage 2	0.169	0.098	0.043
Moderate Damage 3	0.168	0.098	0.043
Extensive Damage 1	0.169	0.099	0.043
Extensive Damage 2	0.173	0.102	0.044
Extensive Damage 3	0.164	0.096	0.042
Collapsed 1	0.167	0.098	0.043
Collapsed 2	0.165	0.097	0.043
Collapsed 3	0.167	0.098	0.043
Collapsed 4	0.168	0.098	0.043
Collapsed 5	0.165	0.097	0.043
Collapsed 6	0.167	0.098	0.043

The static pushover curve in Figure 15 shows that the building has acceptable ductility and stiffness if the building was designed according to the current code minimum requirements. However, six buildings with specifications similar to the sample building collapsed during the earthquake.

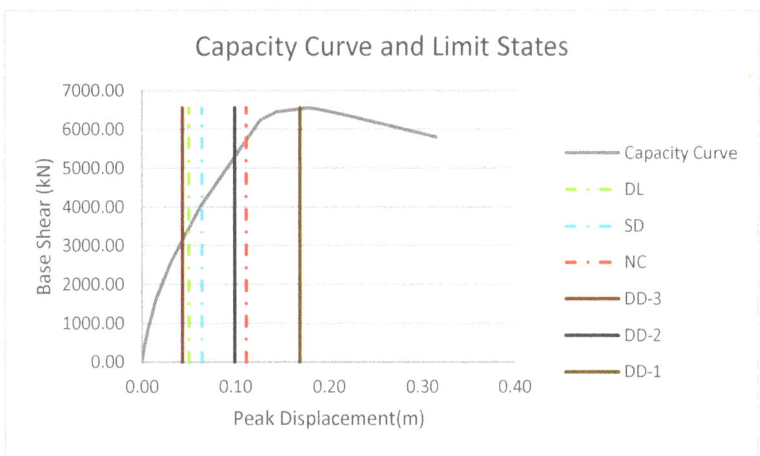

Figure 15. Static pushover curve and limit states of TBEC and Eurocode 8.

Figure 15 illustrates that the DL and SD limits of Eurocode 8 fall between TBEC's DD-2 and DD-3 limit states, with DD-3 being closer than DD-2. The higher limit states, DD-1 and NC, have greater margins than the lower limit states. It can be concluded that the TBEC2018 code is on the safer side.

4. Conclusions

The Pazarcık earthquake was a striking reminder of the seismic risks facing Türkiye. This study focuses on the effects of the devastating earthquake that occurred on 6 February 2023 in Diyarbakır province. It was recorded that six buildings during the earthquake and 17 buildings after the earthquake collapsed; 72 buildings required urgent demolition; 2840 buildings were extensively damaged; 2458 buildings were moderately damaged; and 28,937 buildings were slightly damaged. In the field study conducted in this context, except for the six demolished buildings, three each of extensively damaged, moderately damaged, slightly damaged and undamaged buildings were selected for analysis. For the analyses to be conducted, the current location of each building was visited one by one, and the GPS coordinates were determined on site. Site-specific PGA data were obtained thanks to the earthquake hazard map, which started to be used with the TBEC2018 regulation. Unlike the current regulation, these data were compared with the TSC2007 regulation, which provides fixed PGA values for earthquake zones. In addition to the comparison of the PGA values, the damage limit values were also compared. Eurocode 8 was used for damage limit values. In this context, as a structure that could be found in all selected locations, an eight-storey reinforced concrete building was modelled in SeismoStruct software, and static pushover analysis was performed. The Eurocode 8 limit values were obtained and compared to the Turkish codes to ensure compatibility. It appeared that the 2007 regulation tended to be on the safer side compared to the other regulations in terms of PGA values. In addition, the increases in the PGA values obtained by micro zonation in the current regulation were expected to increase the seismic hazard. It is seen that this may not always be the case because of the local soil conditions and the structural properties that resist earthquake loads.

On the other hand, as shown in Figure 13, it is clearly seen that the spectral values of the earthquake measured in Diyarbakır are far below the design spectra. It is thought-provoking in terms of engineering that this earthquake, which in theory could be expected to have weak effects on a settlement that is so far away, caused such great damage. The outcome from the structural analysis of the selected building has already indicated notable divergences between the projected seismic performance and the actual seismic behaviours. Such discrepancies warrant urgent attention, as they underscore potential inadequacies in the current earthquake design regulations and their implementation.

After evaluating all the data collected in this study, it is concluded that the last two seismic hazard maps and seismic design regulations applied in Türkiye can be considered as successful. However, despite this, significant loss of lives and property still call for further research. A comprehensive analysis of both seismic parameters and structural characteristics is necessary to gain a deeper understanding of the causes of the large-scale losses.

Funding: This research received no external funding.

Data Availability Statement: Most data are included in the manuscript.

Conflicts of Interest: The author declares no conflict of interest.

References

1. Işık, E. Comparative investigation of seismic and structural parameters of earthquakes (M $\geq$ 6) after 1900 in Türkiye. *Arab. J. Geosci.* **2022**, *15*, 971. [CrossRef]
2. Işık, E.; Hadzima-Nyarko, M.; Bilgin, H.; Ademović, N.; Büyüksaraç, A.; Harirchian, E.; Bulajic, B.; Ozmen, H.B.; Aghakouchaki Hosseini, S.E. A comparative study of the effects of earthquakes in different countries on target displacement in mid-rise regular RC structures. *Appl. Sci.* **2022**, *12*, 12495. [CrossRef]
3. Huang, C.; Chen, L.; He, L.; Zhuo, W. Comparative assessment of seismic collapse risk for non-ductile and ductile bridges: A case study in China. *Bull. Earthq. Eng.* **2021**, *19*, 6641–6667. [CrossRef]

4. Yel, N.S.; Arslan, M.H.; Aksoylu, C.; Erkan, İ.H.; Arslan, H.D.; Işık, E. Investigation of the earthquake performance adequacy of low-rise RC structures designed according to the simplified design rules in TBEC-2019. *Buildings* **2022**, *12*, 1722. [CrossRef]

5. Işık, E.; Kutanis, M. Performance based assessment for existing residential buildings in Lake Van basin and seismicity of the region. *Earthq. Struct.* **2015**, *9*, 893–910. [CrossRef]

6. Li, H.; Li, L.; Zhou, G.; Xu, L. Effects of various modeling uncertainty parameters on the seismic response and seismic fragility estimates of the aging highway bridges. *Bull. Earthq. Eng.* **2020**, *18*, 6337–6373. [CrossRef]

7. Ghani, S.; Kumari, S.; Jaiswal, S.; Sawant, V.A. Comparative and parametric study of AI-based models for risk assessment against soil liquefaction for high-intensity earthquakes. *Arab. J. Geosci.* **2022**, *15*, 1262. [CrossRef]

8. Büyüksaraç, A.; Işık, E.; Bektaş, Ö. A comparative evaluation of earthquake code change on seismic parameter and structural analysis; A case of Turkey. *Arab. J. Sci. Eng.* **2022**, *47*, 12301–12321. [CrossRef]

9. Bilgin, H.; Hadzima-Nyarko, M.; Işık, E.; Ozmen, H.B.; Harirchian, E. A comparative study on the seismic provisions of different codes for RC buildings. *Struct. Eng. Mech. Int'l J.* **2022**, *83*, 195–206.

10. Kotoky, N.; Dutta, A.; Deb, S.K. Comparative study on seismic vulnerability of highway bridge with conventional and HyFRC piers. *Bull. Earthq. Eng.* **2019**, *17*, 2281–2306. [CrossRef]

11. Işık, E.; Ademović, N.; Harirchian, E.; Avcil, F.; Büyüksaraç, A.; Hadzima-Nyarko, M.; Akif Bülbül, M.; Işık, M.F.; Antep, B. Determination of natural fundamental period of minarets by using artificial neural network and assess the impact of different materials on their seismic vulnerability. *Appl. Sci.* **2023**, *13*, 809. [CrossRef]

12. Elganzory, A.M.; Novák, B.; Yousry, A.M. Damage assessment and sustainability of RC building in New Cairo City considering probable earthquake scenarios. In *Design and Construction of Smart Cities: Toward Sustainable Community*; Springer International Publishing: Cham, Switzerland, 2021; pp. 47–55.

13. Zhang, Y.; Ouyang, X.; Sun, B.; Shi, Y.; Wang, Z. A comparative study on seismic fragility analysis of RC frame structures with consideration of modeling uncertainty under far-field and near-field ground motion excitation. *Bull. Earthq. Eng.* **2022**, *20*, 1455–1487. [CrossRef]

14. Mertol, H.C.; Tunç, G.; Akış, T.; Kantekin, Y.; Aydın, İ.C. Investigation of RC buildings after 6 February 2023, Kahramanmaraş, Türkiye Earthquakes. *Buildings* **2023**, *13*, 1789. [CrossRef]

15. Wang, X.; Feng, G.; He, L.; An, Q.; Xiong, Z.; Lu, H.; Wang, W.; Li, N.; Zhao, Y.; Wang, Y.; et al. Evaluating urban building damage of 2023 Kahramanmaras, Turkey earthquake sequence using SAR change detection. *Sensors* **2023**, *23*, 6342. [CrossRef] [PubMed]

16. Katsimpini, P.S.; Papagiannopoulos, G.A. Effectiveness of the seesaw system as a means of seismic upgrading in older, non-ductile reinforced concrete buildings. *Vibration* **2023**, *6*, 102–112. [CrossRef]

17. Papagiannopoulos, G.A.; Hatzigeorgiou, G.D.; Beskos, D.E. An assessment of seismic hazard and risk in the islands of Cephalonia and Ithaca, Greece. *Soil Dyn. Earthq. Eng.* **2012**, *32*, 15–25. [CrossRef]

18. Faizah, R.; Amaliah, R.R. Comparative study of Indonesian spectra response parameters for buildings according to 2012 and 2019 seismic codes. *Int. J. Integr. Eng.* **2021**, *13*, 168–175. [CrossRef]

19. Avcil, F.; Işık, E.; Büyüksaraç, A. The effect of local soil conditions on structure target displacements in different seismic zones. *Gümüşhane Üniv. Fen Bilim. Derg.* **2022**, *12*, 1000–1011.

20. Wei, B.; Jia, J.; Bai, Y.; Du, X.; Guo, B.; Guo, H. Seismic resilience assessment of bridges considering both maximum and residual displacements. *Eng. Struct.* **2023**, *291*, 116420. [CrossRef]

21. Khanmohammadi, M.; Eshraghi, M.; Behboodi, S.; Mobarake, A.A.; Nafisifard, M. Dynamic characteristics and target displacement of damaged and retrofitted residential buildings using ambient vibration tests following Sarpol-e Zahab (Iran) Earthquake (MW 7.3). *J. Earthq. Eng.* **2022**, *26*, 6015–6041. [CrossRef]

22. Ghasemi, M.; Zhang, C.; Khorshidi, H.; Zhu, L.; Hsiao, P.C. Seismic upgrading of existing RC frames with displacement-restraint cable bracing. *Eng. Struct.* **2023**, *282*, 115764. [CrossRef]

23. Zhou, P.; Xiong, Z.; Chen, X.; Wang, J. Seismic performance of RC frame structure across the earth fissure based on pushover analysis. *Structures* **2023**, *52*, 1035–1050. [CrossRef]

24. Mazza, F. Damage protection of earthquake resistant structures by means of damped braces. *Adv. Eng. Softw.* **2021**, *160*, 103043. [CrossRef]

25. Harirchian, E.; Jadhav, K.; Mohammad, K.; Hosseini, S.E.A.; Lahmer, T. A comparative study of MCDM methods integrated with rapid visual seismic vulnerability assessment of existing RC structures. *Appl. Sci.* **2020**, *10*, 6411. [CrossRef]

26. Eroglu Azak, T.; Kalafat, D.; Şeşetyan, K.; Demircioğlu, M.B. Effects of seismic declustering on seismic hazard assessment: A sensitivity study using the Turkish earthquake catalogue. *Bull. Earthq. Eng.* **2018**, *16*, 3339–3366. [CrossRef]

27. Bogazici University Kandilli Observatory and Earthquake Monitoring Center (KOERI). Regional Earthquake-Tsunami Monitoring Center (RETMC) Earthquake Data Archive. Available online: http://www.koeri.boun.edu.tr/ (accessed on 7 August 2023).

28. GEOFON Data Centre. GEOFON Seismic Network. Deutsches GeoForschungsZentrum GFZ. Seismic Network. 1993. Available online: https://geofon.gfz-potsdam.de/doi/network/GE (accessed on 7 August 2023).

29. AFAD (Disaster and Emergency Management Presidency, Republic of Türkiye)—For Strong Ground Motion Records. Available online: https://tadas.afad.gov.tr/ (accessed on 7 August 2023).

30. USGS. Earthquake Hazards Program. 2023. Available online: https://earthquake.usgs.gov/ (accessed on 7 August 2023).

31. Hancılar, U.; Şeşetyan, K.; Çaktı, E.; Yenihayat, E.Ş.N.; Malcıoğlu, F.S.; Dönmez, K.; Tetik, T.; Süleyman, H. *Strong Ground Motion and Building Damage Estimations Preliminary Report*; Bogazici University: İstanbul, Turkey, 2023; p. 42.

32. Güllü, H.; Karabekmez, M. Effect of near-fault and far-fault earthquakes on a historical masonry mosque through 3D dynamic soil-structure interaction. *Eng. Struct.* **2017**, *152*, 465–492. [CrossRef]
33. Bedirhanoğlu, İ. *Preliminary Assessment Report for The Kahramanmaraş Pazarcik and Elbistan Earthquakes*; Civil Engineering Department, Dicle University: Diyarbakır, Türkiye, 2023.
34. TBEC. *Turkish Seismic Earthquake Code*; TBEC: Ikeja, Nigeria, 2018; pp. 1–416. Available online: https://www.resmigazete.gov.tr/eskiler/2018/03/20180318M1-2-1.pdf (accessed on 7 August 2023).
35. TSC. *Turkish Earthquake Code*; TSC: Ikeja, Nigeria, 2007; pp. 1–159. Available online: http://www.okangungor.com.tr/wp-content/uploads/2013/05/2007-Turkish-Earthquake-Code.pdf (accessed on 7 August 2023).
36. *EN 1998-1:2004*; Eurocode 8: Design of Structures for Earthquake Resistance. Part 1: General Rules, Seismic Actions and Rules for Buildings. European Committee for Standardization (CEN): Brussels, Belgium, 2004.
37. Chopra, A.K.; Goel, R.K.A. Modal pushover analysis procedure for estimating seismic demands for buildings. *Earthq. Eng. Struct. Dyn.* **2002**, *31*, 561–582. [CrossRef]
38. Oğuz, S. Evaluation of Pushover Analysis Procedures for Frame Structure. Master's Thesis, Middle East Technical University, Ankara, Türkiye, 2005.
39. Krawinkler, H.; Seneviratna, G.D.P.K. Pros and cons of a pushover analysis of seismic performance evaluation. *Eng. Struct.* **1998**, *20*, 452–464. [CrossRef]
40. Papazafeiropoulos, G.; Plevris, V. Kahramanmaraş—Gaziantep, Türkiye Mw 7.8 Earthquake on 6 February 2023: Strong ground motion and building response estimations. *Buildings* **2023**, *13*, 1194. [CrossRef]
41. Güler, K.; Celep, Z. On the general requirements for design of earthquake resistant buildings in the Turkish Building Seismic code of 2018. In *IOP Conference Series: Materials Science and Engineering*; IOP Publishing: İstanbul, Turkey, 2019.

MDPI
St. Alban-Anlage 66
4052 Basel
Switzerland
www.mdpi.com

Buildings Editorial Office
E-mail: buildings@mdpi.com
www.mdpi.com/journal/buildings